NMR SPECTROSCOPY IN LIQUIDS AND SOLIDS

VLADIMIR I. BAKHMUTOV

CRC Press

Taylor & Francis Group

Boca Raton London New York

CRC Press is an imprint of the
Taylor & Francis Group, an **informa** business

CRC Press
Taylor & Francis Group
6000 Broken Sound Parkway NW, Suite 300
Boca Raton, FL 33487-2742

International Standard Book Number-13: 978-1-4822-6270-4 (Paperback)

Library of Congress Cataloging-in-Publication Data

Bakhmutov, Vladimir I.
    NMR spectroscopy in liquids and solids / Vladimir I. Bakhmutov.
        pages cm
    "A CRC title."
    Includes bibliographical references and index.
    ISBN 978-1-4822-6270-4 (alk. paper)
    1. Nuclear magnetic resonance spectroscopy. 2. Spectrum analysis. I. Title. II. Title: Nuclear magnetic resonance spectroscopy in liquids and solids.

QD96.N8B35 2014
543'.66--dc23   2014043706

Visit the Taylor & Francis Web site at
http://www.taylorandfrancis.com

and the CRC Press Web site at
http://www.crcpress.com

Contents

Preface

This book is intended for those who are just beginning research activity in chemistry, biochemistry, geochemistry, chemical engineering, or other science and wish to use NMR as the main research method for investigating solutions and the solid state. This text is not a reference book with a great number of tables and NMR data that can be found elsewhere in the literature; however, it does cover basic NMR concepts that must be understood to perform NMR experiments on liquid and solid samples and to interpret the data collected to determine structures and molecular dynamics. The book has been written by an expert who began his research career using iron permanent electromagnets and continuous-wave irradiation, later moving on to superconducting electromagnets with pulsing NMR. Many NMR experiments mentioned in this text have been taken from the author's own experience, including studies of inorganic and organic molecules in solutions and molecular aggregates and materials in the solid state.

What is new and different in this book?

1. Presents current understanding and applications of solution and solid-state NMR techniques
2. Combines and formulates common principles for NMR experiments in gases, liquids, and solids
3. Pays considerable attention to nuclear relaxation, from the phenomenon itself to applications in solutions and solids
4. Formulates general strategies for studies and demonstrates how to choose the appropriate experiment, what methods and pulse sequences are suitable for particular situations, and how to assign signals in the NMR spectra of simple and complex molecular systems

The book is organized into ten chapters that include numerous illustrations and recommended literature. Chapters 1 to 4 cover the basic principles, and Chapters 5 to 10 describe NMR applications. These chapters have been written in such a way that they can be presented as ten lectures for NMR spectroscopy courses aimed at undergraduate and postgraduate students or young researchers at colleges and universities. Only minimal knowledge of physics and quantum mechanics is assumed.

Preface

Introduction

Nuclear magnetic resonance (NMR) is a powerful instrumental method that rapidly and adequately handles a large number of scientific and applied tasks on both the molecular and atomic levels. Because a NMR signal can be detected in gases, solutions, liquids, and solids, including crystals and amorphous homogeneous or heterogeneous systems, this method can be successfully applied in physics, chemistry, food chemistry, biochemistry, geochemistry, biology, geology, archeology, pharmaceuticals, and materials science fields. Moreover, an interesting version of NMR, magnetic resonance imaging (MRI), has become a reliable diagnostic tool in both human and veterinary science and medicine, particularly for research into the brain. MRI is also applicable to the study of materials chemistry.

The modern arsenal of NMR methods and techniques is extremely large. Due to the increasing complexity of objects involved in NMR studies, this arsenal is constantly growing and being modified. Over its 70-year history, NMR research has stimulated the development of new scientific fields. Successes in radiofrequency technology from 1952 to 1953 led to the appearance of the first commercial NMR spectrometers, which strongly impacted research in chemistry, biochemistry, and biology. In turn, the synthesis and study of complex objects, such as biochemical molecular systems and complex inorganic/organic molecular aggregates, required the availability of more powerful NMR instrumentation offering better spectral resolution and sensitivity. Such improvements would have been impossible in the absence of innovations in magnet technology that created stronger and stronger homogeneous magnetic fields. In addition, the development of computers capable of very fast Fourier transform (FFT) processing has resulted in the routine use of two-, three-, or four-dimensional NMR spectroscopy for structural analyses in solutions and solids to assign signals in the NMR spectra of complex molecular systems.

Due to the impressive development of NMR techniques, NMR applications are expanding from molecular and low-temperature physics to archeology, where, for example, the natural destruction of wood is the focus of researchers. In spite of such developments, it is obvious that without a deep understanding of NMR and nuclear relaxation their successful application cannot be guaranteed. Moreover, such an understanding cannot be gained by any computer search (such as Google, for example), a popular approach among students and young researchers. Searches such as these can provide only fragmented knowledge.

With regard to the history of NMR,[1] after the first successful observation of NMR signals in gases occurred in 1937, C.J. Gorter attempted to detect resonances in the condensed phase in order to detect lithium nuclei in crystalline LiF and protons in crystalline potassium alum. Similar experiments had been performed on ^{19}F nuclei in the crystal solid KF but no NMR signals were found. Today, it is known that the nuclei were invisible due to the exclusively long relaxation times in the very pure crystals. Relaxation times are still an issue for NMR beginners, as the behavior of nuclei in NMR experiments is time dependent, and their relaxation times strongly affect the NMR data obtained.

In contrast, due to the relatively short relaxation times of protons in paraffin wax and water, researchers Purcell and Bloch reported the first NMR signals in the condensed phase in 1945. These experiments led to the application of NMR as an analytical method. Later, Blombergen, Purcell, and Pound[2] formulated the theory of nuclear relaxation now referred to as the BPP theory. It should be noted that the BPP theory is valid for all states of matter—solids, liquids, or gases—where the dipole–dipole internuclear interactions govern nuclear relaxation via molecular motions. Extremely low-temperature relaxation measurements performed in solid H_2 at a temperature of 1 K have emphasized the importance of such motions in the dipole–dipole relaxation process. It should be noted that further reconsiderations of nuclear relaxation processes have led to only small additions to the relaxation theory first published in 1948.

The rate and the character of molecular motions strongly change upon going from liquids to solids. This factor affects the experimental conditions required for the detection of NMR signals. Ice demonstrates the principal differences in NMR experiments on liquids and solids. The line width of the ^1H resonance in liquid water is very small (<1 Hz), whereas the resonance of solid water broadens up to ~10^5 Hz due to strong dipolar proton–proton coupling against a background of very slow molecular motions. It is obvious that different technical solutions are needed to detect NMR signals in solids.

Despite the first early successes in understanding the relaxation behavior of nuclei in liquids and solids, the relaxation mechanism in crystal solids, such as CaF_2, remained unclear. In 1949, Bloembergen[3] reported on the nature of ^{19}F relaxation, termed *spin diffusion*. This relaxation occurs via energy diffusion from nuclear spins to paramagnetic centers present in the crystal as impurities, and it can be effective for the strong dipolar coupling ^{19}F–^{19}F.

The early classical work mentioned above and the later discovery of chemical shifts and spin–spin coupling illustrate important features of NMR: The NMR spectra recorded in liquids and solids show a number of distinguished nuclei in structurally different molecules and/or groups and even their mutual dispositions, whereas nuclear relaxation opens the way to describing molecular dynamics ranging from isotropic or anisotropic reorientations in liquids to phase transitions in rigid solids. These unique features make NMR a powerful analytical method.

The diverse NMR experiments on liquids or solids can be represented by the general scheme shown in Figure I.1. Here, each step, from the preparation of samples to interpretations and conclusions, is equally important. Actions taken at each step can result in errors, experimental or interpretational, because modern NMR spectrometers are complex multiple-pulse devices that operate with low or very high magnetic fields, up to 21.14 T. The modern tendency is to aim for full automation in the performance of experiments, including sample changes and even interpretation of the spectra. It should be noted that such automation exemplifies a black-box approach that can be successful for the routine analysis of reaction mixtures or large series of similar compounds. However, such an approach is quite ineffective in a research laboratory, where precise determination of new structures is an important part of studies. In

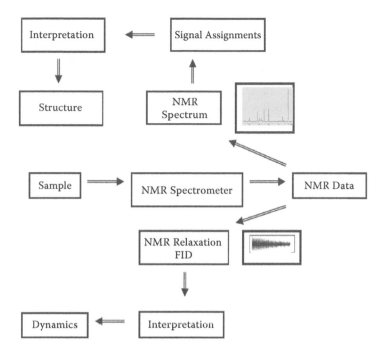

FIGURE I.1 General scheme of NMR experiments performed on solutions and solids to probe their structure and dynamics.

addition, artifact phenomena can appear in NMR experiments, further complicating their interpretation. Their recognition again requires at least minimal understanding of the physical aspects of NMR. Finally, it should be emphasized that NMR experiments resulting in NMR spectra or time-dependent data are revealing only the behavior of nuclear spins and their ensembles in the magnetic field, not groups of atoms and molecules or more complex aggregates. Understanding their nature is a product of our interpretations based on experience and numerous spectral structural relationships.

This book is compressed to ten chapters to minimize its volume. The first three chapters consider the theoretical basis of NMR spectroscopy, the theory of NMR relaxation, and the practice of relaxation measurements. Chapter 4 discusses the general aspects of molecular dynamics and their relationships with NMR. NMR spectroscopy and relaxation studies in solutions are addressed in Chapters 5 and 6, and Chapter 7 considers special issues of NMR in solutions. Chapter 8 leads into the solid-state portion of the book by introducing the general principles and strategies involved in solid-state NMR studies and provides examples of applications of relaxation for the determination of molecular dynamics in diamagnetic solids. Chapter 10 concludes with a discussion on special issues of solid-state NMR, including NMR and NMR relaxation in paramagnetic solids, an area requiring additional knowledge. Finally, the chapters are accompanied by references and recommended literature for further reading.

The book is addressed to undergraduate students, NMR beginners, and young scientists who are planning to work or are already working in the various fields of chemistry, biochemistry, biology, pharmaceutical sciences, and materials science. It provides an introduction to the general concepts of NMR and the principles of its applications, including how to perform adequate NMR experiments and how to interpret the NMR data collected in liquids and solids to characterize molecule systems in terms of their structure and dynamics.

REFERENCES

1. Andrew, E.R. and Szczesniak, E. (1995). *Prog. Nucl. Magn. Reson. Spectr.*, 28: 11.
2. Blombergen, N., Purcell, E.M., and Pound, R.V. (1948). *Phys. Rev.*, 73: 679.
3. Bloembergen, N. (1949). *Physica*, 15: 386.

Author

Vladimir I. Bakhmutov, PhD, works in the Department of Chemistry at Texas A&M University. The 330 scientific articles and 5 books that he has written and published document his outstanding contributions to chemical physics research. Dr. Bakhmutov's research interests include extensive applications of solution and solid-state NMR techniques for chemistry, molecular physics, and materials science. NMR relaxation in liquids and solids applied for structural and dynamic studies is also an area of interest. Dr. Bakhmutov has been granted visiting professorships by Zurich University, Switzerland; Bourgogne University, Dijon, France; Rennes University, France; CINVESTAV, Mexico; and Iberdrola, Zaragosa, Spain.

1 Physical Basis of Nuclear Magnetic Resonance

Nuclear magnetic resonance (NMR) is a complex physical phenomenon, but it can be expressed simply as follows: Magnetically active *nuclei* placed into an external *magnetic field* are excited by radiofrequency irradiation, and the energy absorbed due to the irradiation (*resonance*) is registered as a NMR signal. Nuclei having different environments will correspond to different energies and hence different signals, and the resulting totality of the signals forms a one-dimensional NMR spectrum. Generally, the NMR spectra are recorded as plots of line intensity vs. resonance frequency, where the number of spectral lines corresponds to the number of non-equivalent nuclei dictated by molecular structure and molecular symmetry and depends on spectral resolution.

Figure 1.1 shows one-dimensional NMR spectra recorded in solids where the dispositions of the resonance lines are characterized by chemical shifts measured in universal units (parts per million, ppm). As shown in Figure 1.1A, the number of ^{13}C resonances detected in the solids illustrates well the structural motif of the mercury complex. Resonance lines in spectra can show additional splitting due to spin–spin coupling in solutions and solids or quadrupolar and dipolar coupling in solids. The splitting depends on the nuclei environments. Finally, integral intensities of NMR signals, particularly in solutions, are determined by the numbers of equivalent nuclei in a sample that lead the way to structural analysis. In this context, it should be emphasized that the NMR spectra should be recorded to account for full nuclear relaxation.

Observation of spin–spin coupling and accurate measurement of the spin–spin coupling constants are traditionally used to elucidate molecular structures in solutions. Spin–spin coupling also plays an important role in interpretation of high-resolution solid-state magic-angle spinning (MAS) NMR spectra, as shown in Figure 1.1B. The ^{19}F MAS NMR spectrum of the solid complex $Fe(II)_5(bptz)_4(SbF_6)_{10}(CH_3CN)_{10}$ consists of at least 20 ^{19}F resonance lines. A part of these resonances corresponds to the $^1J(^{19}F-^{121}Sb)$ constant measured as 1940 to 1950 Hz. The other part shows the $^1J(^{19}F-^{123}Sb)$ constant of 1033 Hz. Because the isotropic shift ($^{19}F-^{121,123}Sb$) (see below) is not detectable in solids, the observed number of the resonances is explained by the presence of three non-equivalent ions $[SbF_6]^-$. Because the $J(^{19}F-^{19}F)$ spin–spin coupling constant is not observable, six ^{19}F nuclei in $[SbF_6]^-$ are equivalent due to fast isotropic reorientations of $[SbF_6]^-$ ions in the solid state.

Even these simple examples, showing the logic behind interpreting NMR data, require a clear understanding of how nuclei behave in the external magnetic field, why a single nucleus (e.g., ^1H or ^{13}C) gives several lines in NMR spectra, and why these lines experience additional splitting.

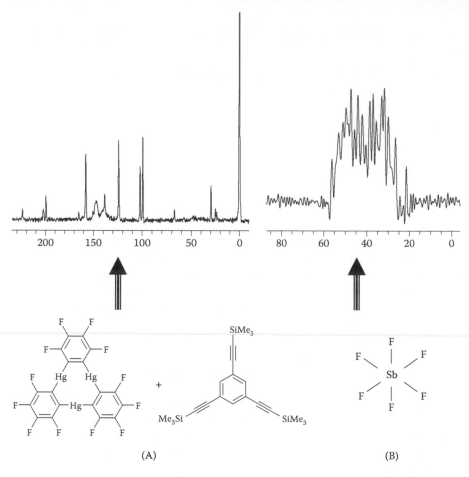

FIGURE 1.1 (A) Solid-state ^{13}C cross-polarization MAS NMR spectrum of mercury complex recorded at a spinning rate of 10 kHz. (B) Single-pulse ^{19}F MAS NMR spectrum of the solid compound $Fe(II)_5(bptz)_4(SbF_6)_{10}(CH_3CN)_{10}$, recorded at a spinning rate of 12 kHz. The high-resolution spectrum was obtained by Gaussian treatment of the free induction decay.

1.1   NUCLEI IN THE EXTERNAL MAGNETIC FIELD

Chemical shift and spin–spin (as well as quadrupolar and dipolar) coupling are fundamental properties of nuclear systems placed into an external magnetic field, noted here and below as B_0. The behavior of the nuclei when B_0 is applied can be rationalized in terms of semiclassical physics or quantum mechanics, both of which can explain NMR. Based on a quantum mechanical description, the central role belongs to nuclear spin, a fundamental property associated with the rotation of a nucleus around its axis, as shown in Figure 1.2A. In terms of quantum mechanics, any system having discrete energy levels can be described by defined sets of quantum numbers and represented by an eigenvalue equation. Nuclear spins in this context can

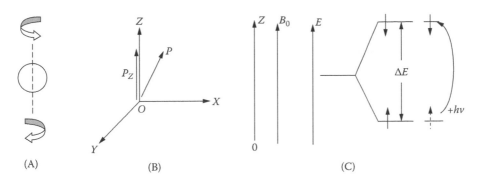

FIGURE 1.2 (A) Nuclear spin associated with rotation of a nucleus around its axis. (B) Projection of the nuclear angular moment P, P_Z, on the OZ axis. External magnetic field B_0 is applied along the OZ direction, whereas the radiofrequency field affecting the P_Z projection is applied along the OX (or OY) axis.

be expressed via wave functions ψ similarly to electron wave functions. These ψ functions, depending on the spatial and spin coordinates, can be used to calculate the energy of spin states on the basis of the Schrödinger equation:

$$\hat{H}\psi = E\psi \tag{1.1}$$

where $\hat{H}$ is an operator (*Hamiltonian*) acting on the wave function. Then, spin operator $\hat{I}$, containing components $\hat{I}_X$, $\hat{I}_Y$, and $\hat{I}_Z$, will correspond to the spin coordinates of the wave functions. As shown below, this quantum mechanical formalism results in energy levels whose appearance in the external magnetic field explains the phenomenon of NMR.

In the semiclassical description of NMR, net magnetization assumes the central role. This magnetization is created in a sample when it is placed into external magnetic field B_0. Any nucleus is considered to be magnetically active and can be observed in NMR experiments if it possesses a non-zero angular momentum (P). In this context, the rule is simple: A nuclei will have zero spin if the number of protons and number of neutrons are equal; for example, ^{16}O or ^{12}C nuclei are not magnetically active. It should be emphasized, however, that most of the elements in the Periodic Table have magnetically active isotopes but their natural abundance is different, as shown in Table 1.1.

Following the formalism of quantum mechanics, nuclear angular momentum P is responsible for the appearance of nuclear magnetic moment μ, which is expressed via Equation 1.2:

$$\mu = \gamma P \tag{1.2}$$

Coefficient γ in this relationship is the nuclear magnetogyric ratio, which is one of the fundamental magnetic nuclear constants strongly dependent on the nature of the nuclei (Table 1.1).

TABLE 1.1

Some NMR Properties of Selected Nuclei

Nucleus	Spin I	Natural Abundance (%)	NMR Frequency v_0 (MHz at $B_0 = 2.3488$ T)	γ (10^7 rad T^{-1} s^{-1})
^{1}H	—	99.98	100	26.752
^{2}H	1	0.016	15.35	4.107
^{11}B	3/2	80.42	32.08	8.584
^{13}C	—	1.108	25.14	6.728
^{14}N	1	99.63	7.22	1.934
^{15}N	−1/2	0.366	10.14	−2.712
^{29}Si	−1/2	4.67	19.87	−5.319
^{19}F	—	100	94.08	25.18
^{31}P	—	100	40.48	10.84
^{17}O	−5/2	0.037	13.557	−3.628
^{93}Nb	9/2	100	24.549	6.567
^{51}V	7/2	99.75	26.3	7.036
^{59}Co	7/2	100	23.7	6.347
^{109}Ag	−1/2	48.2	4.65	−1.083
^{117}Sn	—	7.61	35.63	−9.578
^{119}Sn	_	8.58	37.29	−10.02
^{129}Xe	−1/2	26.4	27.8	−7.399
^{113}Cd	−1/2	12.2	22.2	−5.933
^{199}Hg	—	16.84	17.91	4.815
^{205}Tl	1/2	70.5	57.63	15.589

Nuclear momentum μ, which can be considered to be a small magnet in classical terms, will respond to the presence of external magnetic field B_0, and the response will strongly depend on the γ values. As follows from Table 1.1, which lists the most popular target nuclei, the γ values change in a very large diapason. In turn, this diapason corresponds to a very large set of resonance frequencies upon application of the same magnetic field.

As can be seen in Table 1.1, the γ constant has a sign: negative for ^{15}N, ^{29}Si, ^{17}O, ^{109}Ag, ^{117,199}Sn, and ^{129}Xe nuclei and positive for ^{1}H, ^{2}H, and ^{31}P nuclei. Generally speaking, the sign of γ does not play a role in simple NMR experiments, but it does play a very important role in, for example, indirect detection of nuclei based on heteronuclear multiple quantum coherence (HMQC), heteronuclear single quantum coherence (HSQC), and heteronuclear multiple bond correlation (HMBC) used in solutions, as well as cross-polarization (CP) in solids.

As will be shown in later chapters, an important element in NMR experiments performed in the same external magnetic field is the sensitivity, which fundamentally depends on the magnetogyric constant γ and the natural abundance of nuclei (i.e., on the nuclear nature). In fact, due to the influence of both of these characteristics, NMR sensitivity relative to ^{1}H nuclei with 100% natural abundance decreases by a factor of 10^3, 10^4, and 10^6 for ^{117}Sn, ^{13}C, and ^{2}H nuclei, respectively. This circumstance creates significant problems in detection of these nuclei in solutions and in the solid state. This is particularly important for solids, where resonances can be very broad due to strong dipolar internuclear interactions. Such solid-state NMR experiments require long-term accumulations to achieve good signal-to-noise ratios in the resulting spectra. This is especially problematic for long relaxation times.

Finally, the spin numbers in Table 1.1 also play a very important role. The NMR spectra of nuclei with a spin of 1/2 are remarkably simpler in solutions and in the solid state, as well. It is obvious that, due to the 100% natural abundance of ^{1}H and ^{19}F or ^{31}P nuclei, with $I = 1/2$, they are the most convenient targets in NMR investigations, particularly in solutions, because in the solid state ^{1}H and ^{19}F nuclei experience strong dipolar interactions. These interactions complicate their detection, preventing high spectral resolution and requiring special techniques to collect NMR data.

Going back to the quantum mechanics formalism, the P and μ magnitudes are *quantized*. The projections of the angular nuclear moment P_Z on the OZ coordinate axis, shown in Figure 1.2B as the direction of the external magnetic field B_0, can be written as

$$P_Z = \hbar \times m_I \quad \text{and} \quad \hbar = h / 2\pi \tag{1.3}$$

where h is Planck's constant ($h/2\pi = 1.05457266 \cdot 10^{-34}$ J·s) and m_I is the magnetic quantum number. The m_I number is dictated by nuclear spin I and takes its values from I to $-I$ (i.e., $I, I - 1, I - 2, ..., -I$). In turn, the value of I is a multiple of 1/2.

Spin I of 1/2 is the simplest case, and the angular and magnetic moments can be expressed via Equation 1.4:

$$P_Z = \pm(1/2)\hbar$$
$$\mu_Z = \pm(1/2)\gamma\hbar \tag{1.4}$$

Then, the μ_Z (or P_Z) projections, shown in Figure 1.2C as vectors, can have only two permitted spatial orientations: *parallel* or *anti-parallel* relative to the OZ axis. This agrees with the left- or right-spinning nucleus shown in Figure 1.2A.

In the absence of external magnetic field B_0 (including the Earth's magnetic field), these orientations are energetically equivalent and thus the nuclei occupy a single energy level. When external magnetic field B_0 (strong enough to allow observation of a NMR signal) is applied along the OZ axis, the parallel and anti-parallel μ_Z (or P_Z) orientations become energetically non-equivalent and the initially degenerate energy level undergoes the Zeeman splitting shown in Figure 1.2C. Without going into the theoretical details, it is sufficient to note that the energy difference (ΔE) can be obtained via the Hamiltonian equation shown below, where it is proportional to the magnitude of μ_Z and the strength of the applied magnetic field (B_0):

$$\hat{H} = -\gamma\hbar\hat{I}_Z B_0 \quad \text{and} \quad \Delta E = 2\mu_Z B_0 \tag{1.5}$$

Because the minimal energy difference can be formulated as one quantum (Equation 1.6), Equation 1.5 takes form of Equation 1.7:

$$\Delta E = h\nu \tag{1.6}$$

$$h\nu_0 = 2\mu_z B_0 = \gamma \hbar B_0$$
$$\nu_0 = \gamma B_0, \text{ where } \nu_0 = \omega_0/2\pi \tag{1.7}$$

This equation shows the fundamental condition for the resonance of a single (i.e., uncoupled) nucleus. As seen, a nucleus with a spin of 1/2 will experience a single-quantum transition (the m_I number changes from $-1/2$ to $+1/2$) between low-energy and high-energy Zeeman levels when radiofrequency irradiation is applied at frequency ν_0. The ν_0 frequency, named the Larmor frequency, depends on the nature of the nuclei and the strength of the applied magnetic field (B_0).

An ensemble of nuclei does not change the situation if the nuclei in the ensemble remain uncoupled. A system of coupled nuclei leads to more complex phenomena considered in Chapter 2. Nuclei in an ensemble populate Zeeman energy levels in accordance with heat equilibrium, where the population difference (ΔN) is controlled by Equation 1.8:

$$\Delta N \sim \exp(-\Delta E/kT) \tag{1.8}$$

where T is the temperature in the laboratory coordinate system (or the spin temperature in the presence of forced spin transitions) and k is the Boltzmann constant (1.380658×10^{-23} J/K). Due to their heat equilibrium, the low-energy levels are more populated (Figure 1.3A), resulting in macroscopic magnetization (Figure 1.3B) that can be measured by NMR.

One can show that, due to a low energy difference (ΔE) and the above thermal factor, the population difference (ΔN) is of very small magnitude relative to the total number of the nuclei (N). Even for ^{1}H nuclei, which have the largest γ constant and a 100% natural abundance, the ratio $\Delta N/N$ has been calculated to be as low as 8×10^{-6} for an external magnetic field of 2.35 T and temperature of 300 K. This value highlights the relatively low sensitivity of NMR in comparison with infrared (IR) or ultraviolet (UV) spectroscopy; however, the sensitivity increases at lower temperatures and higher magnetic fields, and it can be enhanced by special techniques or approaches that will be considered in later chapters.

For nuclei having spins of ≥ 1 (i.e., quadrupolar nuclei), the number of Zeeman energy levels increases; thus, it is interesting to consider the number of resonance lines that can be expected in a NMR spectrum. For example, by definition, ^{2}H or ^{14}N nuclei with a spin of 1 correspond to three energy levels in the external magnetic field in accordance with the μ_z values of $+\gamma\hbar(1)$, 0, and $-\gamma\hbar(1)$ (Figure 1.3C). It is obvious that this circumstance cannot lead to a larger set of resonance lines in solutions and liquids. These energy levels are equidistant (i.e., they correspond to the same magnitude of frequency ν_0), and double (or multiple) quantum nuclear transitions where the m_I value changes from +1 to −1 (dashed line) are forbidden. It should

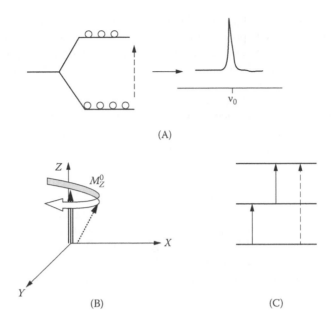

(A)

(B)   (C)

FIGURE 1.3 (A) Zeeman energy levels for spins $I = 1/2$ populating these levels. (B) The macroscopic magnetization M_Z^0 is created in a sample in the presence of an external magnetic field. (C) Zeeman energy levels for spins $I = 1$.

be noted that such multi-quantum transitions can be stimulated by corresponding pulse sequences. As a result, in spite of the presence of three (or more) energy levels, the nuclear transitions between them will be observed as one NMR resonance line.

The situation can change in rigid solids. When fast isotropic molecular motions are absent and the quadrupolar interactions (considered in Chapter 2) are smaller than the Zeeman splitting, the energy levels are no longer equidistant. Therefore, in addition to the strongly intense transitions of $-1/2$ and $+1/2$, other transitions can also be excited to demonstrate increased numbers of resonance lines in a NMR spectrum. For example, in the case of solid-state MAS NMR spectra of quadrupolar nuclei, all of these transitions can be detected, resulting in very wide sideband patterns.

1.2 RADIOFREQUENCY IRRADIATION: CONTINUOUS-WAVE AND RADIOFREQUENCY PULSES

As indicated by Equations 1.5 to 1.7, a NMR signal can be detected when irradiation frequency v approaches the Larmor frequency of v_0. For different nuclei placed in a magnetic field (B_0) of 2.3488 T, it has been shown that their observation requires irradiation with frequencies between 4 and 100 MHz (Table 1.1), corresponding to the radiofrequency diapason. Technically, such an experiment can be realized with continuous-wave (CW) irradiation, applied earlier, or by application of a powerful radiofrequency (RF) pulse to detect the macroscopic magnetization that appears due to the presence of magnetic field B_0.

In terms of semiclassical physics, the unequal populations of the Zeeman energy levels in the magnetic field produce the equilibrium macroscopic magnetization that undergoes the Larmor precession shown in Figure 1.3B. Its frequency is ν_0 or $\omega_0 = 2\pi\nu_0$. Continuous-wave ν_0 irradiation, applied as a RF field rotating in the X,Y plane, reorients the magnetization vector, leading to the appearance of NMR signals.

Figure 1.4A illustrates the process initiated by RF irradiation. The observation of a NMR signal is accompanied by initial absorption of the energy followed by its emission. It is very important to understand that, in the absence of this energy emission, registration of the NMR signal is impossible. In turn, decreasing the energy of an excited nuclear system is possible only through an energy exchange between nuclear spins and their environment. This exchange is directly related to nuclear relaxation, which recovers the initial equilibrium state. Again, in the absence of nuclear relaxation or at its infinitely long time, no NMR signal is observed.

Pulsing irradiation implies application of a powerful radiofrequency pulse at the acting frequency ω_{rf}. This RF pulse, instantly exciting a very large frequency band, causes an oscillating and time-dependent RF field that contributes to external magnetic field B_0. Under these conditions, the total field produced can be written as

$$B_{total} = iB_1 \times \cos(\omega_{rf}t) + kB_0 \tag{1.9}$$

where unit vector i is situated along the space coordinate OX, and the OZ coordinate corresponds to the location of unit vector k. Then, the Hamiltonian in Equation 1.5 transforms to

$$\hat{H} = -\gamma\hbar\left[\hat{I}_zB_0 + \hat{I}_xB_1\cos\left(\omega_{rf}t\right)\right] \tag{1.10}$$

which now contains the rotating component. Due to this component, the wave function ψ' in Equation 1.11, corresponding to the Hamiltonian,

$$\psi' = \exp(-i\psi_{rf}t\,\hat{I}_z) \times \psi \tag{1.11}$$

describes the spin states in the new *rotating* coordinate system (the first term in Equation 1.11 is the rotation operator).

In reality, the accurate quantum-mechanical description of a spin system to solve the Schrödinger equation is not simple. Nevertheless, the quantum mechanical elements, presented above, introduce the new rotating coordinate system required to describe the spin states; thus, the behavior of nuclear magnetization during pulse irradiation can be clearly imagined via the vector model shown in Figure 1.4B, where magnetization vector M, initially orientated in an equilibrium state along the direction OZ, reorients due to the action of the RF pulses.

According to this vector model, the amplitudes and durations of the RF pulses must be sufficient to rotate the magnetization vector by the necessary angle (α). According to the Ernst rule, the angle is defined via Equation 1.12:

$$\alpha = \gamma B_1 t_P \tag{1.12}$$

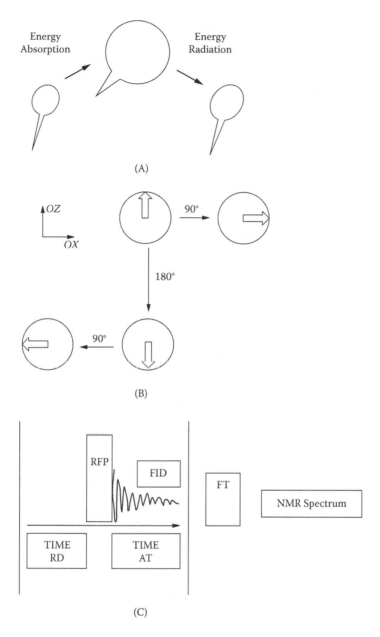

FIGURE 1.4 (A) The change in the energy of a nuclear system initiated by irradiation (shown schematically). (B) Visualization of changes in the orientations of the magnetization vector (in the rotating frame) due to the action of 90° and 180° radiofrequency pulses (vector representation). (C) The simplest one-dimensional NMR experiment, involving three phases: (1) time delay relaxation delay (RD) necessary for full relaxation to recover the initial equilibrium state; (2) time RFP corresponding to the duration of a radiofrequency pulse, and (3) time AT required to collect the NMR data.

where B_1 is the power and t_p is the duration of the pulse. Generally, the duration is adjusted between milliseconds and microseconds.[1] The vector model shown in Figure 1.4B leads to several important conclusions. After a 90° pulse, the new M orientation should be along the positive axis OX, whereas it would be along the negative axis OZ after the action of a 180° pulse. In practice, it is important to note that, for the registering radiofrequency coil along axis OX, the M orientation along the positive axis OX will obviously correspond to the *maximal* intensity of the NMR signal. In contrast, for the same reason, the NMR signal will not be observable after an 180° RF pulse, or the signal will be visibly negative due to the combination of 180° and 90° pulses. The latter observation is valid when the delay time between the 180° and 90° pulses is not sufficient for nuclear relaxation. Generally speaking, this behavior of nuclear magnetization by the pulsing irradiation is common for both liquids and solids. Technically, however, solid-state NMR requires significantly more powerful RF pulses.

The CW irradiation method directly gives NMR spectra recorded as patterns, where the coordinates are intensity and frequency. Thus, these data are expressed as a $F(\nu)$ function. In contrast to CW, NMR data obtained by pulse techniques (free induction decay, or FID) are time dependent and expressed via function $f(t)$. Therefore, regular NMR spectra can be obtained only after the mathematical procedure known as Fourier transform (FT). This procedure treats the FIDs by converting the time domain into the frequency domain and *vice versa* as shown in Equation 1.13:

$$F(\nu) = \int f(t)\exp(-i2\pi\nu t)dt$$
$$f(t) = \int F(\nu)\exp(+i2\pi\nu t)d\nu$$

(1.13)

Thus, NMR spectrometers should be equipped with computers capable of fast numerical Fourier transformation, and the final NMR spectrum is dependent on digital resolution, considered below. Because the majority of current NMR experiments are based on RF pulses acting at the carrier frequency ω, they provide observations of all of the time-evolving signals in the rotating frame. These signals can be characterized by amplitude C, expressed, in turn, via the magnitude of the real and imaginary components:

$$C\exp(i\omega t) = C\cos(\omega t) + iC\sin(i\omega t)$$

(1.14)

Fourier transformation of the term $C\exp(i\omega t)$ produces the complex frequency spectrum where the real part is associated with the so-called absorption line shape (i.e., line shape of detected NMR signals), while the imaginary part represents the dispersive line shape.

The simplest one-pulse NMR experiment, performed with direct excitation of nuclei, involves three steps, as shown in Figure 1.4C: (1) initial time delay (relaxation delay, RD), or cycle delay, to provide nuclear relaxation; (2) the action of a radiofrequency pulse (RFP) to excite the target nuclei; and (3) the collection of NMR data as free induction decays (FIDs) during acquisition time AT. In order to obtain NMR spectra with signal-to-noise ratios sufficient for reliable interpretation of the data, these three steps can be repeated as much as necessary. These cycles dictate the total experimental time, which is particularly important for the analysis of low-concentration solutions or solids, where the lines are usually broad.

Finally, in the experiment shown in Figure 1.4C, the RF pulses can vary. One can apply non-selective pulses, generally short, to excite a very large frequency range and selective pulses, usually long, to excite a limited frequency diapason. In addition, the shape of the pulses can vary. The latter can have a significant effect on the resulting NMR spectra, particularly in advanced NMR experiments on solids.

1.3   FROM NUCLEAR RELAXATION TO SHAPES OF NMR SIGNALS

The shape of an NMR signal is mathematically well defined by an analysis of the behavior of the macroscopic magnetization in the homogeneous magnetic field accounting for the nuclear relaxation. This is a very important advantage of NMR because the line shape observed can strongly affect the interpretation. Figure 1.5A shows a line shape obtained experimentally, which raises several important questions: (1) Does it indicate spin–spin coupling or quadrupolar interaction? (2) Does it indicate the presence of two different structural units? (3) Does it indicate the presence of a gradient of the magnetic field? Only knowledge of the line shape can help provide an answer.

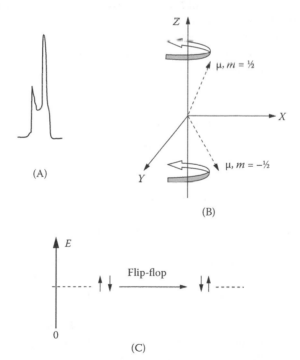

FIGURE 1.5   (A) Shape of a resonance observed experimentally (see the text). (B) Macroscopic magnetization M_Z^0 undergoing Larmor precession. (C) Mutual reorientations of two excited spins leading to spin–spin relaxation.

As noted above, the equilibrium macroscopic magnetization undergoes Larmor precession. Under this condition, the M_Z^0 value is maximal, whereas the M_X^0 and M_Y^0 components are equal to zero. Because the appearance of the macroscopic magnetization itself is directly related to the behavior of nuclear dipoles in the external magnetic field, the dipoles with $m_I = +1/2$ and $-1/2$ undergo Larmor precession (Figure 1.5B). The orientations of the nuclear dipoles with respect to the field B_0 are energetically different. Because the orientation corresponding to $m_I = +1/2$ is preferable, the dipoles are summed to give the magnetization vector along the axis OZ.

When the nuclear spins are irradiated, the M vector reorients. The M_Z^0 vector reduces to the M_Z component, while the M_X and M_Y components become non-zero. Then, the excited spins, capable of an energy exchange with their environments, relax to again reach the initial equilibrium state. Due to the presence of the longitudinal (OZ) and transverse (OX and OY) components in the macroscopic magnetization, their change after the RF irradiation shows two main relaxation mechanisms. The first mechanism, spin–lattice relaxation, corresponds to the energy exchange between nuclear spins and the lattice, which can be defined as a continuum of nuclear and/or electromagnetic moments of any sort, mutually interacting and reducing the absorbed energy. The second mechanism, spin–spin relaxation, corresponds to the situation when two or more nuclear spins experience the so-called flip-flop reorientations shown in Figure 1.5C. It should be emphasized that, in contrast to the spin–lattice relaxation mechanism, these mutual reorientations reduce the lifetimes of the excited spins but the total energy of the system does not change. Therefore, these mutual flips are referred to as *energy-conserving spin transitions*.

The macroscopic magnetization relaxes following the phenomenological equations of Bloch:[2]

$$dM_Z/dt = -\left(M_Z - M_Z^0\right)/T_1$$
$$dM_{X,Y}/dt = -M_{X,Y}/T_2 \qquad (1.15)$$

Here, the longitudinal (or transverse) component of the total nuclear magnetization recovers *exponentially* to an equilibrium magnitude with time constant T_1 (or T_2). These constants or relaxation times play an important role in the detection of NMR signals. In fact, infinitely long T_1 times, corresponding to an ineffective energy exchange between excited spins and the lattice, prevent the observation of NMR signals. In addition, infinitely short T_2 times due to intense spin flips lead to strong line broadenings, complicating the registration of NMR signals.

The shape of NMR signals observed and analyzed in liquids and solids is defined by the dependence of signal intensity on frequency, as a function of $f(\omega)$. According to the Abragam formalism, the shape of a NMR signal corresponding to the $f(\omega)$ dependence can take the form of a Gaussian curve:

$$f(\omega) \sim \left(\Delta v(2\pi)^{1/2}\right)^{-1} \exp\left(-\left(\omega - \omega_0\right)^2/2(\Delta v)^2\right) \qquad (1.16)$$

or a Lorentzian curve:

$$f(\omega) = (\Delta v / \pi)\left[1/\left((\Delta v)^2 + (\omega - \omega_0)^2\right)\right] \tag{1.17}$$

Both of them are shown schematically in Figure 1.6. When nuclear spins are not coupled, the line width of the NMR signal is determined at half-height and calculated through spin–spin (T_2) times:

$$\Delta v \text{ (Hz)} = 1/\pi T_2 \text{ (s)} \tag{1.18}$$

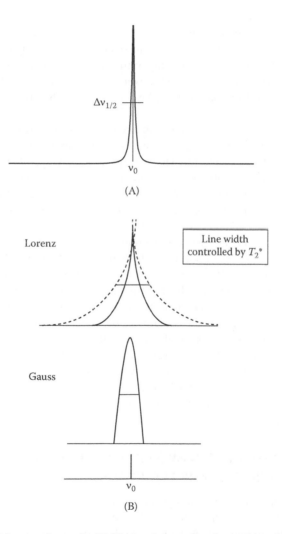

FIGURE 1.6 (A) Lorenz shape of a NMR signal generally observed in solutions. (B) NMR signal centered at frequency v_0 with the Gaussian shape and Lorentzian shape, as well as the influence of a non-homogeneous magnetic field on the line width (top).

or via Equation 1.19:

$$\Delta \nu \, (\text{Hz}) \sim 1/T_2 \, (\text{s}) \qquad\qquad (1.19)$$

for the Lorentzian and Gaussian shapes, respectively. It should be noted that in reality time T_2 can be long, and the homogeneity of the external magnetic field can be imperfect.

The line width in Figure 1.6B can then be determined by the T_2^* time:

$$1/T_2^* = 1/T_2 \, (\text{natural}) + 1/T_2 \, (\text{inhomogeneity}) \qquad\qquad (1.20)$$

Generally speaking, Lorentzian resonance shapes are expected in liquids and solutions, where molecular motions are very fast and high amplitude. It should be noted that such fast isotropic motions can occur in, for example, plastic crystals or solids having the so-called high-temperature phases. Thus, these solids can also show Lorentzian resonance shapes. However, in rigid solids, NMR resonances are usually approximated by Gaussian curves or by a combination of Gaussian and Lorentzian curves.

Because the NMR line shape is of primary importance for researchers from both a practical and a theoretical point of view, several important observations can be made. First, knowledge of the line shape helps to achieve good spectral resolution by shimming coil manipulations in experiments on solutions and solids. Poor shimming procedures can result in an incorrect shape (as shown in Figure 1.5A), which affects all of the observed resonances. Second, the wings of the Lorentzian curves are remarkably stretched compared to those of the Gaussian curves; in other words, two neighboring Lorentzian lines can overlap, while two Gaussian lines can remain resolved. This effect is very important in the context of integrating the signals. Third, the theory of line shapes allows calculation of the signals within the framework of a line-shape analysis to quantitatively characterize spin states experiencing exchanges on the NMR time scale. Such an analysis identifies the lifetimes and energies of the exchanges. The line-shape analysis can also be applied for calculations of dipolar and/or quadrupolar coupling, particularly important for solids.

The concept of the line shape, however, cannot be applied to the so-called power NMR spectra that are built by standard software procedures using the real and imaginary parts (see Equation 1.14). Such power spectra can be useful when the regular NMR spectra show poor signal-to-noise ratios or require non-trivial phase and baseline corrections.

1.4   REGISTRATION OF NMR SIGNALS: GENERAL PRINCIPLES AND NMR EQUIPMENT

Despite the great variety of techniques currently utilized for NMR research, their principal schemes do not differ remarkably and can be represented by the simple schematics shown in Figure 1.7A, which includes a radiofrequency reference generator (or several generators for multi-channel devices), a transmitter, an amplifier, a receiver, a pulse generator, a powerful computer, a magnet, and a NMR probe. The

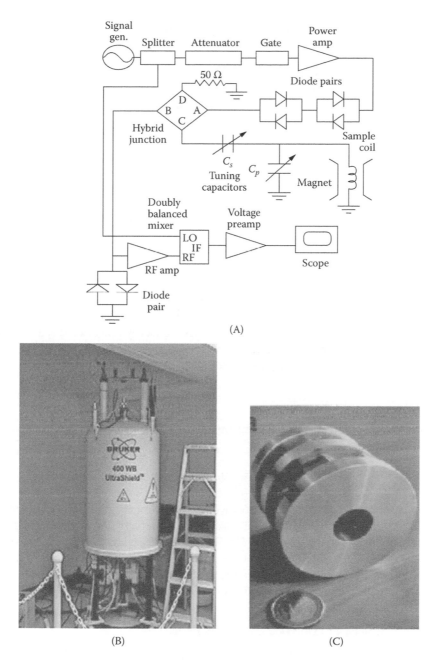

FIGURE 1.7 (A) Scheme of the simplest NMR spectrometer. (B) Regular superconducting magnet. (C) Small magnet used for the NMR-MOUSE® experiments.

macroscopic magnetization in a sample is detected by a RF coil in a NMR probe. Because this signal takes the form $A \exp(i(\omega_0 + \Delta\omega)t)$, where A is signal amplitude, ω_0 is carrier frequency, and $\Delta\omega$ is the desired frequency offset, maximal A values can be reached when the coil in the NMR probe is accurately tuned to carrier frequency ω_0. One should remember that inaccurate tuning can significantly affect the resulting NMR spectra. In fact, the magnetic susceptibility of samples very much depends on the nature of the solutions and solids. It should also be emphasized that, even for a well-tuned NMR probe, the quality of all of the units in Figure 1.7A strongly affects the signal-to-noise ratios in the resulting NMR spectra. When these units are ideal, the NMR sensitivity is dictated by the strength of external magnetic field.

Because the signal in the RF coil is very weak, it must be enhanced by an amplifier and filtered and mixed with an intermediate frequency. Separation of the signal components into real and imaginary parts is the next step. After this separation, the analog-to-digital converter is used to produce a digitized signal, which is processed by computer to provide the final NMR spectrum.

All of the above steps are common for any NMR procedure. Nevertheless, in practice, the requirements for solution and solid-state NMR spectrometers are different.[3] First, in contrast to solutions, spectral windows of 1 MHz or greater would be applied to characterize solids by NMR; therefore, a solid-state NMR spectrometer should be equipped with a 12-bit analog-to-digital converter capable of digitizing spectral windows of up to 5 MHz vs. the use of a 16-bit converter for solution NMR. The latter provides observation of a maximum spectral window of only 125 kHz. Again, because the solid-state NMR spectrum covers a very large frequency range, solid-state NMR spectrometers should be equipped with high-speed digitizers with much faster speeds. Second, the intensity of radiofrequency field B_1 can be defined as follows:

$$2\pi\nu_1 = \gamma B_1 \tag{1.21}$$

In turn, at the radiofrequency pulse corresponding to the 90° acting angle and duration t_p, radiofrequency field ν_1 can be expressed as

$$\nu_1 = 1/4t_p \, (90°) \tag{1.22}$$

Thus, the duration of a 90° pulse of 5 s will correspond only to the radiofrequency field of 50 kHz. This rule should be considered particularly when running NMR experiments in solids. In solid-state NMR, the spectral widths are generally very large, so the intense radiofrequency pulses operating in channels of the spectrometer should *uniformly* excite the corresponding spectral range. Only at this uniform excitation will the NMR spectra recorded not be distorted.

1.4.1 Magnets, Decouplers, and NMR Probes

The NMR spectra of many nuclei—for example, ^{13}C, ^{31}P, or ^{15}N attached to ^{1}H (or ^{19}F) nuclei—can be simplified by the use of decouplers, which are generally applied for both solutions and solids. In solution, a decoupler power of around a few watts is quite sufficient to completely eliminate spin–spin couplings. In solids, however, direct dipolar couplings X–^{1}H (or X–^{19}F) can be completely suppressed only at kilowatt decoupling powers.

A magnet is the source of the strong, stable, and homogeneous magnetic field required for initial nuclear polarization and thus for observation of high-resolution NMR spectra. The earliest NMR experiments were performed with permanent iron electromagnets that generated magnetic fields of 1 to 2 T, corresponding to a working ^1H frequency of ≤ 100 MHz. However, such iron magnets are relatively unstable and very sensitive to temperature. In contrast, superconducting electromagnets (Figure 1.7B) show excellent stability and good homogeneity. Currently, the magnetic fields created are capable of working at a proton frequency of up to 900 MHz, where the stability of the magnetic field necessary for high-resolution NMR experiments in solutions and solids can be improved by field locking. It should be noted, however, that solid-state NMR experiments generally require wide-bore magnets, which can provide various manipulations of samples.

The long-term trend in the development of NMR has been based on the idea that only higher magnetic fields can lead to better and powerful spectrometers because sensitivity and chemical shift dispersion, measured in hertz, increase proportional to field strength.[4,5] This trend implies that a sample must be placed inside the stationary magnet providing the strong magnetic field. However, new developments in the production of small, inexpensive NMR magnets have allowed the creation of downsized magnets (Figure 1.7C). Small NMR magnets with field strengths up to 2 T have been realized, opening the door to highly mobile NMR devices and creating a new domain within NMR studies: low-field NMR spectroscopy.

The NMR probe is a very important component of spectrometers. Generally, a sample is spinning in the NMR probe at a low rate (~20 Hz) (Figure 1.8A, bottom) in studies of solutions to decrease inhomogeneity of the magnetic field. To reduce (or eliminate completely) dipolar interactions in solids, the spinning rate should be much higher (up to 20 to 30 kHz). The design of NMR probes differs considerably between experiments on solutions and those on solids. In addition, it depends on the demands of the research; for example, double and triple resonance or gradient-field experiments are options with specially designed NMR probes.

Variable-temperature measurements in solutions and solids, generally between $-120°$ and $+120°C$, require accurate temperature settings for the heating and cooling equipment and the absence of remarkable temperature gradients through a sample. The corresponding procedures for temperature calibrations are considered in later chapters.

Solid-state experiments on static samples are usually performed with wide-line NMR probes, where a sample is placed in a glass NMR tube located perpendicularly to the external magnetic field. Such wide-line probes can be equipped with flat coils, which are generally needed for biomolecular investigations. As noted above, solid-state NMR probes are generally capable of mechanical rotation of samples at high spinning rates of tens of kHz. Suppressing the strong dipole–dipole interactions in solids at high spinning rates will be maximal when MAS NMR rotors are oriented at the magic angle relative to external magnetic field B_0 (Figure 1.8A). Standard commercial 7-, 4-, 2.5-, and 1.3-mm zirconium oxide rotors can be used for MAS experiments to routinely reach spinning rates of 7, 18, or 25 kHz. Higher spinning rates (up to 60 to 70 kHz) are also available in modern NMR spectrometers.

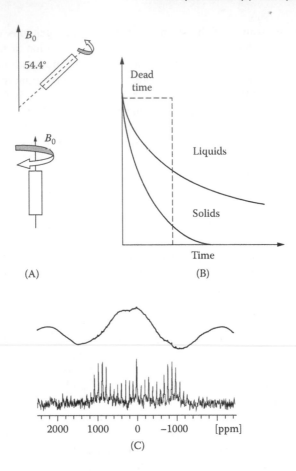

FIGURE 1.8 (A) Sample spinning in a NMR probe. (B) Dead time of a NMR spectrometer affecting the NMR signal intensity. (C) One pulse (top) and Hahn echo (bottom) ^{2}H MAS (5 kHz) NMR spectra recorded during a large sweep.

1.4.2 DEAD TIME IN NMR EXPERIMENTS: INFLUENCE ON NMR DATA

An important element in NMR equipment is dead time because a radiofrequency coil in NMR probes is used for irradiation of a sample and simultaneously for detection of NMR signals. According to the principles of Kisman and Armstrong,[6] the transmitter signal and the probe signal must be distinguished. Technically, a weak signal should be accurately measured immediately after the action of a high-power radiofrequency pulse. This leads to a switching problem and gives rise to the so-called dead time of a NMR spectrometer. Generally, this problem is minimized in modern NMR spectrometers, and their standard components reduce the dead times to only a few microseconds, comparable with radiofrequency pulse lengths.

Because relaxation times and FIDs are relatively long in liquids (Figure 1.8B), dead time does not play a role in solution NMR and the final intensities of signals are not reduced. In contrast, the FIDs in solids are much shorter and the NMR signal loses its intensity if the dead time is relatively long (Figure 1.8B). For obvious reasons, the signal can even become invisible; therefore, the shortest possible dead time is desirable in this situation. In turn, the shortest dead time can result in the problem of *ringing*. After Fourier transformation of the FIDs into NMR spectra, the baselines can be strongly distorted (Figure 1.8C) in the presence of ringing, which is not easy to eliminate in spite of the software currently available. To avoid such distortions, an adjusted dead-time parameter longer than the dwell-time parameter can be applied; however, even in these cases, the baseline distortion problem can still persist. This problem can be minimized by applications of the Hahn echo, spin echo, or solid echo pulse sequences and collection of NMR data immediately after the second registering radiofrequency pulse. It should be added that a procedure to cut first FID points, available in standard NMR software, can also be useful.

1.4.3 Spectral Resolution

One of the major tasks in NMR analysis is determination of a number of magnetically, chemically, or structurally non-equivalent nuclei in solutions and in solids. It is obvious that success in doing so strongly depends on the spectral resolution, which can be formulated as a minimal distinguished distance between two peaks in a NMR spectrum, measured in hertz. In practice, for example, in solution NMR the spectral resolution is characterized by the line widths of NMR signals, measured in hertz, in standard oxygen-free solutions (see Figure 1.6A). The spectral resolution is dictated by the homogeneity of the external magnetic field, which can be improved by standard manipulations with shimming coils (see Chapter 4). This procedure can routinely provide a spectral resolution of 0.1 to 0.2 Hz in solutions.

As noted above, the FIDs collected by NMR are converted to digitized forms by analog-to-digital converters and processed by computer. After Fourier transformation, the final resolution in NMR spectra is dependent on the size of the memory used for this treatment. This size (number of points) defines the digital resolution, which is particularly important in high-resolution NMR, where the digital resolution is obviously a function of sweep widths applied for the data collection and number of points used for the treatments. Figure 1.9 illustrates the situation where a doublet observed for a large memory size transforms to a broad resonance with poor digital resolution. Here it should be emphasized that even careful manipulations with shimming coils giving a good physical resolution will be limited by a small number of points. This number dictates the actual spectral resolution.

Because the nuclear relaxation times (T_2) are long in solutions and the expected resonance lines are generally sharp ($\Delta v = 1/\pi T_2$), the shimming coil manipulations are important for each sample. In addition, collection of NMR data requires a relatively large computer memory. In solids, however, resonance lines are generally broad due to strong dipolar internuclear interactions and short T_2 times. Therefore, the shimming coil manipulations are not critical in solid-state NMR, and a reasonable compromise between sweep width and computer memory size is possible. At the same time, the number of the points should be increased when the FIDs are treated with a Gaussian function to improve the resolution—for example, to resolve overlapped resonances.

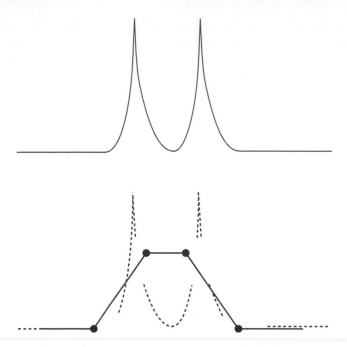

FIGURE 1.9 Schematic illustration of the effect of digital resolution on the observed line shape of the NMR signal, transforming from a doublet to a broad resonance.

Highly resolved NMR spectra can be obtained when the strong external magnetic field is maximally homogeneous and capable of maintaining the resonance conditions in each point through the volume of a sample. This is key to NMR studies of liquids and solids. However, for large heterogeneous samples, the resonance conditions often cannot be satisfied throughout an entire object. At the same time, the external magnetic field can be homogenized in a small volume and a high-resolution NMR signal can still be recorded from this small volume. This small volume can be spatially shifted to record the next signal, and repeating these experiments will allow observation of target nuclei distributed throughout the sample volume. This is the principal idea behind NMR imaging, which was realized in 1973.[7]

1.5 ENHANCEMENT OF SENSITIVITY IN NMR EXPERIMENTS

Due to the low sensitivity of NMR experiments, the characterization of complex molecular systems such as organic molecules in solutions or hybrid materials in solids, or even very low-concentration solutions, is not simple and sometimes impossible. This section considers general principles in pulsed NMR applied to increase the sensitivity in NMR experiments besides the well-known application of stronger initial nuclear polarization in the highest magnetic fields.

The sensitivity can be remarkably enhanced by cryogenic NMR probes. Here, the radiofrequency coil and the associated electronics are cooled to a low temperature—for example, to 25 K or even lower. As a result, the sensitivity can be increased by a factor of 12 to 16 to perform experiments on natural products containing rare nuclei such as ^{13}C, ^{17}O, ^{15}N, etc. Specially developed approaches, based on optical pumping,[8] *para*-hydrogen-induced polarization,[9] and dynamic nuclear polarization (DNP),[10] can increase the sensitivity even more. In gaseous phase studies (e.g., xenon), optical pumping with laser light enhances the initial nuclear polarization by five orders of magnitude. In solutions and solids, DNP plays a more important role.

The DNP effect is based on the presence of species containing unpaired electrons; for example, paramagnetic metal ions or organic groups can be added to samples. Such electron–nucleus systems are commonly described via the following Hamiltonian:

$$H = -\omega_e S_Z - \omega_I I_Z + H_{ee} + H_{en} + H_{nn} \tag{1.23}$$

where the first two terms are Zeeman interactions between spins of electrons and nuclear spins, respectively. The terms H_{ee} and H_{nn} describe electron–electron and nucleus–nucleus coupling, and the H_{en} term corresponds to the hyperfine electron–nucleus interaction. The latter plays a major role in the experiments with DNP generation.

Due to the H_{en} interaction, irradiation of the samples at the electron Larmor frequency ω_e (or near) causes simultaneous electron–nucleus spin flips, which enhance the absolute magnitude of the nuclear polarization via three possible mechanisms: nuclear Overhauser effect, solid effect, or thermal mixing effect. The enhancements expected for the nuclear polarization will depend on the number of unpaired electrons, the line widths in electron spin resonance (ESR), the nuclear relaxation rates in the absence of electrons, and the amplitude of the exciting microwave field applied for the sample. In solids, for example, the enhancement factors can reach 660 and 2600 for ^1H and ^{13}C nuclei, respectively.

Technically, DNP NMR experiments can be performed with specially designed NMR spectrometers and probes. The NMR probes should be equipped with a coil, double-tuned to ^{13}C and ^1H frequencies, for example, and a horn antenna to transmit the microwaves. Generally, the samples hyperpolarized in the solid state at a very low temperature (up to 1.4 K, for example) are rapidly dissolved, heated to a high temperature under pressure, and then mixed with a solution of another reagent in the NMR spectrometer.[11] It has been shown that in solutions signal-to-noise ratios observed by DNP can be increased by a factor of $\gg 10^3$. The disadvantage of this method is obvious in that the samples should be initially treated with a paramagnetic compound distributed homogeneously throughout the sample volume to provide the DNP effect, but this is not always convenient in the chemical sense.

1.6 TWO-DIMENSIONAL AND MULTI-QUANTUM NMR EXPERIMENTS: GENERAL ASPECTS

As shown earlier in Figure 1.4C, the simplest one-dimensional NMR experiment is based on a single-pulse sequence where after the RF pulse the FID is recorded with the acquisition time representing a single time for spin evolution. Then, Fourier transformation gives a regular one-dimensional NMR pattern for intensity vs. frequency.

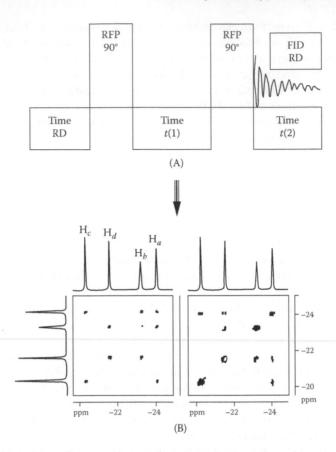

FIGURE 1.10 (A) Pulse sequence and time sections in the simplest 2D NMR experiment, COSY. (B) A high-field region of ^{1}H COSY (left) and NOESY (right) NMR spectrum of the complex $[Ir_2(\mu-H_a)(\mu-Pz)_2(H_b)(H_c)(H_d)(NCCH_3)(PPr_3^i)_2]$ in C_6D_6, where the pairs of nuclei H_a/H_c, H_a/H_b, and H_b/H_d are spin coupled, while the pairs H_d/H_c and H_b/H_d are spatially approximated.

Two-dimensional (2D) NMR adds a second frequency axis. The simplest 2D NMR experiment, correlation spectroscopy (COSY), is shown in Figure 1.10A. Here, due to the action of the first 90° pulse, a nuclear spin system develops during the evolution time, noted as t_1, due to the presence of nucleus–nucleus spin–spin coupling. Thus, the first set of time-dependent data, noted as $f(t_2)$, can be collected after the action of the second radiofrequency pulse during acquisition time t_2. Because the $f(t_1)$ datasets can be recorded for various values of t_1, double Fourier transformation performed for time domains t_1 and t_2 gives two frequency axes. The resulting NMR spectrum can be obtained as a square plot with the diagonal representing the one-dimensional NMR spectrum, while cross peaks appear only in the presence of scalar spin–spin coupling (Figure 1.10B). Similarly, two-dimensional ^{1}H nuclear Overhauser effect spectroscopy (NOESY; see Chapter 5) shows the nuclei that are spatially approximated. When the time evolution in the pulse sequence corresponds to heteronuclear

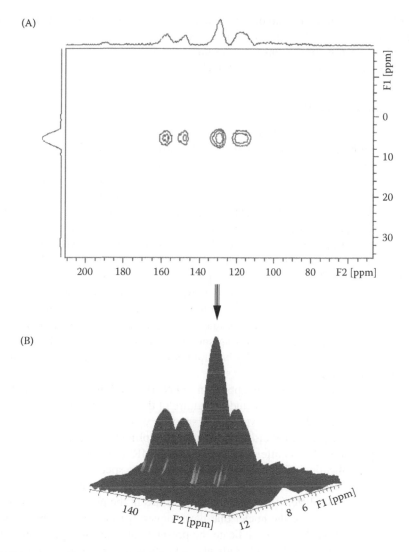

FIGURE 1.11 (A) Correlation ^{13}C–^{1}H solid-state MAS (10 kHz) NMR spectrum for a polymer system containing aromatic rings and carbonyl groups. (B) The same spectrum converted to the pattern with the new intensity axis.

spin–spin coupling, then the 2D NMR spectra show cross peaks (i.e., heteronuclear chemical shift correlations). Figure 1.11A shows an example of the 2D NMR ^{13}C–^{1}H correlation NMR spectrum obtained in a solid, where only the cross peaks are shown in the coordinates F2 vs. F1, corresponding to the ^{13}C and ^{1}H frequencies, respectively. It should be emphasized that standard NMR spectrometer software can convert the spectrum shown at the top of Figure 1.11 into the spectrum shown at the bottom of Figure 1.11, where a third coordinate, intensity, has been added. The NMR spectra show the presence of the distinguished resonances observed for aromatic rings in the

^{13}C NMR. At the same time, all of the signals correlate with a single broad resonance detected in the ^{1}H NMR. This is normal because, due to strong proton–proton dipolar coupling, protons of the rings remain unresolved. Practical applications and the advantages of 2D NMR are considered in Chapter 5.

1.6.1 Artifacts in 2D NMR

Unfortunately, 2D NMR spectroscopy in solutions and solids can suffer from artifacts that often appear as additional (unexpected) artificial peaks in the spectra. There are three common types of artifacts: artifact signals coming from the NMR spectrometer, artifact phenomena caused by the nature of nuclei in a sample, and artifacts due to the choice of incorrect acquisition parameters. The first type can be illustrated by 2D [207]Pb MAS NMR experiments in solids. As will be shown in later chapters, the solid-state MAS NMR spectra of heavy nuclei ([207]Pb) generally show signals that are accompanied by wide-sideband manifolds. All of the signals can be overlapped, and this overlapping complicates interpretation of the spectra. Two-dimensional [207]Pb isotropic/anisotropic correlation NMR spectroscopy allows for their separation. As seen in Figure 1.12A, the corresponding one-dimensional projection shows an isotropic spectrum without sidebands;[12] however, long spin–spin relaxation times, inhomogeneity of the magnetic field, and the bandwidth of the RF pulses produce artifacts in the resulting spectra (top) parallel to the F1 dimension. These artifact signals can potentially be suppressed by the application of composite 180° pulses (bottom).

The second type of artifact can appear in 2D NMR as additional lines due to the specifics of a particular procedure and the nature of the object. Such additional lines are often observed in 2D J-resolved NMR spectra, where artifact signals are produced by effects of the second order in the scalar coupling that are difficult to avoid. The third type of artifact is often observed in 2D NMR spectroscopy performed in solutions by application of double quantum correlation spectroscopy (DQCOSY) pulse sequences. The intense artifacts appear when the spin–lattice relaxation times differ remarkably for different nuclei, and delay times in the pulse sequences do not provide full relaxation for the nuclei that have longer T_1 times. Under these conditions, the final 2D NMR spectra indicate the presence of additional peaks as multiple-quantum diagonals. Only field gradients can help to avoid this artifact problem.

1.6.2 Multi-Quantum NMR

Multi-quantum (MQ) NMR is widely used in solutions and solids, and its practical applications are considered in greater detail in later chapters. Briefly, MQ NMR focuses on spin transitions in systems containing magnetic spin states with coherence orders different from ±1. Such transitions are forbidden but can be stimulated by RF pulses and generated from longitudinal magnetization. This generation is possible when nuclei are quadrupolar or coupled by indirect or direct dipolar interaction. MQ NMR spectroscopy is performed with complex pulse sequences (see Figure 1.12B) and includes the following steps: MQ preparation, MQ evolution, MQ reconversion, Z-filtering, and detection. For example, to generate intramolecular MQ coherences

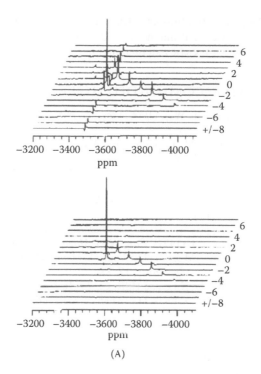

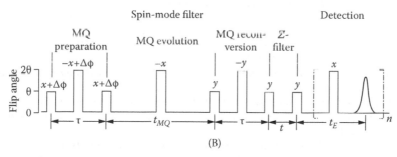

FIGURE 1.12 (A) The 2D PASS [207]Pb NMR spectra at a spinning rate of 8 kHz in a mixture of Pb(NO$_3$)$_2$ and PbSO$_4$. Composite pulses are applied to suppress artifacts. (From Vogt, F. et al., *J. Magn Reson.*, 143, 153, 2000. With permission.) (B) 2D MQ NMR pulse sequence where multi-quantum build-up and decays are obtained with variation in the preparation and reconversion time (τ). (From Blumich, B. et al., *Prog. Nucl. Magn. Reson. Spectrosc.*, 52, 197, 2008. With permission.)

for coupled protons in solutions, at least two 90° pulses are necessary. After the action of the pulses, the MQ coherences evolve and can then be converted back to longitudinal magnetization with the inverse MQ excitation sequence. Then, the next 90° pulse converts the longitudinal magnetization, modulated by the MQ evolution, back in order to observe the signal corresponding to the single-quantum transition. It should be emphasized that the Z-filter is used in combination with appropriate cycling of the RF phases to eliminate stray signals due to imperfect RF pulses and inhomogeneity of the external magnetic field. Finally, details of this technique and the modern theory of MQ NMR can be found in the recommended literature.

REFERENCES AND RECOMMENDED LITERATURE

1. Ernst, R.R. and Anderson, W.A. (1966). *Rev. Sci. Instr.*, 37: 93.
2. Bloch, F. (1946). *Phys. Rev.*, 70: 460.
3. Bryce, D.L., Bernard, G.M., et al. (2001). *J. Anal. Sci. Spectrosc.*, 46: 46.
4. Blumich, B., Casanova, F., and Appelt, S. (2009). *Chem. Phys. Lett.*, 477: 231.
5. Blumich, B., Perlo, J., and Casanova, F. (2008). *Prog. Nucl. Magn. Reson. Spectrosc.*, 52: 197.
6. Kisman, K.E. and Armstrong, R.L. (1974). *Rev. Sci. Instr.*, 45: 1159.
7. Lauterbur, P.C. (1973). *Nature*, 242: 190.
8. Walker, T.G. and Happer, W. (1997). *Rev. Modern Phys.*, 69: 629.
9. Bargon, J. and Natterer, J. (1977). *Prog. Nucl. Magn. Reson. Spectrosc.*, 31: 293.
10. Wind, R.A., Duijvestun, M.J., Van Der Lugt, C., Manenschijn, A., and Vriend, J. (1985). *Prog. Nucl. Magn. Reson. Spectrosc.*, 17: 33.
11. Lee, Y., Zeng, H., Wooly, K.L., and Hilty, C. (2013). *J. Am. Chem. Soc.*, 135: 4636.
12. Vogt, F.G., Gibson, J.M., Aurentz, D.J., Mueller, K.T., and Benesi, A.J. (2000). *J. Magn. Reson.*, 143: 153.

RECOMMENDED LITERATURE

Abragam, A. (1985). *Principles of Nuclear Magnetism*. Oxford: Clarendon Press.
Bakhmutov, V.I. (2009). Magnetic resonance spectrometry. In: *Encyclopedia of Applied Spectroscopy* (Andrew, D.L., Ed.), pp. 933–963. Weinheim: Wiley-VCH.
Blumich B., Perlo, J., and Casanova F. (2008). Mobile single-sided NMR. *Prog. Nucl. Magn., Reson. Spectrosc.*, 52: 197–269.
Duer, M.J., Ed. (2002). *Solid-State NMR Spectroscopy: Principles and Applications*. Oxford: Blackwell Sciences.
Farrar, C.T. (1987). *An Introduction to Pulse NMR Spectroscopy*. Chicago, IL: Farragut Press.
Friebolin, H. (1991). *Basic One- and Two-Dimensional NMR Spectroscopy*. Weinheim: Wiley-VCH.
Harris, R.K. (1983). *Nuclear Magnetic Resonance Spectroscopy*. Avon: Bath Press.
Lambert, J.B. and Riddel, F.G., Eds. (1982). *The Multinuclear Approach to NMR Spectroscopy*. Boston: Springer.
Srivastava D., SubbaRao, R.V., and Ramachandran, R. (2013). Understanding multi-quantum NMR through secular approximation. *Phys. Chem. Chem. Phys.*, 15: 6699–6713.

2 Chemical Shifts and Nuclear Coupling

Theory and Practical Consequences

Commonly, nuclear spins in a sample placed into external magnetic field B_0 for observation of NMR are not isolated. They are capable of interacting with their environments. These interactions can be with the same or different types of nuclei and with electrons, which can be paired or unpaired. Interactions with the closest environments through chemical bonds or through space give rise to the phenomena of chemical shift, spin–spin coupling (homonuclear and heteronuclear), quadrupolar coupling, and dipolar coupling. Independently of the objects and their nature—a solution or a solid (crystalline or amorphous, homogeneous or heterogeneous)— chemical shifts and internuclear couplings are always a focus of researchers as they are the main factors affecting NMR spectra, along with the number of resonances and their line shapes. These parameters should be decoded in structural terms by application of the relevant relationships: spectrum/chemical structure, chemical shift/electron distribution, splitting/neighboring spins, etc., which can be rationalized on the basis of the theoretical considerations presented in this chapter.

2.1 PHENOMENOLOGY OF CHEMICAL SHIFT

In practice, chemical shifts δ (generally implying isotropic chemical shifts) can be defined via dispositions of resonance lines in NMR spectra. For example, carbons in the $CH_3CH_2CH_3$ molecule show two $^{13}C\{^1H\}$ resonances displaced on the frequency scale belonging to group CH_2 and two equivalent groups CH_3. The physical reason for the displacement is a local magnetic field acting on the observed nucleus. When electrons surrounding the nuclei circulate in external magnetic field B_0, they create a new, additional magnetic field that shields the external magnetic field. Thus, this local magnetic field, B_{loc}, acting on the nuclei, shifts the Larmor resonance frequency (ν_0) corresponding to the isolated nuclei to give the so-called isotropic magnetic shielding constant, $\sigma(iso)$, or the isotropic chemical shift, $\delta(iso)$, defined by Equation 2.1:

$$B_{loc} = (1 - \sigma(iso))B_0$$
$$\delta(iso) = -\sigma(iso)$$

(2.1)

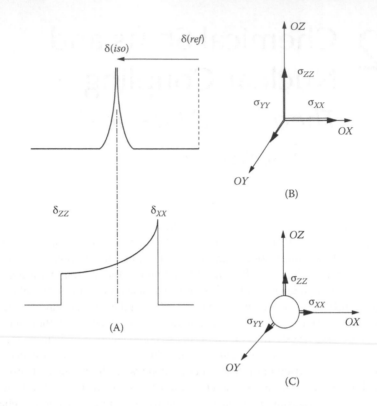

FIGURE 2.1   (A) The line shape of a NMR resonance typical of liquids shows the isotropic chemical shift (top) and the line shape typical of static solids at an axially symmetrical screening tensor, $\sigma_{ZZ} \neq \sigma_{XX} = \sigma_{YY}$ (bottom). (B) Components of the tensor in the absence of symmetry. (C) A spherically symmetrical σ tensor.

The $\delta(iso)$ shift is shown in Figure 2.1A. To avoid the expected dependence of chemical shift values δ, measured experimentally and expressed in frequency units on the strength of applied external magnetic field B_0, the δ values can be expressed in parts per million (ppm) and referred to the internal or external (rather typical of solids) standards via the following relationship:

$$\delta = \left[\nu(sample) - \nu(reference)\right] \times 10^6 / \nu(reference) \tag{2.2}$$

Different compounds can be used as standards—for example, tetramethylsilane (TMS) for ^1H and ^{13}C nuclei, a water solution of H_3PO_4 for ^{31}P nuclei, or a solution of CF_3COOH for ^{19}F nuclei. It should be emphasized that, due to the remarkable influence of magnetic susceptibility, which is different in different samples, an internal reference is obviously preferable. A standard can be directly added to a sample, or any resonance with a well-known chemical shift referred to a standard can be used. One should remember that accuracy in the determinations of chemical shifts reported in the literature, obtained even at very good spectral resolution, depends on digital resolution expressed in ppm or hertz as a distance between two points (see Figure 1.9, Chapter 1).

TABLE 2.1

Components of Screening Tensors and the $\Delta\sigma$ Values (in ppm) for Some Nuclei Measured in Solid-State NMR Spectra

Compound	σ_{ZZ}	σ_{YY}	σ_{XX}	$\Delta\sigma$
$^{14}NH_3$	237.3	278.0	278.0	−44.3
$Me^{14}NC$	370	10	10	360
$HC^{14}N$	348	−215	−215	563
$C_6^{19}F_6$	465.9	310.8	310.8	155.1
$Xe^{19}F_4$	528.5	58.5	58.5	470

In reality, screening constants σ or chemicals shifts δ have three-dimensional magnitudes. They can be expressed via tensors by components σ_{XX}, σ_{YY}, and σ_{ZZ} in the laboratory coordinate system as shown in Figure 2.1B and C. Similarly to the isotropic chemical shifts, these tensor components can be associated with molecular structures.[1] Generally speaking, the laboratory coordinate system does not correspond to the molecular coordinate system; however, in many cases, the Z coordinate is situated along chemical bonds.

The tensor components directly affect resonance shapes via molecular motions. When all three σ (or δ) components differ (i.e., $\sigma_{XX} \neq \sigma_{YY} \neq \sigma_{ZZ}$) and the high-amplitude molecular reorientations averaging these components are absent (e.g., in rigid solids), then the shape of the NMR resonance will be complex. These shapes are represented in later chapters. Figure 2.1A shows a shape corresponding to an axial-symmetrical tensor (i.e., $\sigma_{XX} = \sigma_{YY} \neq \sigma_{ZZ}$). Here, the σ (or δ) components in the laboratory coordinate system are easily determined from a static NMR pattern recorded in rigid solids, as shown in Figure 2.1A; however, in reality, computer simulation procedures are needed for their accurate determinations. Some experimentally determined components of screening tensors for different nuclei are listed in Table 2.1. The unequal σ (or δ) components give rise to chemical shift anisotropy ($\Delta\sigma$), defined in Equation 2.3; the $\Delta\sigma$ values are also shown in Table 2.1.

$$\Delta\sigma = \left[2\sigma_{ZZ} - \left(\sigma_{XX} + \sigma_{YY}\right)\right]/3 \qquad (2.3)$$

In liquids, due to fast intense reorientations, the $\Delta\sigma$ values cannot be observable in NMR spectra; however, they can play an important role in nuclear relaxation (see Chapter 3). Due to the high-amplitude molecular motions in liquids, the σ_{XX}, σ_{YY}, and σ_{ZZ} components are completely averaged to show the Lorenz-shaped resonance (see Figure 2.1A), which is characterized by a $\delta(iso)$ or $\sigma(iso)$ value. The isotropic chemical shifts or the isotropic screening constants are defined via Equation 2.4:

$$\sigma(iso) = 1/3\left(\sigma_{XX} + \sigma_{YY} + \sigma_{ZZ}\right) \qquad (2.4)$$

In other words, the Lorenz shape can be observed, even in rigid solids, when the tensor is spherically symmetrical at $\sigma_{XX} = \sigma_{YY} = \sigma_{ZZ}$ (Figure 2.1C).

2.1.1   Chemical Shift in Diamagnetic Molecular Systems

As generally accepted for diamagnetic samples, where unpaired electrons are absent
the total shielding constant σ for any nucleus can be expressed via the sum of three
main contributors: diamagnetic term $\sigma^{local}(dia)$, paramagnetic term $\sigma^{local}(para)$, and
magnetic anisotropy term σ^*. The first term originates primarily from the circulation
of unperturbed "spherical" electrons (s-electrons) located close to the nuclei. In the
simplest form, this term is well expressed by the Lamb equation:

$$\sigma^{local}(dia) = \left(\mu_0 e^2 / 8\pi m_e\right) \int r\rho_e \, dr \qquad (2.5)$$

where μ_0 is the permeability constant, ρ_e is the charge density, r is the electron–
nucleus distance, and m_e is the mass of the electron. Thus, because the diamagnetic
term, defined via integration over the total s-electron density, shields the external
magnetic field, the resonance frequency should be increased to be observable. In
other words, a larger electronic density produces a larger diamagnetic screening constant and *vice versa*.

   In addition to the local magnetic field, the high-field shift can be caused by electrons in neighboring atoms. This diamagnetic term, or the so-called effect of heavy
atoms, can be expressed simply via Equation 2.6:

$$\sigma^{local}(dia) = 1/4\pi \left(\mu_0 e^2 / m_e\right) \sum Z/r^3 \qquad (2.6)$$

where Z is the atomic number and r is the corresponding interatomic distance. This
heavy atom effect can be particularly large, for example, for ^{13}C nuclei neighboring
such atoms as Br or I. In fact, the compounds CBr_4 and CI_4 show ^{13}C NMR resonances in a very high field at –28.5 and –292.2 ppm (relative to TMS), respectively.

   Another factor is responsible for the highest field ^1H resonances observed in
solutions of transition metal hydrides; for example, the iridium hydride complex
$[HIrCl_2(PMe_3)_2]$ shows the HIr signal at –50 ppm (relative to TMS). The large diamagnetic shifts in such molecules are produced by paramagnetic ring currents within
the incomplete valence d shell of the transition metal atoms and lead to diamagnetic
current in the hydrogen position.[2] This is significantly enhanced by spin–orbital
effects, especially important for 4d- and 5d-electrons.[3] In addition, the hydride protons show extremely large chemical shift anisotropy[4] that is not typical of ^1H nuclei.

   The paramagnetic term is due to the perturbation of non-spherical electrons (i.e.,
p-, d-, and f-electrons).[5] This term can be written in its simplest form as follows:

$$\sigma^{local}(para) = -\left(\mu_0 e^2 \hbar^2 / 6\pi m^2 \, \Delta U\right) < r^{-3} > P_U \qquad (2.7)$$

which includes the average energy of electron excitation ΔU. Function P_U in the equation is the p-electron imbalance, which depends on charge densities and bond orders.
Alternatively, the paramagnetic term can be described in terms of Equation 2.8:

$$\sigma^{local}(para) \sim -(1/\Delta E)(1/r^3)\sum Q \qquad (2.8)$$

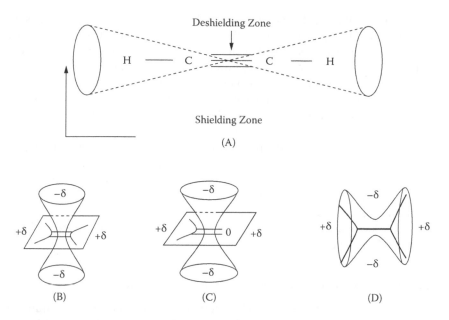

FIGURE 2.2   (A) Shielding and deshielding zones in an acetylene molecule due to the electrons in the triple bond. (B,C,D) The zones with an additional deshielding (+δ, shift toward a lower field) and additional shielding (−δ, shift toward a higher field) expected for the nuclei neighboring with C=C, C=O, and C–C bonds.

which is generally used for analysis of ^{13}C chemical shifts; Q is the bond order parameter. It is obvious that because the excitation energy of s-electrons is too high they do not provide paramagnetic contributions to the chemical shifts of ^{1}H nuclei. In contrast, heavy nuclei (e.g., ^{13}C, ^{19}F, ^{117,119}Sn, ^{199}Hg) can demonstrate very large paramagnetic contributions due to the presence of easily perturbed p-, d-, and f-electrons. Finally, it is pertinent to emphasize that the paramagnetic contribution considered here differs completely from the paramagnetic shift caused by unpaired electrons.

The third term, σ*, which is relatively small on the scale of chemical shifts, is due to electrons of chemical bonds situated at anisotropic neighboring groups (C–C, C=C, C≡C, or aromatic rings) typical of organic molecular systems. Again, the circulation of electrons in such groups creates magnetically anisotropic fields giving rise to the σ* term. Therefore, the chemical shifts observed for nuclei located close to the anisotropic groups will depend on their mutual orientations. Figure 2.2A demonstrates the effect of the field appearing in an acetylene molecule due to the electrons in the triple bond. As can be seen, this field results in shielding and deshielding zones. Because the H–C≡C–H protons are situated in the shielding zone, they show resonance that is high-field shifted to 2 to 3 ppm, in spite of the well-known larger acidic character of the acetylene protons. The magnetically anisotropic effects of the other groups are represented in Figure 2.2B, C, and D, where single C–C bonds cause the smallest effects. Nevertheless, they are large enough to observe different ^{1}H resonances in different conformations of cyclic organic molecules.

Combinations of these three terms explain the large variations in chemical shifts (δ) that are remarkably dependent on the nature of the nuclei. Chemical shifts of ^{1}H nuclei are generally caused by diamagnetic and small magnetic anisotropy contributions. They give the relatively narrow chemical shift range from +20 to –50 ppm. In spite of this, the $\delta(^1\text{H})$ diapason is quite sufficient for a structural analysis of high-resolution NMR spectra in solutions. In contrast, routine applications of ^{1}H solid-state NMR are not simple, as strong dipolar proton–proton interactions reduce spectral resolution, which can be improved only by specific methods.

The chemical shift ranges for other nuclei due to the presence of significant paramagnetic terms are much larger: from –250 to +400 ppm for ^{19}F nuclei, from –125 to +500 ppm for ^{31}P nuclei, from –500 to +850 ppm for ^{15}N nuclei, from +90 to –110 ppm for ^{11}B nuclei, and from +250 to –30 ppm for ^{27}Al nuclei. The δ ranges are particularly large for ^{17}O and ^{195}Pt nuclei, changing from +1120 to –59 ppm or from –1370 to +12000 ppm, respectively. One might assume that such nuclei are preferable in the analytical context.

2.1.2 RELATIONSHIP CHEMICAL SHIFT AND ATOMIC CHARGE

The chemical shifts in NMR spectra in solutions and in solids are key parameters applied for structural analysis, where the relationship between chemical shifts and electronic distributions in molecules plays a central role. As follows from the theory presented in Equations 2.5 to 2.8, both of the main terms depend on charge density, directly via the ρ_e parameter in the diamagnetic contribution and indirectly via the electron–nucleus distance (r) in the paramagnetic term. The latter is valid because this distance decreases upon increasing the positive atomic charge.

Thus, chemical shifts and charges on atoms (or electronic density) can be correlated (often linearly) if the other parameters in the equations do not change or change insignificantly in a set of similar compounds. For example, the average energy of electron excitation ΔE and bond order Q can remain similar in such compounds. This relationship exemplifies the well-known principle often applied by researchers for an analysis of NMR spectra and signal assignments: The low-field displacement of a NMR signal corresponds to decreasing electronic density on an atom.

Generally speaking, the linear correlations $\Delta\delta(^1\text{H})$ vs. $\Delta\rho$ are well established for aromatic compounds, where the $\Delta\rho$ charges are obtained by quantum chemical calculations. In the same context, proton chemical shifts expectedly correlate with the Hammett/Taft constants of groups in benzenes, 2-substituted pyridines, and pyrazines.[6] The correlations for other molecular systems can be found in the recommended literature. One should remember that such correlations show significant deviations upon increasing contributions due to the magnetic anisotropy term σ^*.

The linear correlations of $\delta(^{13}\text{C})$ vs. atomic charge have also been established for various classes of organic compounds. Figure 2.3B illustrates one of them obtained for solutions of polyfluorinated olefins (I) and allyl cations (I–IV) shown in Figure 2.3A.[7] As can be seen, 21 $\delta(^{13}\text{C})$ values measured for sp^2-hybridized carbon atoms linearly correlate with the total MNDO atomic charges. Only two of the $\delta(^{13}\text{C})$ values obtained for *cis*- and *trans*-isomers of cation III (the filled circles) strongly deviate.

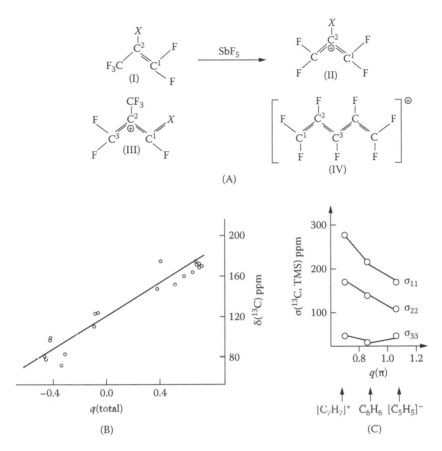

FIGURE 2.3 (A) Polyfluorinated olefins and allylic cations. (B) The dependence of the ^{13}C NMR chemical shifts of sp^2-hybridized carbons of the cations. (C) Shielding tensor principal components for tropylium cation, benzene, and cyclopentadienide anion represented as a function of charge q_π.

It is interesting that the addition of diamagnetic term corrections via Equation 2.6 does not improve the linearity, probably due to the effect of bond order parameter Q in Equation 2.8.

Because the atomic charge values depend on the method of their quantum chemical calculations, these correlations themselves cannot be used for quantitative estimations of atomic charges. However, the tendencies observed can help to assign the signals in ^1H and ^{13}C NMR spectra. Moreover, they open the way to computer analysis of ^{13}C NMR spectra, where assignments of ^{13}C signals are performed on the basis of these tendencies. In the case of heavy nuclei other than ^1H and ^{13}C, the situation is complicated and becomes unpredictable.

The $\delta(^{13}C)$ dependence on π charges (q_π) in aromatic compounds implies changes in the isotropic chemical shifts measured in solutions. However, it is interesting to see how the charge q_π affects three-dimensional ^{13}C shifts, which can be measured experimentally in solids. Figure 2.3C shows the tensor components of chemical shifts obtained in the compounds $C_7H_7^+$, C_6H_6, and $C_5H_5^-$ as a function of charges q_π.[8] Only the two components σ_{11} and σ_{22} correlate with the non-spherically symmetric charges q_π, while the perpendicular component, σ_{33}, does not show any monotonic change. It has been found that the σ_{11} and σ_{22} components are very sensitive to the energies of transition $\sigma - \pi^*$ and $\pi - \sigma^*$. Therefore, their correlation with π charges of the molecules is not surprising. Moreover, the correlations reveal the strong influence of the π charge, mainly on the paramagnetic term.

2.1.3   PREDICTING CHEMICAL SHIFT VALUES

A structural analysis based on one-dimensional NMR (in solutions or in solids) obviously requires the detailed assignment of signals in the NMR spectra, the first and most important step in such an analysis. Generally, the chemical shifts for signals in the NMR spectra of a new compound can be predicted. There are three approaches to prediction of δ values and to the assignment of NMR signals based on (1) database searches, (2) quantum chemical calculations, and (3) calculations with applications of empirical schemes.

A database search based on a structural analogy between known and unknown compounds is probably the simplest and most popular approach in practice because the search utilizes the abundant experimental data establishing spectrum/structure relationships and can demonstrate, for example, the δ values as a function of atomic coordination numbers or atomic oxidative states, hydrogen bonding, complexation, etc. Generally, this approach can be very successful even in the case of heavy nuclei.

The quantum mechanical theory of NMR and the quantum chemical calculations can be used to calculate chemical shift values for known or proposed structures. In turn, these calculations provide rationalization of the experimental NMR spectra. Many of the calculations are very successful or show good linear correlations between the theoretical and experimental values, thus verifying the prediction of chemical shift values. Such calculations can be found in the recommended literature. However, the results of the calculations depend on the basis used. For example, fully relativistic (FR) density functional theory (DFT) calculations better predict the $\delta(^1H)$ values for hydride ligands in transition metal complexes compared to non-relativistic (NR) and scalar relativistic (SR) calculations. According to the calculations, in the rhodium complex $[HRhCl_3(PMe_3)_2]$, the $\delta(^1H)^{NR}$, $\delta(^1H)^{SR}$, and $\delta(^1H)^{FR}$ values are -11.7, -12.6, and -25.6 ppm vs. -31.1 ppm measured experimentally.[3]

In addition, such calculations are not simple; they are extensive and require a deep understanding and much experience. Therefore, in the analytical context, they are not applicable. There are simpler approaches; for example, the semi-empirical MNDO methodology can be applied to calculate the chemical shifts of ^1H, ^{13}C, ^{15}N, ^{17}O, and ^{19}F nuclei. These approaches are particularly successful at parameterization (i.e., at the addition of NMR-specific parameters).[9]

Again, though, in the case of the most popular ^1H and ^{13}C NMR spectra, the prediction of ^1H and ^{13}C chemical shifts on the basis of empirical increment schemes is simpler and more accurate. The complex effect of a neighboring substituent can be replaced for an increment corresponding to this substituent which can be added to the chemical shift of a standard compound in a common equation:

$$\delta(calc) = \delta(st) + \sum Z_1 + \sum Z_2 + \cdots \qquad (2.9)$$

where Z_1, Z_2, etc. represent the first, second, etc. environments. A sum of these increments reproduces well the target chemical shift because the increments are obtained as a product of multiple linear-regression analyses applied for a large number of experimental data. It should be emphasized that the increments are dependent on the structure and class of the compounds; therefore, they can be applied separately for alkanes, alkenes, and aromatic systems, including heterocyclic or condensed systems. Finally, different computer programs can be applied for these calculations.

2.1.4 ISOTROPIC CHEMICAL SHIFT

In addition to the diamagnetic, paramagnetic, and magnetic anisotropy terms contributing to the total chemical shifts, isotropic chemical shifts are also caused by the presence of different isotopes in a molecule. The ^{1}H NMR spectrum of chloroform ($CHCl_3$) best illustrates the isotropic effect on the ^{1}H chemical shift. As seen in Figure 2.4A, the most intense line in the ^{1}H NMR spectrum belongs to ^{1}H nuclei neighboring with the non-magnetically active isotope ^{12}C, which has a natural abundance of 98.9%. In addition to this signal, a less intense doublet appears for ^{1}H neighboring with the isotope ^{13}C, which has a natural abundance of 1.1%. This doublet is explained by the spin–spin coupling ^{1}H–^{13}C; however, the center of the doublet, corresponding to its chemical shift, is detected at a higher field than the intense resonance. Thus, the $\delta(^1$H) difference observed can be explained by the presence of the pairs ^{1}H–^{12}C and ^{1}H–^{13}C in $CHCl_3$. Similarly, H_2 and HD molecules show a singlet and a triplet in the solution NMR spectra at their relative high-field displacement of 0.04 ppm (Figure 2.4B). Theoretically, the isotropic chemical shifts do not originate from electrons in nuclear environments. They are connected with intramolecular dynamics, including the vibration and rotation terms, where the behavior of isotopes is obviously different.

As generally accepted, the above effects (which vary for different nuclei) can be observed by primary and secondary isotope shifts. Primary isotope shifts appear and can be measured in NMR spectra in solutions for different isotopes in the same molecular system. Due to the primary isotope effect, chemical shifts of protons and deuterons in the same molecule will not be identical and show a difference of 0.1 ppm or less.

Secondary isotope shifts are visible in NMR experiments on nuclei with environments containing different isotopes. The typical secondary isotope chemical shifts are listed in Figure 2.4C. As can be seen, the effects depend on the nature of the nuclei. The effects are relatively small for ^{13}C, ^{15}N, and ^{31}P nuclei and moderate or large for heavy nuclei; for example, ^{119}Sn nuclei show the secondary isotope effects reaching 6 or even 9 ppm.

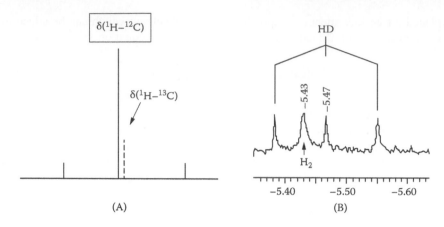

Secondary isotope effects measured for some nuclei in solutions

Detected Nucleus	Isotopomers	δ (ppm)
^{13}C	CH_4	0.0
	CH_3D	−0.2016
	CHD_3	−0.6006
	CD_4	−0.7946
^{31}P	PH_3	0.0
	PH_2D	−0.8045
	PHD_2	−1.6491
	PD_3	−2.5373
^{15}N	NH_3	0.0
	NH_2D	−0.6264
	NHD_2	−1.2491
	ND_3	−1.8687
^{119}Sn	SnH_3^-	0.0
	SnH_2D^-	−3.093
	$SnHD_2^-$	−6.202
	SnD_3^-	−9.325

(C)

FIGURE 2.4 (A) The ^1H NMR spectrum of chloroform ($CHCl_3$). (B) The ^1H NMR spectrum of a mixture of H_2 and HD molecules in a solution. (C) Secondary isotope shifts observed for some nuclei.

Even very small values of the secondary isotope effect play a very important role for two reasons. First, they allow determination of the isotope content and distribution in molecules by NMR when necessary. Second, the isotropic chemical shifts observed—for example, for ^1H nuclei at ^{12}C isotopes vs. ^{13}C, for ^{19}F nuclei at ^{12}C isotopes vs. ^{13}C, or for ^{19}F nuclei at ^{121}Sb isotopes vs. ^{123}Sb nuclei—can strongly affect the NMR spectra of symmetrical molecular systems. Due to perturbation of the shifts, these systems become non-symmetrical structures in a NMR sense and their NMR spectra can become extraordinarily complex. In contrast, the isotropic chemical shifts remain unresolved in magic-angle spinning (MAS) NMR spectra of solids because of strong line broadening.

2.2 CHEMICAL SHIFTS IN THE PRESENCE OF UNPAIRED ELECTRONS

The magnetic moment of electrons is known to be larger by 662 times than that of protons. Therefore, unpaired electrons located close to detected nuclei produce a strong local magnetic field that masks the natural NMR spectroscopic properties of nuclei (i.e., their isotropic chemical shifts or the components of chemical shift tensors). Theoretically, the large magnetic moment of an electron located at a paramagnetic center (e.g., paramagnetic metal cations or organic radicals) affects the nuclear Larmor resonance frequency via effective magnetic field B_{EF}, expressed in Equation 2.10:

$$B_{EF} = \left(2\pi A\mu_B g B_0 S(S+1)/3\gamma_N kT\right)$$
$$+\left(\mu_0 \mu_B^2 g^2 B_0 S(S+1)(3\cos^2\theta - 1/3kTr^3\right) + B_{BMS} \tag{2.10}$$

Here, A is the isotropic electron–nucleus hyperfine coupling constant (observable by electron paramagnetic resonance), μ_B is the Bohr magneton measured as 9.2741×10^{-24} J/T, r is the electron–nucleus distance, θ is the angle between the electron–nucleus vector and external magnetic field B_0, and g is the fundamental constant of electrons, referred to as the electron g factor (see the recommended literature). The first term in Equation 2.10 can be expressed as follows:

$$H = A \times S \times I \tag{2.11}$$

written for nuclear and electron spins I and S. This term produces the so-called Fermi contact (FC) paramagnetic chemical shift, which originates from direct delocalization of the unpaired electron density on an atom with an observed nucleus. Because of this delocalization, the FC shifts strongly depend on hyperfine coupling constant A. In addition, the sign of this constant dictates a low or high field displacement of signals in the NMR spectra. It is remarkable that the FC shifts are independent of distances between electrons and nuclei due to the above-mentioned electron delocalization. On the whole, the FC shift can reach extremely large magnitudes, up to 3000 ppm (or higher), depending on the nature of the nuclei and paramagnetic centers.

The second term in Equation 2.10, corresponding to a magnetic field created in a sample by an unpaired electron, shows direct dipolar electron–nucleus interaction. This interaction strongly depends on the electron–nucleus distance as a factor of r^3. In addition, this term is angle dependent via $3\cos^2\theta - 1$. Here, θ is defined as an angle formed by the r vector and the direction of external magnetic field B_0. In rigid static solids, where fast high-amplitude molecular reorientations are absent, the spectral effect of this field is obvious: Nuclei resonances will experience strong broadenings that are similar to dipole–dipole internuclear interactions. In solutions, this large dipolar electron–nucleus term can manifest as residual dipolar couplings. Finally, the effect can be seen when molecules containing paramagnetic centers experience partial alignment in the external magnetic field.[10]

In theory, the dipolar term in Equation 2.10 can be reduced to zero when $3\cos^2\theta - 1$ = 0 due to fast and isotropic molecular rotations, particularly in solutions. In reality, however, the electron–nucleus dipolar interaction is not completely quenched even in the presence of these fast motions. It follows from the formalism describing any dipolar interactions between vectors. Therefore, even for fast-moving molecules, the anisotropy of the static magnetic moment will be responsible for a residual dipolar interaction. The latter produces the so-called pseudocontact paramagnetic shift. According to the theory, the value of a pseudocontact chemical shift (PCS) is given by

$$\delta(PCS) = (1/12\pi)r^{-3}\left[\Delta\chi_{AX}(3\cos^2\theta - 1) + (3/2)\Delta\chi_{RH}\sin^2\theta\cos 2\phi\right] \quad (2.12)$$

Here, r is the distance between the paramagnetic center and the nucleus, θ and ϕ are the angles describing the location of the nuclear spin relative to the principal axes (where the paramagnetic center is located) and the axes of magnetic susceptibility tensor χ with the axial (AX) and rhombic (RH) components, respectively.[11] This expression shows that the PCS term depends on the electron–nucleus distance and atomic coordinates. Thus, molecular geometry can be potentially determined even in solutions. In solids, the pseudocontact chemical shifts can be observed in systems containing paramagnetic centers with anisotropic unpaired electrons, such as the metal ions Dy^{3+}, Tb^{3+}, or Fe^{3+}.

The last term in Equation 2.10, B_{BMS}, indicates the presence of a demagnetization field that appears in a sample due to bulk magnetic susceptibility (BMS). In general, this effect is insignificant for diamagnetic compounds in solutions and in the solid state but it can play a very important role in NMR experiments on paramagnetic molecular systems (see Chapter 10).

2.2.1 KNIGHT SHIFTS

Knight shifts, measured experimentally in metals and alloys, and their interpretation represent a specific field of NMR applications requiring experience in and knowledge of solid-state physics. Some examples will be considered in Chapter 10, but here it should be emphasized that Knight shifts and FC shifts demonstrate the fundamental difference in their origin. The FC shifts are measured in non-metallic systems containing unpaired localized electrons. In contrast, the conducting electrons in metals or alloys are not localized. This feature leads to the situation where each observed nuclear spin "sees" the magnetic field produced by all of the electrons. The general expression for the total shift (δ) observed in metals and alloys takes the form of Equation 2.13:

$$\delta = \delta_C + K_S \quad (2.13)$$

Here, the observed chemical shift (δ_C) is obtained by summing all of the chemical shift contributions from the orbital motion of the core electrons, while the Knight shift (K_S) is a sum reflecting all of the conducting electron spin- and orbital-dependent contributions. Because these magnitudes are tensors, they can be diagonalized to give the principal components $\delta(iso)$ and $\delta(aniso)$ in Equation 2.14:

$$\delta(iso) = \delta_{C(iso)} + K_{(iso)}^{SPIN} + K_{(iso)}^{ORB}$$
$$\delta(aniso) = \delta_{C(aniso)} + K_{(aniso)}^{SPIN} + K_{(aniso)}^{ORB}$$

$$(2.14)$$

where the orbital contributions $K_{(iso)}^{ORB}$ and $K_{(aniso)}^{ORB}$ are connected with the field-induced Van Vleck orbital susceptibility of the partially filled electron levels. The practical measurement and interpretation of the shifts for ^{27}Al or ^{29}Si nuclei in URu_2Si_2 and UNiAl systems can be found in References 12 and 13 and will be discussed further in Chapter 10.

2.3 SPIN–SPIN COUPLING

In accordance with the quantum mechanical theory of NMR, the Hamiltonian describing the spin states in isotropic media takes the form

$$h^{-1}\hat{H} = -\sum_{j}(2\pi)^{-1}\gamma_j B_0\left(1-\sigma_j\right)\hat{I}_j^z + \sum_{j<k}J_{jk}\hat{I}_j\hat{I}_k$$

$$(2.15)$$

The first part of the equation shows the nuclear spins, which are placed into external magnetic field B_0 and isolated (i.e., uncoupled). Thus, this part corresponds to the screening constant σ (or chemical shift δ). The second term in the equation represents interactions between spins that become coupled and can be characterized by the spin–spin coupling constant J_{jk}. These J_{jk} constants are relatively small and typically take values from 1–2 to 1200 Hz, as a maximum. Therefore, the spin–spin coupling is always considered as perturbation of the first term. This perturbation of the Zeeman energy levels by spin–spin coupling can be easily represented as shown in Figure 2.5A. Here, the symbols α and β represent the parallel and anti-parallel mutual orientations valid for nuclear spins of 1/2. The spin states for a system consisting of two nuclei can be written as $\beta\beta$, giving total magnetic quantum number $(m_I) = -1$; as $\alpha\beta$, giving $m_I = 0$; as $\beta\alpha$, giving $m_I = 0$; and as $\alpha\alpha$, giving $m_I = 1$. In turn, these spin states correspond to the energy diagram in the Figure 2.5A. Thus, one can say that the first nuclear spin can "see" the second spin in two different states so its energy levels will undergo an additional splitting by energy J measured in Hz. Because double quantum transitions (e.g., $\alpha\alpha \leftrightarrow \beta\beta$) are forbidden, the energy diagram describing two nuclei will correspond to four resonance lines instead of the two lines expected for uncoupled nuclei.

It should be emphasized that the J constant in Equation 2.15 measured in Hz does not depend on external magnetic field B_0. In other words, the constant characterizes only the energy of spin–spin coupling. By definition, the J constant is of scalar magnitude and has a sign: positive or negative. As generally accepted, for example, the J constant for two nuclei will be positive if the anti-parallel spin orientations are more stable energetically and *vice versa*. The sign of the J constant depends on the nature of the coupled nuclei and various structural factors. For example, the $^1J(^1H-^{13}C)$ constant through one chemical bond is always positive, whereas $^1J(^{19}F-^{13}C)$ is negative. The $^nJ(^1H-^1H)$ constants, as well as $^nJ(^{19}F-^{19}F)$, can be positive and negative, where the magnitude and the sign depend on the number of chemical bonds (n) and the

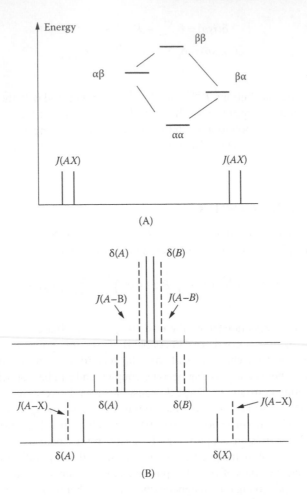

FIGURE 2.5 (A) Perturbation of the Zeeman energy levels by spin–spin coupling in a simple spin system. (B) Evolution of the simple spin system AX (bottom) to the strongly coupled spin system AB (top).

molecular geometry. One should remember that the absolute sign of the J constant (which can be determined by special NMR experiments) does not play an important role in the analytical context. At the same time, the relative J constant signs are extremely important. In fact, the different relative signs of the J constants result in quite different spectral patterns observed in NMR spectra, when spin systems are strongly coupled.

2.3.1 STRONGLY COUPLED SPIN SYSTEMS

Figure 2.5A shows the NMR spectrum of two magnetically different and coupled nuclei, where line intensities are distributed as 1:1:1:1. Such a pattern corresponds to a simple spin system, named AX, where the chemical shift difference $\Delta\delta^{AX}$ (measured in Hz) is significantly larger than the spin–spin coupling constant $J(A - X)$. In practice, a spin system is simple when the $J(A - X)$ corresponds to Equation 2.16:

$$J(A - X): [\Delta\delta^{AX}(Hz)]/[J^{AX}(Hz)] > 10 \tag{2.16}$$

The main feature of the spectra of simple spin systems is that the chemical shift values and the spin–spin coupling constants are measured directly from the spectrum as shown in Figure 2.5B. Here, dashed lines correspond to the chemical shifts A and X, while the distance between solid lines is the spin–spin coupling constant $J(A - X)$.

Generally speaking, the condition in Equation 2.16 provides a simple analysis of the line multiplicity (N) observed for the A resonance:

$$N = 2nI + 1 \tag{2.17}$$

where n is the number of the X nuclei having spin I. Even relative intensities within multiplets can be predicted in accordance with the rule of the so-called Pascal's triangle. For example, protons in group CH_2–CH_3 should exhibit two resonances appearing as a quartet and a triplet, respectively. In other words, these quartet and triplet lines are a spectral signature of group CH_2–CH_3.

The situation changes completely for a decreasing $\Delta\delta^{AX}$ (Hz)/J^{AX} (Hz) ratio. Then, the initial AX spin system transforms to a strongly coupled system named AB using the alphabetic principle to emphasize the small chemical shift difference $\Delta\delta^{AB}$. Figure 2.5B demonstrates the evolution of the AX spectral pattern to the AB spectral pattern when the chemical shift difference reduces. The solid lines observed in the spectrum and the chemical shifts do not coincide, whereas the $J(AB)$ constant is still determined directly as the distance between the lines.

Upon an increase in the number of nuclei in strongly coupled spin systems—such as ABX, AMX, AB_2, ABC, or $AA'BB'X$ (where term AA' corresponds to two coupled nuclei with the $J(A - A')$ constant at the same chemical shifts), the spectral patterns become very complex due to the appearance of combinational transitions. In this situation, the number of observed lines and their intensities are difficult to predict *a priori*. Moreover, the dispositions of the lines in the spectrum and the distances between them do not correspond to chemical shift values and spin–spin coupling constants. In addition, the combinational transitions can show low-intensity lines. They play an important role in recognizing the type of spin systems. The NMR spectra of such strongly coupled spin systems can be analyzed only via simulation or by fitting to the experimental data with the help of convenient computer programs based on the quantum-mechanical formalism. Figure 2.6 gives an example of a strongly coupled seven-spin system formed by ^{19}F nuclei in allyl cation A. The ^{19}F NMR spectrum of this species, recorded in a SO_2ClF–SbF_5 medium, consists of more than 140 lines, which can be rationalized only by computer fitting to the experimental pattern, leading to the negative $J(^{19}F$–$^{19}F)$ constants shown in Figure 2.6B.

It should be emphasized that in the cases of strongly coupled spin systems, observed experimentally, computer treatment of NMR spectra is necessary to at least determine the structure of the compounds under investigation. This can be illustrated by the ^{19}F and ^{13}C NMR spectra recorded for allyl cation II in Figure 2.6C.

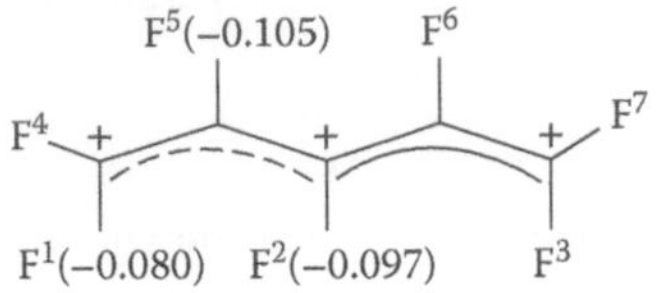

$F^5(-0.105)$ F^6

F^4 +   +   + F^7

$F^1(-0.080)$ $F^2(-0.097)$ F^3

(A)

The ^{19}F – ^{19}F spin–spin constants in the allyl cation

$J(F^1F^2) = -136.6$ Hz
$J(F^5F^6) = -52.7$ Hz
$J(F^1F^3) = -10.4$ Hz
$J(F^1F^7) = -12.0$ Hz
$J(F^4F^7) = -9.9$ Hz
$R(1 \cdots 2) = 2.750$ Å
$R(5 \cdots 6) = 2.660$ Å

(B)

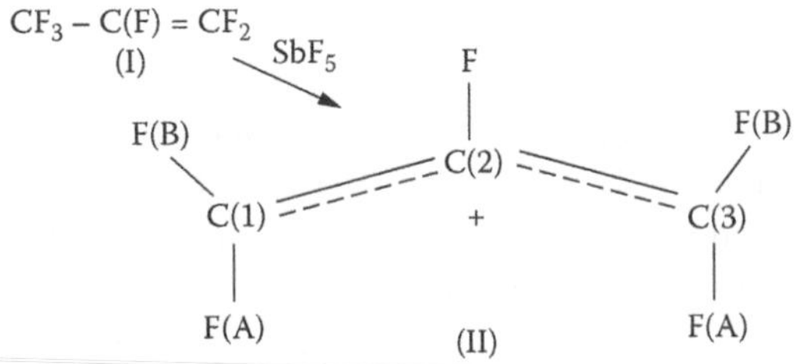

$CF_3 - C(F) = CF_2$
(I) $\xrightarrow{SbF_5}$

F(B)   F(B)

C(1) ===== C(2) ===== C(3)

F(A)   +   F(A)

(II)

(C)

FIGURE 2.6 (A) Structure of the allyl cation with the distribution of charges. (B) The spin–spin constants obtained by computer analysis of the experimental ^{19}F NMR spectrum recorded for A. (C) Reaction between olefin and SbF_5 in a SO_2ClF–SbF_5 medium at low temperatures.

In fact, this cation might be formed in the reaction between olefin I and SbF_5 in a SO_2ClF–SbF_5 medium at low temperatures.[14] After mixing the reagents, the new and very complex pattern appears in the ^{19}F NMR spectrum shown in Figure 2.7A, where the resonances are expectedly low-field shifted in accordance with the distribution of the positive charge. However, this observation does not prove the cation formation. This pattern can be reproduced only by computer simulations performed on the basis of a spin system $AA'BB'K$, where K is the ^{19}F nucleus at the central carbon atom. The calculations in terms of this spin system give reasonable J values in accordance with the symmetry of allyl cation II. The situation becomes more complicated for the ^{13}C NMR spectrum recorded for the same product. Again, the spectrum shows a very complex pattern requiring computer simulation. However, as it follows from Figure 2.7B, the resonance of the C(1,3) nuclei can be calculated as the X part of spin system $ABCDKX$ in contrast to the above-mentioned system, $AA'BB'K$, formed by ^{19}F nuclei. This change can be explained by a loss of the symmetry due to the isotope shift of ^{19}F nuclei located at ^{13}C and ^{12}C. The isotropic shift is determined as 0.11 ppm. Finally, because the J constant does not depend on external magnetic field B_0, the strongly coupled spin systems can be simplified (but not always) by experiments at the highest magnetic fields.

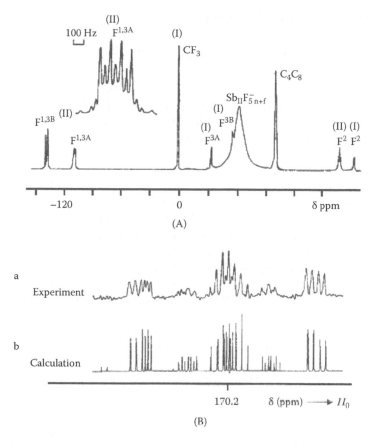

FIGURE 2.7 (A) Low-temperature ^{19}F NMR spectrum recorded for a product of mixing olefin (I) and SbF$_5$ in SO$_2$ClF–SbF$_5$ medium. (B) The experimental (a) and theoretical (b) ^{13}C NMR spectrum of the same mixture.

2.3.2 SPIN–SPIN COUPLING VIA CHEMICAL BONDS

The spin–spin coupling through chemical bonds is transmitted via electrons surrounding nuclei. The bonding electron orbitals play a major role; therefore, the values of ^nJ constants depend on molecular structure (i.e., electron distribution), atomic hybridization, the number of chemical bonds (n) separating the nuclei, and interbond angles. As generally accepted, the total J coupling energy is comprised of three contributions coming from the spin–orbital (SO):

$$H_{SO} = (\mu_0/4\pi) g_L \mu_B \gamma_N \hbar \sum 2L_e I_N r_{eN}^{-3} \tag{2.18}$$

from the spin–dipolar (SD):

$$H_{SD} = -(\mu_0/4\pi) g_S \mu_B \gamma_N \hbar \sum \left[r_{eN}^{-3} I_N S_e - 3 r_{eN}^{-5} (I_N r_{eN})(S_e r_{eN}) \right] \tag{2.19}$$

and from the Fermi contact (FC) interactions:

$$H_{FC} = (2/3)\mu_0 g_S \mu_B \gamma_N \hbar \sum \delta(r_{eN}) I_N S_e \qquad (2.20)$$

where e and N are the electron and nucleus, respectively; L_e is the orbital angular moment operator, r_{eN} is the distance between the electron and nucleus, and $\delta(r_{eN})$ is the Dirac delta function. Because the Dirac delta function takes a value of 1 at $r_{eN} = 0$ or a value of 0 otherwise (thus demonstrating the electron density directly on the nuclei), only spherical s-electrons will play a major role in the FC term.

Relative contributions of the SO, SD, and FC terms to the J constants obtained for some nuclei by quantum chemical calculations are listed in Figure 2.8A. The data show that the SO, SD, and FC contributions and the spin–spin constants or even their signs change with the nature of nuclei and the number of chemical bonds. The interesting case is the $^2J(F–F)$ constant in CF_4 molecules, where the total J value is caused by spin–dipolar interaction; however, the FC contribution generally dominates in spin–spin coupling through chemical bonds. This fact allows simple chemical interpretations in terms of s-electron densities and their changes due to the action of various factors. Figure 2.8B demonstrates such an interpretation, based on

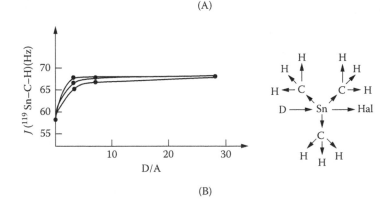

Theoretically calculated contributions to total spin–spin coupling constants (given in Hz)				
Coupling Constant	Molecule	FC	SO	SD
$^1J(C–C)$	C_2H_6	11.99	−0.55	0.3
	C_2H_2	69.43	4.13	2.84
$^1J(F–C)$	CH_3F	−126.30	−4.15	6.92
$^2J(F–F)$	CF_4	−1.09	6.20	16.73

(A)

(B)

FIGURE 2.8 (A) Relative contributions of the FC, SO, and SD terms to the J constants obtained by quantum chemical calculations for some nuclei. (B) The $^2J(^{119}Sn–C–^1H)$ constants measured for $(CH_3)_3SnCl$ in CD_2Cl_2 at the addition of pyridine at 28°, −30°, and −50°C (from bottom to top).

a remarkable increase in the $^2J(^{119}Sn-C-^1H)$ constants of methyltin halides (acceptor A) observed in different electron-donating solvents (donor D). These donors (acetone, dioxane, pyridine, etc. added to CD_2Cl_2) have an effect dependent on the amounts of added solvents and temperature. As can be seen, the J curves reach a plateau at a D/A ratio of 10, corresponding to complexation. The scheme in Figure 2.8B shows the electron redistribution resulting from donor–acceptor interactions that increase the s-electron density on the H atoms. It is obvious that quantitative treatments of these curves can give the corresponding equilibrium constants.[15]

Among the several factors changing the J coupling constant values, the pronounced influence of bond angles is of great interest for researchers in the context of a structural analysis by NMR; for example, the geminal constants $^2J(^1H-^1H)$ in CH_2 groups increase from 0 to 40 Hz as a function of the bond angle. In turn, the bond angle depends on carbon hybridization. The vicinal $^3J(^1H-^1H)$ constants increase from 0 to 12–14 Hz and follow the well-known Karplus curve. This curve changes with dihedral angle φ. Thus, the ^3J constant is a powerful instrument for a conformation analysis of organic molecules (considered in Chapter 5).

Figure 2.9A illustrates the generally accepted W rule. The maximal spin–spin coupling is transmitted via four chemical bonds. In agreement with this rule, in the allyl cation the $^4J(^{19}F-^{19}F)$ constant between nuclei F(3) and F(4) is larger than that between nuclei F(2) and F(3) by a factor of 3.

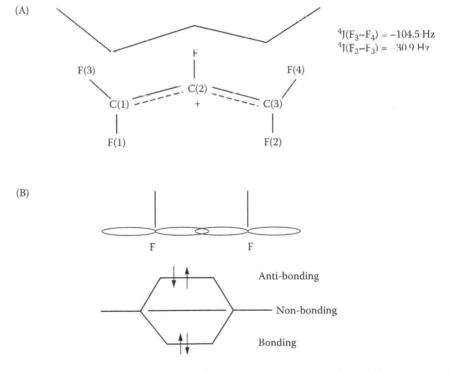

FIGURE 2.9 (A) The W rule corresponding to the maximal spin–spin coupling transmitted via four chemical bonds. (B) Spin–spin coupling through space by non-bonding interactions.

2.3.3 SPIN–SPIN COUPLING THROUGH SPACE

Potentially spatial proximity of some nuclei can lead to the phenomenon known as through-space (t-s) spin–spin coupling, which can be seen in Figure 2.6. The $^4J(^{19}F–^{19}F)$ constants for nuclei F(1) and F(2) are maximal (–136.6 Hz) in spite of the four chemical bonds separating the nuclei. It has been shown that the through-space (t-s) contribution, evaluated from the total J constant of different F-allyl cations, varies between –100 and –144 Hz and correlates with internuclear distance r (Å) in Equation 2.21:

$$^4J(t\text{-}s) = (2600/r^3) + 3.6 \tag{2.21}$$

The $^4J(t\text{-}s)$ value goes to zero when distance $r \rightarrow \infty$.[16] According to quantum chemical calculations, the t-s coupling is again dominated by the FC interaction. However, the interaction occurs via non-bonding electron orbitals. This mechanism is shown in Figure 2.9B for closely located F atoms. Similarly, the fluoro- and trifluoromethyl-substituted mono- and di-arylthallium derivatives also exhibit the $J(^{205}Tl–^{19}F)$ coupling constants through space, when the fluoro- or trifluoromethyl groups are located *ortho* to the thallium atoms.[17] The $^{19}F–^1H$ t-s coupling constants are also significant; for example, they can be seen in fluorine-substituted aryl vinyl selenides and sulfides.[18] Further discussion regarding the recognition of $J(^{19}F–^{19}F)$ and $J(^{31}P–^{31}P)$ coupling through space in organic and inorganic molecules and their detailed theoretical treatments can be found in the recommended literature.

In the context of the mechanisms considered above for spin–spin coupling, it should be noted that through-space coupling differs completely from direct dipolar interactions between nuclear spins (dipolar coupling), which play an important role in solid-state NMR factors affecting the line shapes of NMR signals in solids. This coupling is transmitted directly through space but not via electrons surrounding the nuclei.

2.3.4 PROTON–PROTON EXCHANGE COUPLING

This surprising phenomenon, related to exchange interactions in electron paramagnetic resonance (EPR) spectroscopy, can be observed, for example, in the low-temperature ^1H NMR spectra in solutions of transition metal polyhydrides, one of which is shown in Figure 2.10A. The hydride regions of such polyhydrides exhibit ^1H complex spectral patterns that are difficult to rationalize. These patterns can be formally calculated as the strongly coupled spin systems *ABC* or *ABX*, where the $J(^1H–^1H)$ coupling constants take abnormally large values, reaching 1500 Hz.[19]

Such extremely large $J(^1H–^1H)$ magnitudes cannot be predicted by theory in terms of the strongest FC or other interactions in Equations 2.18 through 2.20. In addition, these $J(^1H–^1H)$ constants are strongly temperature and solvent dependent, in contrast to the regular spin–spin coupling constants. It has been found that the coupling is quenched when the protons are replaced for deuterons or tritium atoms. All of these features reveal the quantum mechanical nature of the coupling.

Currently, there are several models revealing the nature of this phenomenon within the framework of a quantum mechanical proton–proton exchange. A full description of the quantum mechanical exchange is complex but begins from the

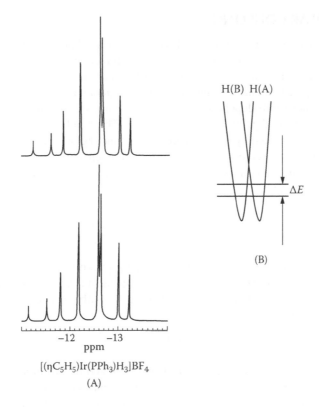

FIGURE 2.10   (A) For the solution ^{1}H{^{31}P}, the experimental (bottom) and theoretical (top) NMR spectrum of the hydride complex (hydride region). (B) An additional splitting ΔE due to quantum mechanical H(B)–H(A) exchange.

single-particle wave functions corresponding to the ground state of two particles, localized in two potential wells. When the particles (the H atoms) are separated by a large distance, then their vibrational motions are independent; however, at distances ≤1.65 Å, these single-particle wave functions can overlap, resulting in the additional splitting (ΔE_{exc}) depicted in Figure 2.10B. Thus, the splitting observed in NMR experiments as the $J(^1\mathrm{H}-^1\mathrm{H})$ constant $J(exch)$ is not of a magnetic nature but rather corresponds to the rate of the quantum mechanical exchange. A simplified expressions for the $J(exch)$ constant is

$$J(exch) = \left[-3\hbar a/\left(8\pi m\delta^3\right)\right](3/\pi)^{1/2}\exp\left[-3/4\left(a^2+\lambda^2\right)/\delta^2\right] \qquad (2.22)$$

This equation includes internuclear distance a, characteristic distance λ, and three-dimensional harmonic oscillator δ, thus demonstrating delocalization of the individual particles and predicting the $J(exch)$ quench in the case of deuterons or tritium atoms.

2.4   DIPOLAR COUPLING

Nuclear spins can be coupled directly through the space that is typical of solids (but not via electrons surrounding the nuclei), where molecular reorientations are very slow or completely absent. Nuclei placed into external magnetic field B_0 can experience dipole–dipole homonuclear and heteronuclear interactions. Resonance frequencies of these nuclei will depend on these interactions, which can be quantitatively characterized via the dipolar coupling constant. The dipolar coupling shows how the magnetic field created by neighboring spins in the region of an observed nucleus varies when inter-spin vectors reorient in external magnetic field B_0.

The dipolar coupling constant (D) can be written as follows, which is valid for two different nuclei, I and S:

$$D = (\mu_0/4\pi)\hbar\gamma_I\gamma_S/r(I-S)^3 \qquad (2.23)$$

Here $r(I-S)$ is the internuclear distance and factor r^3 exhibits the very strong influence of distance on the D value. In turn, the dipolar interaction in pairs of nuclei depends on the orientation of the internuclear vector with respect to the direction of static magnetic field B_0 via

$$(3\cos^2\theta - 1)/r^3 \qquad (2.24)$$

where θ is the angle between the internuclear vector and the B_0 direction.

As has been demonstrated theoretically, the dipolar coupling constant is typically about 1000 times larger than the scalar spin–spin coupling constant. In fact, the dipolar coupling constant, expressed in frequency units, has been calculated for two protons at a H–H distance of 2 Å to be as large as 30,000 Hz compared to only several hertz (20 Hz sometimes) for spin–spin scalar coupling.

In spite of the very large D magnitudes due to the relation shown in Equation 2.24, dipolar coupling goes to zero when molecules move rapidly and isotropically in solutions or liquids. Thus, the disposition of resonances and their shapes in solutions are not influenced by dipolar coupling, although it can play a major role in nuclear relaxation (Chapter 3). In contrast, in rigid solids the dipolar interactions are not averaged and strongly affect the line shape. In other words, the dipolar coupling constant can be determined experimentally in NMR spectra of static solids.

Figure 2.11A shows the classical Pake powder pattern expected for two ^1H nuclei in a static sample. Here, the shape of the resonance is formed by superposition of two sub-spectra corresponding to a ^1H axially symmetrical screening tensor (dashed lines), and the distance between the two singularities is equal to the dipolar coupling constant, shown as ω_D. At a uniform distribution of internuclear orientations in powdered samples placed in the external magnetic field, the static NMR spectra will show, even for one sort of nuclei, a superposition of many lines characterizing the various dipolar couplings. Because of this distribution, the resonance is anisotropically broadened. This broadening effect, inversely proportional to internuclear distances, depends on the nature of the nuclei. For example, a ^1H resonance of static ice is broadened up to ~10^5 Hz. At the same time, ^{29}Si–^{29}Si dipolar coupling is weak due

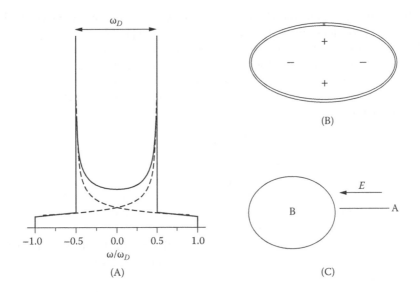

FIGURE 2.11 (A) Classical Pake powder spectrum observed for two ^{1}H nuclei in a static sample. (B) Non-spherical distribution of charges at nuclei having spin numbers > 1/2. (C) Non-homogeneous electric field E changing along chemical bond A–B.

to the low γ value and low natural abundance of ^{29}Si nuclei. As a result, line widths will be around only 2 to 3 kHz. As mentioned above, MAS NMR experiments (see Figure 1.8A in Chapter 1) partially or completely reduce this broadening.

2.5 QUADRUPOLAR COUPLING

Distributions of charges for many nuclei with spin numbers > 1/2 are non-spherical (Figure 2.11B). Such distributions are responsible for the appearance of nuclear quadrupole moments Q, which are also fundamental properties of nuclei. Spins of quadrupolar nuclei (e.g., ^{2}H, ^{14}N, ^{27}Al, ^{81}Br, ^{127}I, ^{35}Cl) can interact not only with the external and local magnetic fields in samples but also with electric field gradients (EFGs) for these nuclei. Some of the quadrupolar nuclei (^{81}Br and ^{79}Br, or ^{35}Cl and ^{37}Cl), having extremely large quadrupole moments (25.4 and 30.5 or –8.16 or –6.44 × 10^{-30} C·m^2, respectively) can be objects of nuclear quadrupole resonance (NQR) in contrast to deuterium, which has a small quadrupole moment of 0.286 × 10^{-30}.

Analogous to NMR, NQR is based on quantized energy levels corresponding to different orientations of quadrupole moment Q with respect to the electric field gradient (EFG). In turn, the electric field gradient (e.g., its eq_{ZZ} component) is connected with chemical structure and is generally directed along a chemical bond. This component is expressed via the electrostatic potential (V) as shown in Equation 2.25:

$$eq_{ZZ} = \partial^2 V / \partial^2{}_Z \tag{2.25}$$

Figure 2.11C shows that the eq_{ZZ} component characterizes a non-homogeneous electric field (E), which changes along chemical bond A–B.

The electrostatic potential (V) is of scalar magnitude, whereas the electric field gradient is tensor. This tensor includes the non-zero components eq_{XX}, eq_{YY}, and eq_{ZZ} and the zero off-diagonal elements. By convention, the largest element of the EFG tensor, eq_{ZZ}, is always oriented along the A–B bond (i.e., along the Z-axis).

Interactions between the nuclear quadrupole moments Q and the electric field gradients can be energetically characterized via the nuclear quadrupolar coupling constant (QCC):

$$QCC = e^2 q_{zz} Q / h \qquad (2.26)$$

where the eq_{ZZ} term corresponds to the principal component of the EFG tensor and e is the elementary charge, measured as $1.6021773 \times 10^{-19}$ C. The QCC values vary and can reach extremely large magnitudes of the order of 10^6 to 10^9 Hz. Some QCC values that are strongly dependent on the nature of the nuclei are presented in Figure 2.12A, where the QCC changes from 0 to 8 MHz.

Generally speaking, the QCC value measures the size of the electric field gradient, whereas its spatial extension or shape is defined by the asymmetry parameter η shown in Equation 2.27:

$$\eta = |eq_{XX} - eq_{YY}| / eq_{ZZ} \qquad (2.27)$$

It is obvious that at $eq_{XX} = eq_{YY}$ and hence at $\eta = 0$, the electric field gradient is an axially symmetric tensor.

It should be emphasized that the shapes and sizes of the electric field gradients and QCC values, respectively, depend strongly on the symmetry of the charge distributions around the nuclei, even if they have large quadrupolar moments. In some sense, the QCC can be used as a measure of the charge distributions. For example, a symmetrical charge distribution for ^{14}N nuclei in $NH_4^+Cl^-$ results in a small QCC value of 0.016 MHz vs. 0.9 or even 3.98 MHz in the compounds $EtONO_2$ or $MeNH_2$, respectively, where charge distributions for the ^{14}N nuclei are not symmetrical. The same tendency can clearly be seen for ^{17}O nuclei in the compounds MoO_4^{2-} and CO.

Theoretically, the electric field gradient for any quadrupolar nucleus can be written at a semi-quantitative level via the sum of nuclear and electronic terms:

$$eq_{ZZ} = + \sum_n K_n \left(3z_n^2 - r_n^2\right) / r_n^5 - e < \psi * \left| \sum_i \left(3z_i^2 - r_i^2\right) / r_i^5 \right| \psi \qquad (2.28)$$

where K and e are charges of the neighboring nuclei and electrons, respectively, and r_n and r_i are the corresponding distances. Thus, the electric field gradient opens the way to describing bonding modes in molecules when the QCC values (Equation 2.26) are accurately determined by NMR or NQR methods.

In spite of the very large quadrupolar coupling constants for some nuclei, they do not affect, by themselves, the chemical shifts and line shapes of signals in NMR spectra in solutions and liquids due to fast isotropic molecular motions in the external magnetic field averaging the quadrupolar interactions. Nevertheless, their presence manifests in nuclear relaxation. In contrast, quadrupolar interactions play a major role in solid-state NMR, particularly in rigid solids. The line shape in rigid

Compound	Nucleus	QCC (MHz)
O^2	^{17}O	−8.42
CO	^{17}O	4.43
MoO_4^{2-}	^{17}O	0.7
$NaNO_3$	^{14}N	0.745
CH_3CN	^{14}N	4.00
$MeNH_2$	^{14}N	3.98
$EtONO_2$	^{14}N	0.9
NH_2Cl	^{14}N	0.016
Solid PdD	^2H	0.000
BD_3NH_3	^2H	0.105
BH_4^-	^{11}B	0.0

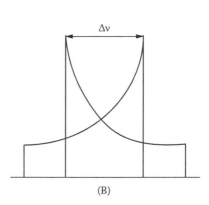

(A)   (B)

FIGURE 2.12 (A) QCC values and their variations for ^{17}O, ^{14}N, ^2H, and ^{11}B nuclei. (B) A ^2H resonance in a static powder sample, where Δν is quadrupolar splitting.

solids depends directly on QCC values. Figure 2.12B shows a ^2H resonance typical of static powder samples. As in the case of the dipolar coupling in Figure 2.11A, the shape of the resonance consists of two sub-spectra to show the quadrupolar splitting, Δν. Again, similar to the case in Figure 2.11A, the axially symmetric electric field gradients are responsible for this shape.

As follows from the theory, in the absence of fast molecular reorientations the Δν splitting gives the static QCC value via Equation 2.29:

$$\Delta v = 3/4(e^2 q_{ZZ} Q/h) \tag{2.29}$$

One can now formulate the following important statement: If a static known QCC value and an experimental QCC value, determined from the NMR spectrum via Equation 2.29, are similar, then molecular (or group) reorientations are absent. Moreover, as will be shown in further chapters, the shape of the NMR signal for quadrupolar nuclei in solids provides quantitative probing molecular mobility.

REFERENCES AND RECOMMENDED LITERATURE

1. Strub, H., Beeler, A.J., Grant, D.M., Michel, J., Cutts, P.W., and Zilm, K.W. (1983). *J. Am. Chem. Soc.*, 105: 3333.
2. Buckingham, A.D. and Stephens, P.J. (1964). *J. Chem. Soc.*, 2747.

3. Hrobárik, P., Hrobáriková, V., Meier, F., Repisky, M., Komorovsky, S., and Kaupp, M. (2011). *J. Phys. Chem. A*, 115: 5654.
4. Garbacz, P., Terskikh, V.V., Ferguson, M.J., Bernard, G.M., Kodziorek, M., and Wasylishen, R.E. (2014). *J. Phys. Chem. A*, 118: 1203.
5. McConnel, H.M. and Chesnut, D.B. (1958). *J. Chem. Phys.*, 28: 107.
6. Syrova, G.P. and Sheinker, Yu.N. (1972). *Chem. Heterocycl. Comp.*, 3: 345.
7. Bakhmutov, V.I., Galakhov, M.V., and Fedin, E.I. (1987). *Izv. Akad. Nauk SSSR, Ser. Khim.*, 3: 675.
8. Facelli, J.C., Grant, D.M., and Michl, J. (1987). *Acc. Chem. Res.*, 20: 152.
9. Williams, D.E., Peters, M.B., Wang, B., and Merz, K.M. (2008). *J. Phys. Chem. A*, 112: 8829.
10. Clore, G.M. and Iwahara, J. (2009). *Chem. Rev.*, 109: 4108.
11. Bertini, I., Luchinat, C., and Parigi, G. (2002). *Prog. Nucl. Magn. Reson. Spectrosc.*, 40: 249.
12. Bernal, O.O., Rodrigues, C., and Martinez, A. (2001). *Phys. Rev. Lett.*, 87: 196402.
13. Nowak, B. and Troc, R. (2000). *Solid State NMR*, 18: 53.
14. Galakhov, M.V., Petrov, V.A., Bakhmutov, V.I., Belen'kii, G.G., Kvasov, B.A., German, L.S., and Fedin, E.I. (1985). *Izv. Akad. Nauk SSSR, Ser. Khim.*, 306.
15. Petrosyan, V.S., Yashina, N.S., Bakmutov, V.I., Permin, A.B., and Reutov, O.A. (1974). *J. Organomet. Chem.*, 72: 71.
16. Bakhmutov, V.I., Galakhov, M.V., and Fedin, E.I. (1985). *Magn. Reson. Chem.*, 23: 11.
17. Pecksen, G.O. and White, R.F. (1989). *Can. J. Chem.*, 67: 1847.
18. Afonin, A.V. (2010). *Russ. J. Org. Chem.*, 46: 1313.
19. Zilm, K.W., Heinekey, D.M., Millar, J.M., Payne, N.G., Neshyba, S.P., Duchamp, J.C., and Szczyrba, J. (1990). *J. Am. Chem. Soc.*, 112: 920.

RECOMMENDED LITERATURE

Abragam, A. (1985). *Principles of Nuclear Magnetism*. Oxford: Clarendon Press.
Barone, V., Contreras, R.H., and Snyder, J.P. (2002). DFT calculation of NMR J_{FF} spin–spin coupling constants in fluorinated pyridines. *J. Phys. Chem. A*, 106: 5607.
Duer, M.J., Ed. (2002). *Solid-State NMR Spectroscopy: Principles and Applications*. Oxford: Blackwell Sciences.
Günther, H. (2013). *NMR Spectroscopy: Basic Principles, Concepts and Applications in Chemistry*. Weinheim: Wiley–VCH.
Harris, R.K. (1983). *Nuclear Magnetic Resonance Spectroscopy*. Avon: Bath Press.
Hierso, J.C. (2014). Indirect nonbonded nuclear spin–spin coupling: a guide for the recognition and understanding of "through-space" NMR *J* constants in small organic, organometallic, and coordination compounds. *Chem. Rev.*, 114: 4838.
Lambert, J.B. and Riddel, F.G., Eds. (1982). *The Multinuclear Approach to NMR Spectroscopy*. Boston: Springer.
Lee, J-A. and Khitrin, A.K. (2008). High-precision measurement of internuclear distances using solid-state NMR. *Concepts Magn. Reson. Part A*, 32A: 56.
McConnel, H.M. and Chesnut, D.B. (1958). Theory of isotropic hyperfine interactions in p-electron radicals. *J. Chem. Phys.*, 28: 107.
Poole, C.P. (1983). *Electron Spin Resonance*. New York: Wiley.
Sadoc, A., Biswal, M., Body M., Legein, C., Boucher, F., Massiot, D., and Fayon, F. (2014). NMR parameters in column 13 metal fluoride compounds (AlF_3, GaF_3, InF_3 and TlF) from first principle calculations. *Solid State Nucl. Magn. Reson.*, 59–60: 1.
Zilm, K.W., Heinekey, D.M., Millar, J.M., Payne, N.G., Neshyba, S.P., Duchamp, J.C., and Szczyrba, J. (1990). Quantum mechanical exchange of hydrides in solution: proton–proton exchange couplings in transition-metal polyhydrides. *J. Am. Chem. Soc.*, 112: 920.

3 Nuclear Relaxation
Theory and Measurements

Nuclear relaxation recovers the initial equilibrium state of spins excited by radio-frequency (RF) irradiation, thus allowing observation of a NMR signal. In other words, the frequency- and time-dependent data collected by pulse NMR experiments represent the same phenomenon. In spite of this, NMR applications are grouped under two separate categories: NMR spectrometry and NMR relaxation. The former is most popular among researchers who use NMR for structural analyses; however, even in this case, knowledge on the theory of nuclear relaxation is needed to perform even the simplest NMR experiments and to interpret the results. To emphasize the importance of this statement, it should be noted that the first attempt, in 1936, to detect ^1H and ^7Li nuclei was unsuccessful because of the extremely long relaxation times. In addition, because nuclear relaxation affects the shapes of resonances and line widths in solutions and solids, an understanding of these effects is also necessary to avoid misinterpretations of NMR spectra. Relaxation studies can open the way not only to a deeper understanding of molecular dynamics via identification and characterization of motions on the large frequency scale but also to structural diagnostics via determination of interatomic distances and special relaxation criteria.

According to the general concept presented in Chapter 1, nuclear relaxation occurs via an energetic exchange between nuclear spins excited by radiofrequency energy and their environment. In turn, the exchange is possible only in the presence of fluctuating magnetic fields created in the lattice. These magnetic fields are generated by nuclear and/or electron magnetic moments, and the nuclei and the electrons are physically located in molecules. Therefore, the thermal molecular motions result in the oscillating magnetic fields and they are responsible for nuclear relaxation.

The frequency scale of molecular motions (motions of groups or atoms) can be very large—from slow translational motions and rotational reorientations to very fast molecular librations and vibrations with a frequency range of 10^{12} to 10^{14} Hz. In accordance with the general principles of NMR, spin–lattice relaxation can be effective when the frequencies of fluctuating are close to the Larmor resonance frequency, v_0 (or ω_0).

3.1 MOLECULAR MOTIONS: COMMON CHARACTERISTICS

Figure 3.1A shows a CH_2 group with protons H(A) and H(B), which are coupled by dipole–dipole interaction (DDI, see below). This interaction is given as vector oriented—for example, along the direction perpendicular to the applied external

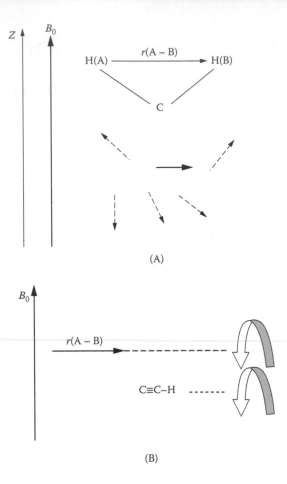

FIGURE 3.1   (A) Representation of a CH_2 molecular fragment that experiences rotational and translational reorientations in external magnetic field B_0. (B) The free rotation around the C–H axis in a molecular acetylene fragment.

magnetic field, B_0. The strength of a local magnetic field, created by proton A in the place of proton B, will then depend on the internuclear distance $r(A - B)$ and the angle formed by this $r(A - B)$ vector and the direction of the external magnetic field.

In the presence of rotational and/or translational motions, this angle will change, thus changing the local magnetic field. As a result, these motions will result in proton–proton dipole–dipole relaxation. Figure 3.1B illustrates another case, where the carbon and the proton (see bond C–H) experience dipolar coupling in a molecular acetylene fragment. However, the free rotation around the C–H axis does not change the spatial orientation of the dipolar vector, and this rotation motion does not contribute to the ^{13}C relaxation, at least not by the dipole–dipole mechanism. In reality, the situation is more complex because molecular reorientations in the external magnetic field can have a composite character, including low-amplitude vibrations and librations.

Any noncomposite molecular motion can quantitatively be characterized at a given temperature by a single motion correlation time, τ_C. The correct definition of the molecular motion correlation time, τ_C, comes from the so-called autocorrelation function in the theory of nuclear relaxation. Here, the τ_C value represents an average time necessary for a molecule to progress through one radian. However, in its simplest form, the τ_C can be formulated as a time between two molecular reorientations, or a tumbling time. Because any molecular reorientations in the solid state or in solutions require structural changes in the closest environments, the moving molecule must overcome an energy barrier (E_a), defined in Equation 3.1 as the activation energy:

$$\tau_C = \tau_0 \exp(E_a/RT) \tag{3.1}$$

where τ_0 is the correlation time constant. It is generally accepted that magnitude τ_0^{-1} is the frequency of attempts to overcome energy barrier E_a. In common cases, the τ_0 values are in a diapason of $\sim 10^{-13}$ to 10^{-14} s. According to Equation 3.1, upon cooling, molecular reorientations slow down following an exponential law, and a larger correlation time corresponds to a slower motion.

Molecular motions in liquids and in solids vary widely. Molecules can experience fast thermal rotational reorientations, slow and fast translational motions, or a combination of these. In addition, there are various intramolecular motions such as rotations around the simple (or partially double) chemical bonds and segmental motions in polymeric molecules. Therefore, these motions make their own contributions to nuclear relaxation as a function of motional frequency. In spite of the variation in motions, in many cases the molecules can be accepted as spheres, and the motions can be described by a single correlation time.

In accordance with the Stokes–Einstein–Debye theory regarding liquids and solutions, the molecular motion correlation times depend on molecular sizes (or molecular volumes) and the bulky viscosity of medium η (solvents). For spherical molecules behaving as totally rigid bodies, the rotational correlation time (τ_C) is written as follows:

$$\tau_C = (4\pi a^3 \eta)/3kT \tag{3.2}$$

where a is the molecular radius. Thus, in solutions, the molecular motion correlation times increase upon going from small organic molecules (typically, $\tau_C \leq 10^{-11}$ s) to polymeric systems (typically, $\tau_C \geq 10^{-9}$ s). In the case of nonspherical molecules (when the symmetry of molecules insignificantly deviates from the sphere), the rotational correlation time can generally be expressed via Equation 3.3:

$$\tau_C = [(4\pi a^3 \eta)/3kT]\phi C + \tau_0 \tag{3.3}$$

Here, ϕ is a shape parameter (accepted as 1 for a sphere); C, the slip coefficient, is a measure of the hindrance to rotation experienced by a molecule in dense liquids; and τ_0 is the inertia contribution to the overall rerientational time (this term is usually ignored in the case of liquids).[1] The equation shows that deviation from the spherical symmetry affects the τ_C time, which in turn will change relaxation times, as shown below.

In some cases, molecular motions in liquids are characterized by the rotational diffusion constant D, which is used instead of the correlation times (τ_C). The relationship between both of the magnitudes is shown in Equation 3.4.

$$D = 1/6\tau_C \tag{3.4}$$

The rotational molecular reorientations play a major role in solution NMR. Because the correlation times (τ_C) of these reorientations are between 10^{-11} and 10^{-9} s and the τ_0 values are 10^{-13} to 10^{-14} s, the motions lead to narrowing NMR signals, which usually have a Lorentz shape. One can show that such correlation times correspond to relatively low energy barriers between 0.8 and 5 kcal/mol. Thus, relaxation methods can be successfully applied for quantitative studies of molecular motions in solutions.

In general, molecular motions in solids are strongly slower. Their correlation times, corresponding to motions of whole molecules or only some groups within molecules or molecular aggregates, have been estimated to be between 10^{-4} and 10^{-6} s in a temperature region between $-150°$ and $+250°C$. Thus, the activation energies of these motions in solids will grow to 18 to 25 kcal/mol or even more. Nevertheless, small molecules can show fast reorientations even in the solid state. For example, the solid compound BD_3ND_3 in the high-temperature tetragonal phase (above 225 K) shows reorientations of BD_3 and ND_3 groups, which correspond to the E_a values of 1.4 kcal/mol at $\tau_0 = 1.1 \times 10^{-13}$ s and of 1.7 kcal/mol at $\tau_0 = 4.4 \times 10^{-14}$ s, respectively. In the low-temperature orthorhombic phase, these motions require higher energies: $E_a = 3.2$ to 6.2 kcal/mol.[2] Some of the fast motions in solids and their studies will be considered in Chapter 9.

Finally, solids can sometimes manifest surprising motions that look like motions in liquids. This is particularly the case for plastic crystals consisting of small weakly interacting molecules. These molecules possess an orientational degree of freedom and move quickly and isotropically to give sharp resonances in the NMR spectra. Another example is the solid complexes $[M_2(-HC^1C^2C^3Me_2)(CO)_4(C_5H_5)_2]^+[BF_4]^-$ (where M = Mo, W), which show a rotation about the C–C$^+$ bond on the NMR time scale with a free energy $\Delta G^{\neq}$ of 13 to 14 kcal/mol.[3]

3.1.1 ISOTROPIC AND ANISOTROPIC MOLECULAR REORIENTATIONS

High- or low-amplitude molecular motions can be described as isotropic or anisotropic. This classification is very important because isotropic and anisotropic motions have different effects on NMR spectra and line shapes of resonances, particularly in solids, as well as on relaxation times in solutions and solids. As noted above, a molecular motion can be accepted as isotropic when a single τ_C value (at a given temperature) and a single E_a value completely describe the motion. For geometric reasons, the fully symmetrical rigid spherical molecules shown in Figure 3.2A undergo isotropic rotational reorientations, particularly in dilute solutions. Again, due to the symmetry, for example, motions of the octahedral rhenium complexes $Re(CO)_6$ or $Mn(CO)_6$ in solutions can be satisfactorily described as isotropic.

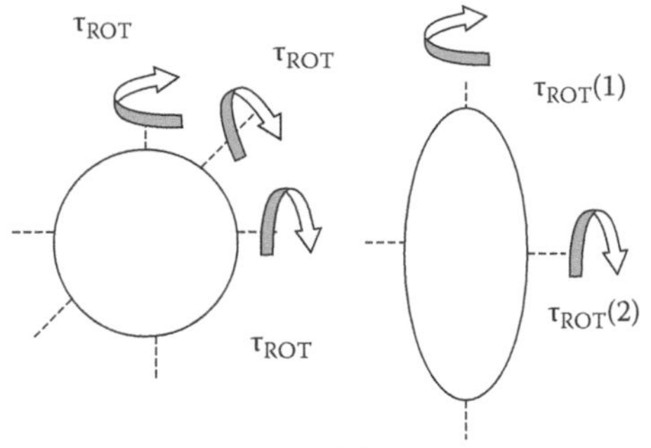

(A)

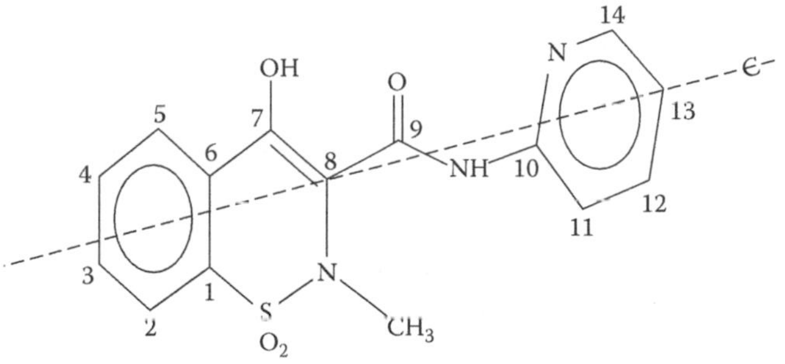

(B)

Carbon No.	T_1 (s)	τ_c (s)
14	0.476	$9.1 \; 10^{-11}$
12	0.476	$9.1 \; 10^{-11}$
4	0.312	$1.42 \; 10^{-10}$
3	0.25	$1.8 \; 10^{-10}$
5	0.385	$1.14 \; 10^{-10}$
2	0.385	$1.14 \; 10^{-10}$
13	0.196	$2.40 \; 10^{-10}$
11	0.500	$8.70 \; 10^{-10}$

FIGURE 3.2 (A) A fully symmetrical sphere and the simple symmetrical ellipsoid experiencing isotropic and anisotropic reorientations, respectively. (B) The ^{13}C T_1 times and the effective correlation times obtained for the organic molecule in DMSO-d_6 (295 K).

Even a simple symmetrical ellipsoid, however, is characterized by two different directions of spatial extension (Figure 3.2A). It is obvious that molecules of such a shape (or close to ellipsoid) will have at least two different inertia moments. As a result, rotational reorientations around these directions will be characterized by two correlation times, $\tau_c(1)$ and $\tau_c(2)$, and by two values of activation energy. The difference between them will depend on the geometries of real objects. In some sense, the organic molecule depicted in Figure 3.2B can be accepted as an ellipsoid in molecular

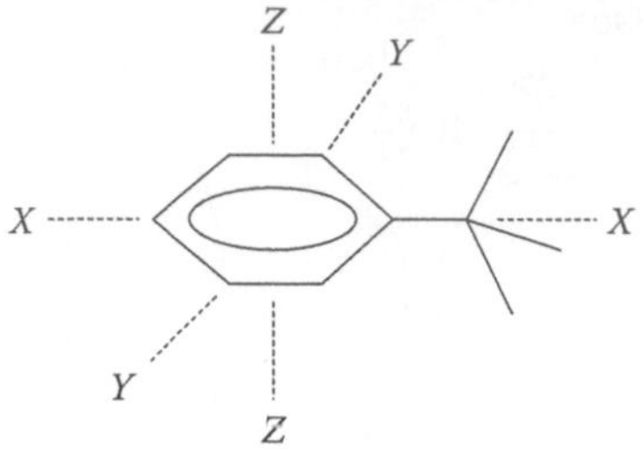

Z

Y

X ---------- X

Y

Z

Axis	τ_C (295 k) (s)	τ_0 (s)	E_a kcal/mol
XX	$0.98\ 10^{-11}$	$2.5\ 10^{-12}$	0.8
YY	$2.2\ 10^{-11}$	$2.9\ 10^{-12}$	1.2
ZZ	$1.1\ 10^{-11}$	$0.25\ 10^{-12}$	2.2

FIGURE 3.3 Molecular reorientations around axes ZZ, YY, and XX in toluene and the parameters of anisotropic motions obtained in net liquid.

rotation around the given axis. The table in this figure lists effective molecular motion correlation times that were calculated from the $^{13}C\ T_1$ times on the basis of an isotropic model.[4] As follows from the data, the correlation times obtained for different labels of the same molecule are quite different, and the difference can reach a factor of 9.7. Similarly, the difference in the τ_C values is still significant for toluene molecules in net liquid for anisotropic motions. Figure 3.3 illustrates these toluene motions as reorientations around the axes ZZ, YY, and XX, where the τ_C value and the activation energy change strongly for ZZ-axis reorientations. In the case of water, the difference is significantly smaller. The rotational correlation time determined for motions of the O–D bond vector in D_2O changes between 5.8 ps at 275 K and 0.86 ps at 350 K, while the out-of-plane vector correlation time ranges from 4.4 ps at 275 K to 0.64 ps at 350 K. Thus, even the rotational motions of water are anisotropic.[5]

Generally speaking, molecular motions of complex molecular systems such as bulky inorganic aggregates or peptides are always anisotropic, even in solutions. They should be described by a number of motional parameters corresponding to the correlation times of overall tumbling and the correlation times of slower and faster internal motions to account for their weighting coefficients.[6]

3.2 MECHANISMS OF SPIN–SPIN AND SPIN–LATTICE NUCLEAR RELAXATION

In general, molecular motions modulate the fluctuating magnetic fields around nuclei via time-dependent dipole–dipole, quadrupole, spin–rotation, scalar, and chemical shift anisotropy interactions, resulting in nuclear relaxation by various mechanisms. Depending on the magnetic properties of the nuclei and the nature of the compounds, different relaxation mechanisms can operate simultaneously, providing the corresponding independent contribution to a total relaxation rate. In some cases, two "independent" mechanisms can experience interference to give cross-relaxation contributions. This is valid for liquids and solids; however, rigid solids can show a unique relaxation mechanism referred to as *spin diffusion*.

By definition, nuclear relaxation is effective when frequencies of molecular reorientations or molecular groups are close to the Larmor frequency. Therefore, independently of the nature of the condensed phase, the relaxation rates $1/T_1$ and $1/T_2$ (or $1/T_{1\rho}$, considered below) can be commonly expressed via Equation 3.5:

$$1/T_{1,2} = CJ(\omega_0, \tau_C) \tag{3.5}$$

where C can be taken as the force constant (dipolar or quadrupolar coupling and other) and $J(\omega_0, \tau_C)$ is a function of the spectral density. This function shows how a combination of resonance frequency ω_0 ($\omega_0 = 2\pi\nu_0$, where ν_0 is the working frequency of a spectrometer expressed in Hz) and molecular motion correlation time τ_C weakens the influence of the C constant on the relaxation rate. Thus, the C constant depends on the nature of nuclei and molecular structures, while the type of the $J(\omega_0, \tau_C)$ function is dictated by molecular mobility (τ_C) and the type of molecular motions. Here one should remember that Equation 3.5 implies only isotropic molecular reorientations.

3.2.1 Intramolecular Dipole–Dipole Relaxation

Intramolecular dipole–dipole interactions between two nuclei of one sort (e.g., protons) or two nuclei of different sorts, one of which is an observed proton, give rise to homonuclear or heteronuclear dipolar coupling, respectively (see Equation 2.22 in Chapter 2). The force dipolar constants, DC_{H-H} and DC_{H-B}, measured in Hz, are given by Equations 3.6 and 3.7:

$$DC_{H-H} = 0.3(\mu_0/4\pi)^2 \gamma_H^4 \hbar^2 r(H-H)^{-6} \tag{3.6}$$

$$DC_{H-B} = (4/30)(\mu_0/4\pi)^2 r(H-B)^{-6} \gamma_H^2 \gamma_B^2 \hbar^2 I_B (I_B + 1) \tag{3.7}$$

where γ_H and γ_B are the nuclear magnetogyric constants of ^1H and B nuclei, respectively; I_B is the spin of the B nucleus; μ_0 is the permeability of the vacuum; and $r(H-H)$ and $r(H-B)$ are the internuclear distances.[7] Equation 3.7 is written for 100% natural abundance of the B nucleus; therefore, it should be modified when the natural abundance of the B nucleus is smaller (e.g., ^{13}C, ^2H, ^{11}B). One can show that, in general, the DC constants will have an influence on dipolar nuclear relaxation when their values are between 10^4 and 10^5 Hz.

Two important consequences from the theory can be deduced. First, the equations show the force constants to be proportional to the inverse sixth power of the internuclear distance. This strong dependence demonstrates, for example, that the force of dipolar coupling reduces by 244 times when the internuclear distance decreases from 1 to 2.5 Å. Thus, theoretically, intramolecular dipole–dipole interactions are always more effective than intermolecular dipolar contacts. At the same time, in practice, this statement is only valid for solutions. In fact, solid-state intermolecular dipolar interactions can also be strong because distances between molecules are reduced and motions are limited.

Equations 3.6 and 3.7 are written for a couple of nuclei. In reality, all of the possible dipole–dipole contacts should be taken into account for interpretation of relaxation data. Due to the large γ factors and the high natural abundance, the dipolar coupling strength will be highest for a pair of protons and ^{19}F nuclei or ^{31}P nuclei and minimal if neighboring nuclei have low γ and low natural abundance. Comparing the properties of ^{1}H and ^{2}H isotopes shows that dipolar coupling will be remarkably reduced upon the replacement of protons for deuterons. That is why the use of isotropic displacement is popular in relaxation experiments, as it eliminates undesirable dipole–dipole contributions, such as intermolecular dipolar interactions with solvents. In fact, dipole–dipole interactions of protons in metal–H bonds and deuterons of solvent C_6D_6 contribute only 3% to the ^{1}H T_1 relaxation rate in the rhenium complex $HRe(CO)_5$.

According to the Bloembergen–Purcell–Pound (BPP) theory, spectral density functions $J(\omega_0,\tau_C)$ are given in the forms of Equations 3.8 and 3.9 for homonuclear dipolar spin–lattice (T_1) and spin–spin (T_2) relaxation, respectively:

$$1/T_1^{DD} = (2/5)\gamma^4\hbar^2 r^{-6}I(I+1)\left[\tau/\left(1+\omega_i^2\tau^2\right)+4\tau/\left(1+4\omega_i^2\tau^2\right)\right] \qquad (3.8)$$

$$1/T_2^{DD} = (1/5)\gamma^4\hbar^2 r^{-6}I(I+1)\left[3\tau+5\tau/\left(1+\omega_i^2\tau^2\right)+2\tau/\left(1+4\omega_i^2\tau^2\right)\right] \qquad (3.9)$$

where the influence of the τ_C values on T_1 and T_2 times differs significantly. Because the molecular motion correlation times τ_C are temperature dependent (see Equation 3.1), the T_1 plots in semilogarithmic coordinates in Figure 3.4 will be symmetrical and V-shaped, passing upon cooling through minima. The minima are observed at the τ_C value in Equation 3.10:

$$\tau_C = 0.62/\omega_0 \qquad (3.10)$$

In contrast, upon cooling, the T_2 time in Equation 3.9 reduces consistently, corresponding to an increase in line widths in NMR spectra after the T_1 minimum in the so-called zone of wide lines. The effect is particularly important in solids. In fact, due to fast molecular reorientations in solutions, T_1 minima are generally observed at the lowest temperatures, especially for relatively small molecules. However, in solids, where molecular motions are slow, the T_1 minima can be reached only in high-temperature zones.

As a result, at room temperature and lower, the resonances will be strongly broadened due to short T_2 times. These times will be quite a bit smaller than T_1 times. For example, ^{19}F nuclei in F^- ions and ^{29}Si nuclei in the silicate octadecasil show $T_1(^{19}F) = 2.6$ s vs. $T_2(^{19}F) = 3.5$ ms and $T_1(^{29}Si) = 22$ s vs. $T_2(^{29}Si) = 18$ ms at room temperature.

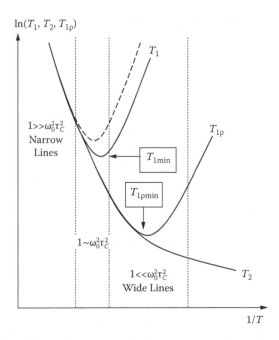

FIGURE 3.4 Temperature dependences (in $1/T$ units) of the dipole–dipole relaxation times T_1, T_2, and $T_{1\rho}$, presented in the semilogarithmic coordinates. The dashed T_1 curve corresponds to an experiment performed at a higher magnetic field. The regions $1 \gg \omega_0^2\tau_C^2$, $1 \sim \omega_0^2\tau_C^2$, and $1 \ll \omega_0^2\tau_C^2$ correspond to fast, intermediate, and slow molecular motions, respectively, on the frequency scale of NMR.

Similarly, the spectral density functions $J(\omega_0,\tau_C)$ take the form of Equations 3.11 and 3.12 for the heteronuclear dipole–dipole relaxation times T_1 and T_2, respectively:

$$1/T_1 = (4/30)\left(\mu_0/4\pi\right)^2 r(A-B)^{-6}\gamma_A^2\gamma_B^2\hbar^2 I_B\left(I_B+1\right)\times \tag{3.11}$$

$$\left\{\left[3\tau/\left(1+\omega_A^2\tau^2\right)+6\tau/\left(1+\left(\omega_A+\omega_B\right)^2\tau^2\right)+\tau/\left(1+\left(\omega_A-\omega_B\right)^2\tau^2\right)\right]\right\}$$

$$1/T_2 = (4/30)\left(\mu_0/4\pi\right)^2 r(A-B)^{-6}\gamma_A^2\gamma_B^2\hbar^2 I_B\left(I_B+1\right)\times \tag{3.12}$$

$$\left[\begin{array}{l} 4\tau+3\tau/\left(1+\omega_A^2\tau^2\right)+6\tau/\left(1+\omega_B^2\tau^2\right)+\tau/\left(1+\left(\omega_A-\omega_B\right)^2\tau^2\right) \\ +6\tau/\left(1+\left(\omega_A+\omega_B\right)^2\tau^2\right) \end{array}\right]$$

where ω_A and ω_B are the resonance frequencies of the A and B nuclei.

The important element of the BPP theory is the dependence of disposition of the T_1 minimum on the curve shown in Figure 3.4 on the strength of the external magnetic field. The T_1 minima are shifted toward high-temperature zones with increasing magnetic field strength, opening the way to measurements of T_{1min} times for small molecules, even in solutions.

Analogous to spin–lattice relaxation time T_1, relaxation time $T_{1\rho}$ measured in the rotating coordinate frame for homonuclear dipolar coupling is given in Equation 3.13:

$$1/T_{1p} = (1/5)\gamma^4\hbar^2 r^{-6} I(I+1)\left[3J^D(2\omega_I) + 5J^D(\omega_I) + 2J^D(2\omega_I)\right] \quad (3.13)$$

Thus, the $T_{1\rho}$ plots in semilogarithmic coordinates also show minima. Because an effective radiofrequency field operating in spin-locking NMR experiments for $T_{1\rho}$ measurements is significantly weaker than magnetic field B_0, the $T_{1\rho}$ minima in solutions can be observed only at the lowest temperatures relative to those for observation of the T_1 minima. Generally, T_1 minima can be easily detected only in solutions of large molecules or in solids due to slow molecular motions.

Equations 3.8 to 3.13 account for a single dipole–dipole contact. Because the relaxation rates are of additive magnitudes, in the presence of a large number of such contacts all of the dipolar contributions should be summed as shown in Equation 3.14:

$$1/T_1^{OBS} = \sum 1/T_1 \quad (3.14)$$

Finally, consideration of Figure 3.4 and comparison of Equations 3.8 and 3.9, or Equations 3.11 and 3.12, lead to a very important conclusion: Relaxation experiments performed for T_1/T_2 measurements in the wide-line zone (i.e., at $T_1 \neq T_2$) give directly (via T_1/T_2 ratios) the correlation times τ_C without knowledge of the dipolar constants. This approach is often used for solids. Another conclusion that can be drawn from T_1 and T_2 temperature dependencies is that at lower temperatures, even in solutions, the spectral resolution can be controlled by short T_2 times but not homogeneity of the magnetic field.

3.2.2 INTERMOLECULAR DIPOLE–DIPOLE RELAXATION

Fast translational molecular motions in solutions modulate the oscillating local magnetic fields, which lead to intermolecular dipolar relaxation of nuclear spins, located in different molecules. When molecules are spherical and translational, molecular motions are fast (see the zone of fast reorientation at $1 \gg \omega_0^2\tau_C^2$ in Figure 3.4), and the spin–lattice relaxation rate of spins I ($1/T_1(I)$) by translational molecular motions can be given by

$$1/T_1(I) = \left(\mu_0/4\pi\right)^2 (8/45) N_S \gamma_I^2 \gamma_S^2 \hbar^2 \left[S(S+1)/D_{IS}\right] r(I-S) \quad (3.15)$$

where N_S is the concentration of spins S, $r(I-S)$ is the closest interatomic distance, and D_{IS} is the translation self-diffusion constant. When the N_S and $r(I-S)$ values are known, then the diffusion constants can be determined by accurate evaluation of the intermolecular relaxation contribution. In general, this evaluation is not simple. Therefore, diffusion coefficients in liquids are measured more often by applications of gradient pulse techniques, which are considered Chapter 7.

McConnell[8] gives a more attractive (from a practical point of view) expression for the intermolecular dipolar relaxation of identical spins in Equation 3.16. This equation shows the relationship among the T_1 time, viscosity η, and the number of independent magnetic spins per unit volume:

$$1/T_1 = 32\pi^2\gamma^2 N\,I(I+1)\eta/15kT \tag{3.16}$$

Finally, elimination, or at least minimization, of the relaxation contribution caused by intermolecular interactions can be easily realized by simply decreasing the concentrations of solutions. In fact, this contribution is often considered to be undesirable, particularly for the application of structural relaxation criteria or for studies of molecular dynamics.

3.2.3 QUADRUPOLAR RELAXATION

Because spins I of a quadrupolar nucleus can interact with the electric field gradient eq_{ZZ} at this nucleus, generally oriented along a chemical bond, random rotational reorientations of this bond result in a fluctuating magnetic field, which, in turn, causes spin–lattice relaxation with time constant $T_1(Q)$ in Equation 3.17:

$$1/T_1(Q) = (3/50)\pi^2\,(2I+3)\big(I^2\,(2I-1)\big)^{-1}\big(e^2 q_{ZZ}Q/h\big)^2\big(1+\eta^2/3\big)\times \tag{3.17}$$
$$\left[\tau_C/\big(1+\omega_Q^2\tau_C^2\big)+4\tau_C/\big(1+4\omega_Q^2\tau_C^2\big)\right]$$

where η is the asymmetry parameter of the electric field gradient and I is the nuclear spin. The spectral density function $J(\omega_0,\tau_C)$ takes the same form as in the case of homonuclear dipole–dipole interactions.

Potentially, this mechanism can be very powerful to give very short relaxation times and very broad resonances even in solutions because the electric field gradient evolves directly at the nucleus. In fact, generally, T_1 times of quadrupolar nuclei in solutions and the solid state are remarkably shorter than those measured for nuclei with a spin of 1/2. However, one should remember that $T_1(Q)$ values can vary within large limits depending on the asymmetry of an environment. For example, the $T_1(^{127}I)$ time in a solution of SnI_4 is measured at room temperature as 0.15×10^{-3} ms vs. 1 ms in IF_6^+ ions, where the ^{127}I environment is symmetrical. The $T_1(^{14}N)$ time reduces dramatically from 1220 ms in MeNC to 0.3 ms in $(Co(NH_3)_6)(ClO_4)_3$. Similarly, the $T_1(^{35}Cl)$ value decreases upon going from symmetrical ClO_4^- (270 ms) to nonsymmetrical $CFCl_3$ (38.3×10^{-3} ms). It is also apparent that with an ineffective quadrupolar mechanism quadrupolar nuclei will relax via, for example, the dipolar mechanism.

As mentioned above, the spectral density functions $J(\omega_0,\tau_C)$ in the quadrupolar and dipolar relaxation are identical. Therefore, the temperature $T_1(Q)$ dependencies are again V-shaped and symmetrical in their semilogarithmic coordinates (the case of isotropic molecular motions), and they pass again through minima at $\tau_C = 0.62/\omega_Q$. In the high-temperature zone (the zone of fast molecular motions), Equation 3.17 converts to Equation 3.18:

$$1/T_1(Q) = 0.3\pi^2 (2I+3)\left(I^2 (2I-1)\right)^{-1} \left(e^2 q_{ZZ} Q/h\right)^2 \left(1+\eta^2/3\right)\tau_C \qquad (3.18)$$

This form is convenient for experimental determinations of nuclear quadrupolar coupling constants (QCCs) when correlation time τ_C is found independently or *vice versa*. It should be noted that for the τ_C values of 10^{-11} to 10^{-12} s, the $T_1(Q)$ times are primarily dictated by quadrupolar coupling constant values.

Some uncertainty in Equation 3.18 is connected with the asymmetry parameter of the electric field gradient, η, which is unknown *a priori*. It can be found independently by a line-shape analysis performed for NMR signals of quadrupolar nuclei in static solids. However, because the η value changes between 0 and 1 (see Chapter 2), the factor $(1 + \eta^2/3)$ in Equation 3.18 can increase between 1 and 1.33, respectively. Thus, this uncertainty does not seem to be large in the context of the τ_C values determined by T_1 measurements.

Finally, an important question for researchers is can a nucleus (S) with spin of 1/2, neighboring with a quadrupolar nucleus, be involved in the quadrupolar relaxation mechanism? Theoretically, this event is possible, if each spin flip of a quadrupolar nucleus relaxing via the quadrupolar mechanism is accompanied by the corresponding spin flip of nucleus S. However, the probability of this phenomenon is negligible due to large differences in the energies of corresponding levels.

3.2.4 Relaxation by Chemical Shift Anisotropy

In the presence of chemical shift anisotropy (CSA) as a manifestation of three-dimensional chemical shift, random molecular motions in solutions or the solid state modulate the local fluctuating magnetic fields. In turn, these oscillating fields cause nuclear relaxation of nucleus I via the CSA mechanism. The corresponding T_1 and T_2 times and the spectral density functions $J(\omega_I, \tau_C)$ can be represented in Equations 3.19 and 3.20, respectively:

$$1/T_1(\text{CSA}) = (1/15)\gamma_I^2 B_0^2 (\Delta\sigma)^2 \left[2\tau_C / \left(1+4\omega_I^2 \tau_C^2\right) \right] \qquad (3.19)$$

$$1/T_2(\text{CSA}) = (1/90)\gamma_I^2 B_0^2 (\Delta\sigma)^2 \left[8\tau_C + 6\tau_C / \left(1+4\omega_I^2 \tau_C^2\right) \right] \qquad (3.20)$$

This mechanism is obviously effective for nuclei showing the large CSA ($\Delta\sigma$) values. The $\Delta\sigma$ values can be large, such as for ^{13}C, ^{15}N, ^{19}F, ^{31}P, or ^{195}Pt nuclei. Sometimes, the CSA contribution can be significant even for protons. For example, the H–Ir hydride ligand in the complex $[HIrCl_2(PMe_3)_2]$ shows a CSA value of 100 ppm, thus providing a significant CSA contribution to the total relaxation rate.[9]

The important feature of the CSA mechanism is that the relaxation time is field dependent even for $\omega_I^2 \tau_C^2 \ll 1$ (i.e., in the zone of fast molecular motions), in contrast to dipolar and quadrupolar relaxation. As follows from Equations 3.19 and 3.20, $T_{1,2}(\text{CSA})$ times decrease with increasing strength of the external magnetic field proportionally to B_0^2. First, this feature reveals the presence of CSA relaxation. Second,

NMR experiments performed in the highest magnetic fields B_0 can potentially increase contributions of CSA relaxation to total relaxation times even for protons with relatively low $\Delta\sigma$ values (~20 ppm). This effect should be taken into account for studies of quadrupolar nuclei probed at the highest magnetic fields.

Generally speaking, complete domination of the CSA relaxation mechanism in solutions and in solids is a rather rare phenomenon, although it is possible when other relaxation channels are absent or minimized. For example, ^{13}C nuclei in proton-free fullerene molecules (C_{60}) show field-dependent T_1 times (chlorobenzene-d_5) corresponding to a 67% contribution of the CSA mechanism. The domination of CSA relaxation can be revealed in practice by measurements of T_1 and T_2 times. Equation 3.21 can be obtained from Equations 3.19 and 3.20:

$$T_1(CSA)/T_2(CSA) = 7/6 \qquad (3.21)$$

which shows a T_1/T_2 ratio of 7/6. This ratio corresponds to the situation when nuclei relax only via CSA interactions.

3.2.5 SPIN–ROTATION AND SCALAR RELAXATION MECHANISMS

The spin–rotation (SR) mechanism is generally effective in solutions for nuclear relaxation of relatively small-sized molecules that show fast rotations in non-viscous media. Magnetic moments of electrons in molecules create magnetic fields even in the absence of external magnetic field B_0. Rotational molecular reorientations cause fluctuation of these fields, resulting in spin–rotation relaxation with time constant T_1, as shown in Equation 3.22:

$$1/T_1(SR) = I_r^2 C^2/9\hbar^2 \, \tau_C \qquad (3.22)$$

where I_r is the molecular inertia moment; C is the spin–rotational constant, measured in frequency units; and τ_C is the rotational correlation time.

It should be noted that the simple form of Equation 3.22 is valid only for spherical molecules, where the effectiveness of SR relaxation is governed by force constant C. For some nuclei, such as ^{31}P or ^{19}F, the C values are large enough to provide the SR contributions to total relaxation times; for example, the C constant for ^{19}F nuclei can reach 2000 Hz.

Equation 3.22 shows the unique features of the spin–rotational mechanism. In contrast to other mechanisms, $T_1(SR)$ times, measured in the high-temperature range, decrease upon heating. This feature can be used as a good test for the presence of SR relaxation. For example, this behavior is often observed for ^{31}P relaxation of phosphorus-containing compounds in solutions. Finally if the C constant is known and the SR mechanism dominates, then the rotational correlation times can be easily determined from T_1/T_2 measurements.

Figure 3.5A shows nucleus A coupled to nucleus X. The A–X coupling becomes time dependent due to X spin flips. As a result, the fluctuating scalar interaction affects the spin–spin relaxation of nucleus A and line shape of its resonance. This

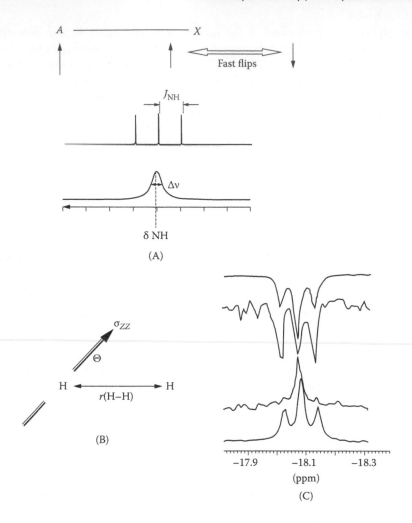

FIGURE 3.5 (A) Scalar relaxation of the second kind shown schematically for nucleus A. (B) The vector $r(H\cdots H)$ experiences a fast tumbling, where Θ is the angle between the $r(H-H)$ vector and the principal axis of the proton chemical shift tensor. (C) The ^1H NMR inversion–recovery spectra obtained for H^A ligand in complex $(H^X)_2$–Ta(Cp)$_2$–H^A–W(CO)$_5$ at increasing delay time from top to bottom. (From Bakhmutov, V.I. et al., *Inorg. Chem.*, 36, 4055, 1997. With permission.)

so-called scalar relaxation of the second kind corresponds to broadening effects, as has been shown for the ^1H resonance of groups HNR$_2$ or H$_2$NR. For slow relaxation of ^{14}N nuclei, the ^1H resonance exhibits a 1:1:1 triplet with a $^1J(^1H-^{14}N)$ constant of ~60 Hz, which transforms to a broad singlet due to fast flip-flop motions of ^{14}N spins during a detection period at the proton frequency. The line width, Δv_{obs}, now corresponds to time T_2(Scalar) in Equation 3.23:

$$\Delta v_{obs} = 1/\pi T_2 \text{ (Scalar)} \tag{3.23}$$

where the T_2(Scalar) time is given by Equation 3.24:

$$1/T_2 \text{ (Scalar)} = (8/3)\pi^2 J^2 (AX) * I_X \left(I_X + 1\right) T_{1X} \Big/ \left[1 + \left(\omega_X - \omega_A\right)^2 T_{1X}^2\right] \quad (3.24)$$

Thus, the T_2(Scalar) time determined experimentally allows calculation of the T_1 time for X nuclei when the spin–spin coupling constant $J(AX)$ is known and *vice versa*. In practice, this type of nuclear relaxation plays an important role only in high-resolution NMR spectra in solutions, where the broadening effects can be easily observed. Moreover, the broadening effect observed for nuclei neighboring with quadrupolar nuclei can be used for assignments of signals in NMR spectra.

3.2.6 CROSS-RELAXATION AND COUPLED RELAXATION

Nuclei commonly can relax in solutions and solids by different mechanisms simultaneously. For example, when relaxation includes dipole–dipole interaction, chemical shift anisotropy, and spin–rotational pathways, the total relaxation rate $1/T_1(tot)$ can be written as

$$1/T_1(tot) = 1/T_1(DDI) + 1/T_1(CSA) + 1/T_1(SR) \quad (3.25)$$

where all of the contributions are independent. They can be evaluated by convenient spectral or chemical procedures (e.g., replacement of protons for deuterons can reduce dipole–dipole interactions). However, two independent mechanisms of different natures can experience interference to give a cross-correlation mechanism. The interference is often observed for the dipole–dipole and CSA interactions in solutions. In fact, cross-correlations between the ^{15}N CSA and ^{15}N–^{1}H dipole–dipole interactions lead to differential transverse relaxation times, measured for two components of ^{15}N–H doublets.[10]

A pair of protons separated by the distance $r(\text{H}\cdots\text{H})$ and participating in a fast molecular tumbling (Figure 3.5B) provides an example of a quantitative consideration of the interference for dipolar and CSA mechanisms. The cross-relaxation term, $1/T_1(DDI,CSA)$ can be written in Equation 3.26 as a function of the dipolar relaxation rate, $R(DDI)$ (i.e., $1/T_1(DDI)$):

$$1/T_1 \text{ (DDI,CSA)} = -\left(4\pi/5\sqrt{3}\right)\left(3\cos^2\Theta - 1\right)\left(v_0 \,\Delta\sigma\right)* \quad (3.26)$$
$$R(DDI)\Big/\left[\left(\mu_0/4\pi\right)\gamma_H^2\hbar/r(\text{H}\cdots\text{H})^3\right]$$

where Θ is the angle between the $r(\text{H}-\text{H})$ vector and the principal axis of the proton chemical shift tensor.[11] Theoretically, one can show that at $r(\text{H}-\text{H}) = 2.4$ Å, $\Delta\sigma = 26$ ppm, and $v_0 = 400$ MHz, Equation 3.26 gives the ratio shown below when the principal axis of the proton chemical shift tensor is perpendicular to the $r(\text{H}-\text{H})$ vector:

$$1/T_1(DDI,CSA) = 0.3R(DDI) \quad (3.27)$$

Thus, according to Equation 3.27, the cross DDI/CSA mechanism can provide up to 30% of the measured relaxation rate, particularly for strong external magnetic fields and therefore large $\Delta\sigma$ values. In other words, even for protons this interference is not negligible. For solution NMR, the presence of the interference DDI/CSA terms can be determined experimentally by application of the pulse sequence $180° \to \tau \to 20°$ instead the standard inversion–recovery experiments performed for T_1 determinations (see the recommended literature).

Coupled relaxation is a special phenomenon in NMR that can potentially appear during the measurement of relaxation times in molecules with strongly coupled spin systems. As has been shown theoretically for a coupled spin system, the second hard 90° pulse, acting as the registering pulse in the inversion–recovery pulse sequence (see below), gives mixing the *eigenstate* spin populations.[12] Due to this mixing, which is theoretically possible even in the case of simple spin system AX_2, the distribution of line intensities for A nuclei will not be an equilibrium distribution (i.e., 1/2/1). A remarkable perturbation of equilibrium line intensities can be seen experimentally when the inverted spin system goes to equilibrium. Figure 3.5C demonstrates this effect for the bimetallic trihydride complex $(H^X)_2-Ta(Cp)_2-H^A-W(CO)_5$ observed in the ^1H inversion–recovery NMR spectra in solutions.[13] Based on the data, one of the partially relaxed ^1H NMR spectra shows only the central component of triplet H^A, while the other components are close to zero.

At a quantitative level, treatment of the data gives a ^1H T_1 difference between the central and side lines approaching 40%. The central line corresponds to the T_1 times of 1.68 s vs. 2.33 and 2.16 s obtained for the side lines. Generally speaking, an accurate analysis of the coupled relaxation is not simple. First, the coupled relaxation can be multi-exponential and, second, the analysis requires treatments based on a density matrix. At the same time, in practice, the coupled relaxation can be successfully approximated by a single effective T_1 time if the line intensities are minimally perturbed in the collected data.

3.3   SPIN DIFFUSION IN SOLIDS

Due to the very large electron magnetic moment, unpaired electrons in paramagnetic centers in rigid solids (often present as impurities) give rise to the phenomenon of spin diffusion. Generally speaking, nuclear spin–lattice relaxation in rigid solids should be extremely slow because molecular motions are strongly restricted. However, even very small concentrations of paramagnetic centers can increase the spin–spin relaxation rate, $1/T_1$, due to mutual spin flips (see Figure 3.6). This mechanism, Bloembergen spin diffusion,[14] is based on flips of spins belonging to dipolar-coupled nuclei that lead to energy-conserving spin transitions. The phenomenon consists of spin transfer from the initial location in a sample to a paramagnetic center, where it relaxes very rapidly due to powerful electron–nucleus dipolar interactions (see below). The measured relaxation times, T_1^{SD}, are controlled by spin diffusion coefficient D in Equation 3.28. In turn, coefficient D is defined by

$$1/T_1^{SD} = (1/3)8\pi NpC^{1/4}D^{3/4}, \text{ where } C = (2/5)\gamma_I^2\gamma_S^2\hbar^2 S(S+1) \qquad (3.28)$$

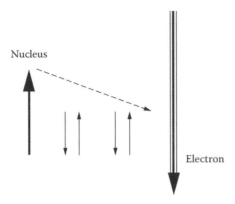

FIGURE 3.6 The relaxation mechanism caused by diffusion of spins. These spins are dipolar coupled and transfer to a paramagnetic center due to the energy-conserving spin transitions.

and

$$D = (M^{1/2}/30)a^2 \tag{3.29}$$

In these equations, Np is the paramagnetic center density, M is a second moment of the dipolar internuclear interaction, and a is the internuclear distance.

Equations 3.28 and 3.29 show that the effectiveness of the spin-diffusion mechanism depends strongly on internuclear dipolar coupling constant C and nuclear constant γ. In addition, the effectiveness depends on the natural abundance of nuclei increasing or reducing the strength of the dipolar coupling. Therefore, spin diffusion can dominate (or at least provide a significant contribution to) spin–lattice relaxation of protons or ^{27}Al nuclei with 100% natural abundance. For ^{13}C nuclei (e.g., in polymer systems) or for ^{29}Si nuclei (e.g., in kaolinite), this mechanism is practically negligible.

In contrast to nuclear relaxation by dipolar, quadrupolar, or CSA interactions directly connected with molecular mobility and structurally (or spectrally) important parameters, the T_1^{SD} measurements result in spin-diffusion constants D. These constants characterize domain sizes in heterogeneous organic or inorganic solids rather than their structure. For protons, the D values in various polymers are typically around 0.6 to 0.8 nm^2/ms.[15] The values of 9.7×10^{-9} to 1.5×10^{-6} m^2/s have been obtained for acidic protons as a function of number x in solids $[(NH_4)_{1-x}Rb_x]_3H(SO_4)_2$ $(0 < x < 1)$.[16]

Finally, in addition to the spin-diffusion mechanism unique for solids, heavy nuclei with spins of 1/2 can show unusual relaxation mechanisms in solids. For example, ^{207}Pb nuclei can relax via a spin–phonon Raman scattering mechanism, where the phonons create a local magnetic field by modulating the valence electron shell motion relative to the nuclear/electron core. The T_1 times are independent of the magnetic field strength but are inversely proportional to the square of the temperature. Both of the features can be used to detect this unusual relaxation process.[17]

3.4   PARAMAGNETIC RELAXATION

Magnetic fields created by a large magnetic moment of an unpaired electron, μ_S $= -g_e\beta_e S$, can cause nuclear relaxation via direct dipolar and/or contact interactions between nuclei and electrons located in paramagnetic centers when their concentrations are relatively large. This paramagnetic relaxation can be observed in solutions and the solid state. The paramagnetic dipolar mechanism is controlled by reorientations of electron–nucleus vectors r in accordance with Equations 3.30 and 3.31 written for spin $S = 1/2$ and the T_1 and T_2 times, respectively.

$$1/T_1^{DD} = 0.1\gamma_I^2\gamma_S^2\hbar^2 r^{-6}\left[\begin{matrix}\tau_e\Big/\left(1+\left(\omega_I-\omega_S\right)^2\tau_e^2\right)+\tau_e\Big/\left(1+\left(\omega_I^2\tau_e^2\right)^2\right) \\ +\tau_e\Big/\left(1+\left(\omega_I+\omega_S\right)^2\tau_e^2\right)\end{matrix}\right] \tag{3.30}$$

$$1/T_2^{DD} = (1/20)\gamma_I^2\gamma_S^2\hbar^2 r^{-6}\left[\begin{matrix}\left(1+\left(\omega_I-\omega_S\right)^2\tau_e^2\right)+3\tau_e\Big/\left(1+\omega_I^2\tau_e^2\right) \\ +6\tau_e\Big/\left(1+\omega_S^2\tau_e^2\right) \\ +6\tau_e\Big/\left(1+\left(\omega_I+\omega_S\right)^2\tau_e^2\right)\end{matrix}\right] \tag{3.31}$$

where the resonance frequencies of nucleus I and electron S are ω_I and ω_S. Correlation magnitude τ_e in these expressions takes the form

$$\tau_e = \left(\tau_C^{-1} + \tau_S^{-1}\right)^{-1} \tag{3.32}$$

where τ_C is the rotational molecular correlation time and τ_S is the electron relaxation time. It should be noted that the latter is valid only for solutions, where molecular reorientations are fast. In solids, molecular motions are slower and τ_e corresponds to the electron relaxation time.

Because nuclei situated closer to paramagnetic centers should relax much faster (in accordance with the r^{-6} factor), the paramagnetic dipolar $T_{1,2}$ times can be rationalized in structural terms if the other parameters in the equations are known independently.

When Fermi contact electron–nucleus coupling is strong, the flips of electron spins can be simultaneously accompanied by flips of nuclear spins.[19] This Fermi contact relaxation mechanism, arising due to delocalization of the unpaired spin density in a nucleus under investigation, is given by Equations 3.33 and 3.34, showing T_1^{CON} and T_2^{CON} times, respectively.

$$1/T_1^{CON} = (2/3)S(S+1)(A/\hbar)^2\left[\tau_{E2}\Big/\left(1+\omega_I^2\tau_{E2}^2\right)\right] \tag{3.33}$$

$$1/T_2^{CON} = (1/3)S(S+1)(A/\hbar)^2\left[\tau_{E1}+\left(\tau_{E2}\Big/\left(1+\omega_I^2\tau_{E2}^2\right)\right)\right] \tag{3.34}$$

Here, τ_{E1} and τ_{E2} are the longitudinal and transverse electron spin relaxation times, respectively. It should be emphasized that the Fermi contact relaxation mechanism generally dominates under the condition $\omega_S\tau > 1 > \omega_I\tau$. When this mechanism is predominant, T_2 times will be much shorter than T_1 times and resonance lines will be particularly broad or can become spectrally invisible.

For strong dipole–dipole proton interactions (e.g., proton–proton interactions) and during time-averaged magnetization of the electrons, spin–spin relaxation of ^1H nuclei can show the Curie mechanism:

$$1/T_2^{Curie} = (1/5)\left(\mu_0/4\pi\right)^2 \omega_H^2 g^4 \mu_B^4 S^2 (S+1)^2 (3KT)^{-2} r^{-6} \tag{3.35}$$

$$\left\{\left[4\tau_r + \left(3\tau_r/\left(1+\omega_H^2\tau_r^2\right)\right)\right] - \left[4\tau_C + \left(3\tau_C/\left(1+\omega_H^2\tau_C^2\right)\right)\right]\right\}$$

where g is the electron g factor; μ_0 is the permeability of free space; μ_B is the magnetic moment of the free electron; and S is the electron spin quantum number. Correlation times in Equation 3.35 are defined as $1/\tau_C = 1/\tau_r + 1/\tau_E$, where τ_E is the electron spin relaxation time and τ_r is the molecular rotational correlation time. In practice, the Curie mechanism can provide a remarkable relaxation contribution only in solutions of macromolecules; this mechanism is negligible even for medium-size molecular systems.[20]

3.5   RELAXATION TIME MEASUREMENTS

Recognition of the relaxation mechanisms and evaluation of the relaxation contribution required for further interpretation in terms of structure or molecular dynamics can obviously be based only on the reliable and accurate measurement of relaxation times. Currently, the hardware and software of NMR spectrometers for both solutions and solids provide simple and convenient procedures for the accurate measurement of relaxation times T_1, T_2, and $T_{1\rho}$, including non-selective, selective, and bi-selective time constants. Nevertheless, the quantitative treatment of collected relaxation curves (exponential or non-exponential) still requires some experience.

3.5.1   NON-SELECTIVE, SELECTIVE, AND BI-SELECTIVE T_1 TIMES

Non-selective (or regular) T_1 times are determined by inversion–recovery, saturation–recovery, or progressive saturation NMR experiments, which are described in the recommended literature. The first method is most popular and is considered in Figure 3.7A. The inversion–recovery method is based on applications of two powerful radiofrequency pulses to register the corresponding NMR signal. The equilibrium macroscopic magnetization is shown in Figure 3.7A as the vector located along the OZ direction, and a registering coil can be seen along the axis OX. A 180° pulse converts the magnetization from OZ to $-OZ$, and its projection on the OX axis will be zero. Therefore, a second 90° pulse is needed to register

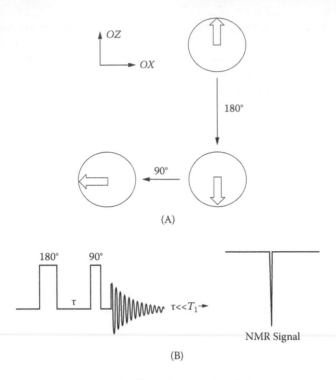

(A)

(B)

FIGURE 3.7   (A) Evolution of magnetization by the action of 180° and 90° radiofrequency pulses. (B) The NMR signal in anti-phase, appearing at a short delay time τ in the inversion–recovery NMR experiment.

any NMR signal. It is obvious that for a very short τ delay ($\tau < 5T_1$), as shown in Figure 3.7B, the observed signal will be maximal at a negative intensity. Generally speaking, this intensity should be equal to that in an equilibrium state at a very short τ delay. This is a good test for pulse calibrations. Increasing the τ value leads to an evolution of intensity from negative to positive values as a function of time τ; these can be treated by a standard nonlinear three-parameter fitting routine of NMR spectrometers to calculate T_1 values. The delay time τ typically varies from values of $\tau \ll T_1$ to $\tau \approx 3T_1$.

In order to achieve good signal-to-noise ratios in NMR spectra, the following cycle can be repeated (n is the number of scans):

$$(\text{RD} \rightarrow 180° \rightarrow \tau \rightarrow 90° \rightarrow \text{AT})_n \tag{3.36}$$

When both of the pulses in Equation 3.36 are hard and short enough for uniform excitation of a large frequency range (typically, powerful pulses with a duration of ~5 to 10 μs excite a range of frequencies on the order of 10^5 Hz), the inversion–recovery experiments result in the determination of non-selective T_1 times.

As follows from Figure 3.7A, spin inversion plays a critical role in inversion–recovery experiments. The quality of the inversion can be improved by application of a composite pulse cluster, $90^\circ_\phi 240^\circ_\phi 90^\circ_\phi$, rather than the simple 180° pulse. To perform saturation–recovery experiments, the 90° saturating pulse replaces the inverting 180° pulse. Finally the progressive saturation method is based on the pulse sequence shown below:

$$\text{Dummy pulses} \to \text{RD} \to 90^\circ \to \text{AT} \qquad (3.37)$$

where the saturation effect is realized with the help of the dummy pulses while T_1 measurements are carried out with the varied delay times, $\tau = \text{RD} + \text{AT}$.

The powers and durations of 180° pulses can be adjusted to reach excitation for a very limited frequency region. These pulses (typically their durations are close to 20 to 30 ms) are often used in solutions and solids. In solutions, they can even excite a single line in NMR spectra to probe its relaxation. Then, the selective (or soft) 180° pulses applied in the inverse–recovery experiments:

$$\text{RD} \to 180^\circ_{sel} \to \tau \to 90^\circ \to \text{AT} \qquad (3.38)$$

$$\text{RD} \to 180^\circ_{sel} \to 180^\circ_{sel} \to \tau \to 90^\circ \to \text{AT} \qquad (3.39)$$

result in determinations of the selective relaxation times, T_{1sel}. Finally, two soft inverting 180° pulses in the pulse sequence shown in Equation 3.39 will excite two frequency regions, thus allowing determination of the bi-selective relaxation times T_{1bis}. All of the experiments (T_1, T_{1sel}, and T_{1bis}) performed for nuclei with dipolar coupling open the way to evaluating dipole–dipole contributions. In practice, it should be emphasized that the relaxation times generally change according to Equation 3.40:

$$T_{1sel} > T_{1bis} > T_1 \qquad (3.40)$$

3.5.2 Measuring $T_{1\rho}$ and T_2 Times

Figure 3.8A shows a spin-locking NMR experiment aimed at determination of $T_{1\rho}$ times. In this experiment, the irradiating radiofrequency field, B_{1Y}, is applied as long as time τ at the phase shifted by 90°. Due to this pulse, the magnetization is locked along the OY direction. Because this radiofrequency field is weaker by several orders than external magnetic field B_0, the magnetization will decay at specific time constant $T_{1\rho}$ due to the lower frequency molecular motions. Again, variation in the τ values shown in Figure 3.8A leads to magnetization $M(\tau)$ in accordance with Equation 3.41. The relaxation curves obtained give the spin–lattice relaxation times measured in the rotating coordinate system:

$$M(\tau) = M_0 \exp(-\tau/T_{1\rho}) \qquad (3.41)$$

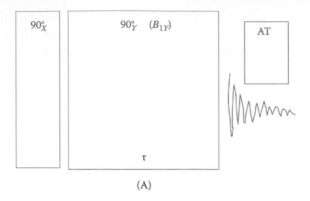

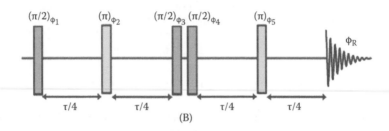

FIGURE 3.8 (A) Spin-locking NMR experiment performed for determination of the T_1 time. (B) Multiple quantum-filtered pulse sequence used for suppression of echo modulations appearing at strong homonuclear scalar couplings.

The spin-locking experiments are particularly important for solids, where molecular motions are strongly restricted. In fact, due to the slow motions in solids, the $T_{1\rho}$ values will differ strongly from T_1 and T_2 times. This difference provides additional information on low-frequency molecular reorientations. In contrast, liquids have a very high molecular mobility and show $T_{1\rho} \sim T_2$.

As mentioned in Chapter 1, time constant T_2 describes the recovery of transverse magnetization components. Therefore, by analogy with Figure 3.7A, it can be determined by the Hahn echo sequence (Equation 3.42) or Carr–Purcell pulse sequence (Equation 3.43):

$$90° \to \tau \to 180° \tag{3.42}$$

$$(90°x' \to \tau \to 180°y' \to \tau \text{ (first echo)} \to \tau \to 180°y' \to \tau \text{ (second echo)} \dots \tag{3.43}$$

where, again, τ variations result in the T_2 time. Making these measurements with modern NMR spectrometers is simple and convenient; however, in some cases, the homonuclear scalar couplings can complicate T_2 determinations by measuring echo signals. In fact, these couplings can lead to the appearance of echo modulations. These modulations can be successfully suppressed even in the case of strongly coupled two-spin systems by advanced NMR methodology, based on a linear combination of multiple-quantum filtered experiments.[21] The pulse sequence applied in Figure 3.8B is not simple. Here, the first 90° pulse is followed by two spin echo blocks. Each spin echo block includes two variable evolution delays, noted as $\tau/4$, separated by a strong 180° pulse. Then, multiplex phase cycling is applied in order to filter the experimental NMR signals via zero quantum or double quantum coherences. Finally, the T_2 times can be accurately determined for variations in the total delays.

3.6   EXPERIMENTS AND MEASUREMENTS: ERRORS AND PROBLEMS

It is generally accepted that quantitative or semiquantitative interpretations of relaxation times in the context of structural studies or characterizations of molecular dynamics will be reliable when experimental errors in determining T_1, $T_{1\rho}$, and T_2 are minimized as much as possible. Following the phenomenological equations of Bloch in Chapter 1, the longitudinal and transverse components of the total nuclear magnetization recover exponentially to an equilibrium state with time constants T_1 and T_2. This simplest exponential type of nuclear relaxation is typical of liquids due to the fast and anisotropic molecular reorientations. However, even in liquids, spin systems described by more than two energy levels (i.e., coupled relaxation) can show non-exponential behavior. It should be added that non-exponential relaxation is often observed in solids. Therefore, the appearance of unusual relaxation behavior in NMR experiments requires a clear understanding of its nature and the reason for the non-exponentiality, as well as knowledge of how to treat the data and calculate T_1 or T_2 times.

3.6.1   INSTRUMENTAL ERRORS

Several factors can significantly affect experimental measurements and reduce the potential accuracy in relaxation time determinations; further details can be found in the recommended literature, but some of these factors are considered in this section.

- *Temperature control in the relaxation experiments.* Because the duration of relaxation experiments is generally long and T_1/T_2 times are temperature dependent, the temperature should be stable within ±0.5°C. This is particularly important for solids probed by MAS NMR, where the high spinning rates increase the temperature. In addition, the temperature in a sample should be calibrated before the experiment (e.g., using a methanol thermometer in a liquid or a conventional method in a solid).
- *Calibrations of radiofrequency pulses.* The 90° and 180° radiofrequency pulses should be accurately calibrated using any of several available procedures. This calibration should be done for each temperature because temperature is known to influence pulse length. Badly calibrated pulses produce effectively non-exponential magnetization recovery curves when in reality they are actually exponential. Treatment of such curves with the standard fitting routine of NMR spectrometers can result in large errors in T_1 and T_2 calculations.

- *Signal-to-noise ratio.* The influence of the signal-to-noise ratio is obvious. In fact, poor signal-to-noise ratios reduce accuracy in the determination of integral (or pick) intensity. This is particularly important in τ regions where the line intensities are close to zero. Thus, in the presence of spectral noise, errors in determining intensities depend on the τ values. In practice, such points should be weighted less in final calculations.[22]

- *Inhomogeneity of radiofrequency pulses.* This is rather a technical problem. It is well known that the radiofrequency field is homogeneous only at the center of the radiofrequency coil. Away from the center, the pulses become imperfect, potentially leading to large deviations from exponential relaxation behavior. This effect, shown in Figure 3.9A, obviously increases errors in relaxation time calculations, particularly in solid-state NMR.

- *Choice of setting parameters.* Total relaxation delays applied in pulse sequences should obviously correspond to complete relaxation of nuclei in each cycle of the measurements. When the delays are too short, the signal intensities will be strongly distorted. One should remember that, for example, the delay close to a T_1 value recovers only ~63% of an equilibrium magnetization while a 99% recovery of the equilibrium magnetization can be achieved for a delay of $5T_1$. Because the type of nuclear relaxation (exponential or non-exponential) is generally unknown *a priori*, the experimental relaxation curves should be properly characterized. For example, the τ values used for inversion–recovery experiments would cover a range from $0.1T_1$ to $3T_1$ to discriminate the exponential and non-exponential processes. It should be noted that the appearance of overshoots in the inversion–recovery curves, particularly at τ values close to $T_1 \div 2T_1$, is a good test for the presence of complex nuclear relaxation. Finally, standard NMR adjustments and some experience in relaxation experiments regularly give errors in T_1 and T_2 determinations of less than 3 to 5%. However, this is only valid for properly chosen relaxation models.

3.6.2 TREATMENT OF RELAXATION CURVES: APPROACHES AND PROBLEMS

Treatment of an experiment is a very important step that plays a major role in interpreting relaxation data, which depends on the quality of the experimental results and the relaxation model used. In solutions, magnetization recovery can be exponential or non-exponential, such as in the case of coupled relaxation. Moreover, the process can become bi-exponential or multi-exponential, depending on the number of relaxation components. NMR relaxation is not simple in solids. In fact, even if each site in inhomogeneous solids relaxes exponentially with its own relaxation time, the total recovery curves or echo decays can significantly deviate from an exponential function. The exception is relaxation via spin diffusion. This relaxation is generally exponential because all of the spins in a sample relax identically. However, one should remember that spin diffusion is ineffective for the short τ delays applied, for example, in inversion–recovery experiments. In such short τ regions, magnetization recovery is proportional to time ($\tau^{1/2}$). Then, with increasing τ values, magnetization recovery again becomes exponential.

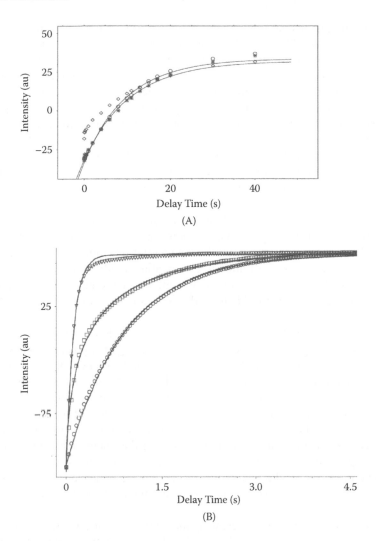

FIGURE 3.9 (A) The ^1H T_1 curves of static D_2O in a 4-mm MAS NMR rotor filled at 100% (O), 50% (*), and 23% (◇). (B) The bi-exponential experimental curves: (O) at $T_1(short) = 0.1$ s, at $T_1(long) = 1$ s, $a = 7\%$, treated by the exponential (solid line); (□) at $a = 50\%$, treated by the stretched exponential (solid line); (∇) at $a = 93\%$, treated by the exponential (solid line).

Commonly, non-exponential NMR relaxation can be approximated by the stretched exponential function:

$$f(t) = \exp(-t/T_1)^\beta)$$ (3.44)

where T_1 is a representative time of the whole spin and the β parameter takes a value between 0 and 1. Applying Equation 3.44 for spin diffusion, it has been shown that spin diffusion is completely absent for $\beta = 1/2$, whereas the case of $1/2 < \beta < 1$ can be explained by the limited spin diffusion. On the other hand, the stretched exponential is theoretically justified as a smooth distribution of relaxation times, figuring largely for short delay times. The generally accepted approach to the treatment of relaxation curves is based on best fittings of signal intensities measured experimentally to a model function in order to determine the fitted parameters. One of these parameters is the relaxation time. Mathematically, this treatment represents an incorrect inverse problem suffering from numerical instability. In addition, solutions to this problem (also known as the inverse Laplace transform problem) are sensitive to the algorithms applied.

In practice, researchers use convenient computing programs based on the Levenberg–Marquardt algorithm to reach the best fit, such that a larger number of varied parameters will obviously provide better fittings. This situation can lead to uncertainty, particularly when the model function used for calculations is not straightforward. In fact, the model can be a sum of two or more exponentials, a sum of the Gaussian and exponential functions, or a stretched exponential. It is easy to show that, for example, the initial bi-exponential relaxation curves having short and long components are well fitted to exponentials (with good statistics) for short components with fractions $a \leq 5$ to 7%, in spite of a large number of experimental points. This effect is shown in Figure 3.9B. Only for fractions $a > 10\%$ can the best fit actually be reached bi-exponentially.

Similarly, some combinations of the T_1(short) and T_1(long) components and their fractions can give equally good fits to experiments using bi-exponential and stretched exponential models. When short and long relaxation T_1 components are commensurable, the treatments are not completely single valued; the curves can be well fitted to exponentials, bi-exponentials, or stretched exponentials. In other words, the experimental relaxation curves can look exponential in spite of their more complex character or they can be equally well treated by bi- or stretched exponentials. This statement shows that the model function chosen for the calculations should be established independently. In practice, representation of signal intensity as a function of the τ time in semilogarithmic coordinates can help to better identify the character of the relaxation. In fact, exponential processes should show straight lines in these coordinates.

It is generally accepted that increasing the number of varied parameters provides better fittings; however, one should remember that for variations within three parameters, for example, their reproducibility in the presence of noise contributions to the same signal intensity is preferable to application of a four-parameter fitting function.

3.7 ARTIFACTS IN RELAXATION TIME MEASUREMENTS

Generally, artifacts in NMR spectra are referred to as artifacts because their appearance is difficult to predict or explain. Artifact phenomena in relaxation experiments complicate their interpretation. Artifacts caused by inhomogeneity of the radiofrequency field or imperfections in the radiofrequency pulses have been mentioned above. The so-called radiation damping effect due to the interaction of a RF coil with the bulk magnetization of a sample can also affect the relaxation behavior of nuclei, particularly protons. In fact, for relatively short damping times, the measured ^{1}H T_1 values will be shorter than real ones and the exponential relaxation itself can become

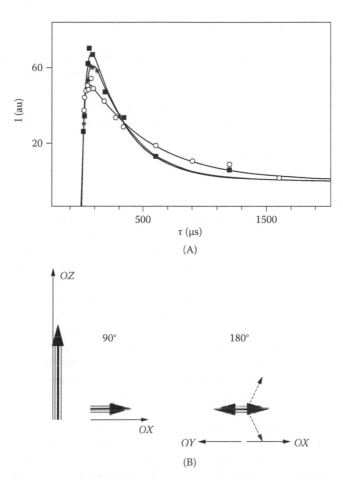

FIGURE 3.10 (A) Hahn echo experiments carried out in the static solid SiO_2–Al_2O_3–NiO (13 wt%): (O) and (■) with a well-tuned and strongly detuned NMR probe, respectively; (*) in the presence of the inhomogeneous RF field. (B) Behavior of the magnetization in the presence of an imperfect 180° pulse.

non-exponential.[23] In practice, the effect can be strongly reduced by isotope dilution in order to decrease the ^1H concentration in a sample or by experiments performed with detuned NMR probes.

Measurements of the spin–spin relaxation times, T_2, by echo pulse sequences can show artificial attenuation of echo signals.[24,25] The attenuation can be observed in experiments on samples with very short T_2 times for short initial τ delays applied in the pulse sequence. Figure 3.10A shows one such experiment performed on the solid SiO_2–Al_2O_3–NiO (13 wt%). The initial echo intensity (obtained by the Hahn echo pulse sequence) for short τ delays increases with increasing τ values and then normally decreases close to an exponent. This abnormal effect can be accepted as an artifact and ruled out to treat only the tail of the curve with an exponent giving a T_2 time (see the first term in Equation 3.45):

$$I(\tau) = I_0(a \exp(-\tau/T_2) - (1 - a) \exp(-\tau/T^A)) \tag{3.45}$$

On the other hand, the experimental curves shown in Figure 3.10A can be approximated by a function in Equation 3.45, where T_2 is the real spin–spin relaxation time, and the time constant T^A characterizes the initial loss of the echo intensity. Because experiments carried out with a detuned NMR probe and increased inhomogeneity of the radiofrequency field show the same behavior, there are only two physical reasons for these artificial attenuations. The first reason is that the dead time of a resonance circuit (receiver is closed; see Chapter 1) is commensurable with τ delays or opening the receiver is not synchronized with pulse actions. This effect can actually lead to a natural loss in intensity.

The second reason, particularly important for solids, has to do with the very large line widths generally observed for resonances. In amorphous (or paramagnetic) solids, resonance frequencies are widely distributed, and after a 90° pulse in Equation 3.42 a uniform 180° rotation for all of the excited spins becomes difficult or even impossible because the amplitude strongly decreases for larger offsets. Under this circumstance, the refocusing 180° pulse can act as a selective one, creating conditions similar to those in NMR experiments in inhomogeneous magnetic fields. Magnetization dephasing in the X,Y plane during the time between the first and second radiofrequency pulses normally refocuses to produce the echo signal. However, if the 180° pulse is not perfect, part of the dephasing magnetization continues to dephase after the action of this pulse. As a result, the destructive interference, shown in Figure 3.10B, results in a loss of intensity. Under these conditions, time T^A in Equation 3.45 will characterize the attenuation of this destructive interference in time.[26]

The so-called vortex effect does not play a role in NMR relaxation experiments on solutions due to the low rates of spinning (around 20 Hz); however, it can produce artifact phenomena in solids spinning at high rates. Figure 3.11 shows a standard 4-mm MAS NMR rotor (it may be a 7- or 2.5-mm rotor), where a powder solid fills in only 20% of the rotor space. Thus, the static powder is out of the center of the RF coil. A lack of perfection in the RF pulses can result in distorted relaxation time values; that is, the relaxation process can become non-exponential. This undesired phenomenon can appear in MAS NMR experiments with 50% (or even greater) filled rotors. In fact, due to spinning at high rates, a powder takes on a solid-vortex form, the RF field is again inhomogeneous, and the pulses are again imperfect. Generally speaking, such an effect can be important even for other NMR experiments requiring good calibrations of RF pulses.

Another vortex mechanism can appear in relaxation NMR experiments performed on high-spinning samples containing liquids or liquid components, such as in experiments on suspensions of heterophase samples. Upon spinning, the initial exponential spin–lattice relaxation observed in the static samples can transform to a bi-exponential process demonstrating the presence of a fast relaxing component. The appearance of this artificial fast relaxing component can be explained by the centrifugal force causing an accumulation of oxygen at an inverted conical liquid surface during spinning.[27]

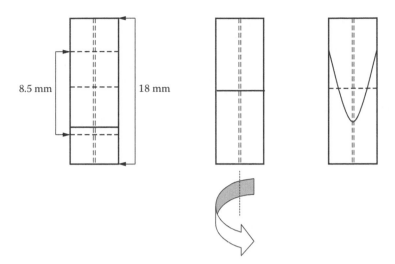

FIGURE 3.11   MAS NMR rotor with a size of 4 mm × 18 mm and RF coil (5 mm × 8.5 mm). (Left) 23% filled static rotor. (Middle) 50% filled static rotor. (Right) 50% filled rotor spinning around the vertical axis for simplicity.

REFERENCES AND RECOMMENDED LITERATURE

1. Martin, N.D., Issa, M.H., McIntyre, R.A., and Rodriguez, A.A. (2000). *J. Phys. Chem. A.*, 104: 11278.
2. Penner, G.H., Chang, Y.C., and Hutzal, J. (1999). *J. Inorg. Chem.*, 38: 2868.
3. Galakhov, M.V., Bakhmutov, V.I., and Barinov, I.V. (1991). *J. Magn. Reson. Chem.*, 29: 506.
4. Rossi, C. (1992). *Chem. Phys. Lett.*, 193: 553.
5. Ropp, J., Lowrence, C., Farrar, T.C., and Skinner, J.L. (2001). *J. Am. Chem. Soc.*, 123: 8047.
6. Mayo, K.H., Daragan, V.A., Idyatullin, D., and Nesmelova, I. (2000). *J. Magn. Reson.*, 146: 188.
7. Bloembergen, N., Purcell, E.M., and Pound, R.V. (1948). *Phys. Rev.*, 73: 679.
8. McConnell, J. (1987). *The Theory of Nuclear Magnetic Relaxation Time in Liquids.* New York: Cambridge University Press.
9. Garbacz, P., Terskikh, V.V., Ferguson, M.J., Bernard, G.M., Kodziorek, M., and Wasylishen, R.E. (2014). *J. Phys. Chem. A*, 118: 1203.
10. Wang, C. and Palmer, A.G. (2003). *Magn. Reson. Chem.*, 41: 866.
11. Aime, S., Dastru, W., Gobetto, R., and Viale, A. (2001). NMR relaxation studies of polynuclear hydride derivatives. In: *Recent Advances in Hydride Chemistry* (Poli, R. and Peruzzini, M., Eds.), p. 351. Amsterdam: Elsevier.
12. Schaublin, S., Hohener, A., and Ernst, R.R. (1974). *J. Magn. Reson.*, 13: 196.
13. Bakhmutov, V.I., Vorontsov, E.V., Boni, G., and Moise, C. (1997). *Inorg. Chem.*, 36: 4055.
14. Bloembergen, N. (1949). *Physica*, 15: 386.
15. Chen, Q. and Schmidt-Rohr, K. (2006). *Solid State Nucl. Magn. Reson.*, 29: 146.
16. Hayashi, S. and Omi, H. (2010). *Solid State NMR*, 37: 69.
17. Grutzner, J.B., Stewart, K.W., Wasylishen, R.E., Lumsden, M.D., Dybowski, C., and Beckmann, P.A. (2001). *J. Am. Chem. Soc.*, 123: 7094.

18. Solomon, I. (1955). *Phys. Rev.*, 99, 559.
19. Bloembergen, N. (1957). *J. Chem. Phys.*, 27: 572.
20. Clore, G.M. and Iwahara, J. (2009). *Chem. Rev.*, 109: 4108.
21. Barrere, C., Thureau, P., Thevand, A., and Viel, S. (2011). *Chem. Commun.*, 47: 9209.
22. Daragan, V.A., Loczewiak, M.A., and Mayo, K.H. (1993). *Biochemistry*, 32: 10580.
23. Krishnan, V.V. and Murali, N. (2013). *Prog. Nucl. Magn. Reson. Spectrosc.*, 68: 41.
24. Bakhmutov, V.I. (2009). *Solid State NMR*, 36: 164.
25. Viel, S., Ziarelli, F., Pagès, G., Carrara, C., and Caldarelli, S. (2008). *J. Magn. Reson.*, 190: 113.
26. Bakhmutov, V.I. and Silber, S. (2012). *J. Spectrosc. Dyn.*, 2: 14.
27. Bakhmutov, V.I. (2012). *Concepts Magn. Reson.*, 40A: 186.

RECOMMENDED LITERATURE

Bakhmutov, V.I. (2005). *Practical NMR Relaxation for Chemists*. Chichester: Wiley.
Duer, M.J., Ed. (2002). *Solid-State NMR Spectroscopy: Principles and Applications*. Oxford: Blackwell Sciences.
Harris, R.K. (1983). *Nuclear Magnetic Resonance Spectroscopy*. Avon: Bath Press.
Keeler, J. (2010). *Understanding NMR Spectroscopy*, 2nd ed. Chichester: Wiley.
Levitt, M. (2008). *Spin Dynamics: Basics of Nuclear Magnetic Resonance*, 2nd ed. Chichester: John Wiley & Sons.
Poole, C.P. (1983). *Electron Spin Resonance*. New York: Wiley.
Smerald, A. (2013). *Theory of the Nuclear Magnetic $1/T_1$ Relaxation Rate in Conventional and Unconventional Magnets*. Heidelberg: Springer.

4 NMR and Molecular Dynamics
General Principles

Molecular dynamics has attracted the attention of researchers because it strongly affects the physical and chemical properties of various systems, from simple and complex molecules in solutions to materials in the solid state. Motions of molecules or molecular groups can affect the detected NMR parameters as a function of frequencies of motions and their character. Depending on the rate and type of intra- and intermolecular reorientations, they can change a number of observed resonances, their line widths and shapes, chemical shifts, spin–spin coupling constants, quadrupolar splitting, and dipolar couplings or relaxation times in solutions and the solid state. Therefore, all of these factors must be taken into account for quantitative analysis of molecular mobility by NMR spectroscopy in order to determine the types, frequencies, and activation energies of molecular reorientations.

The types of molecular motions and their rates vary widely. They change from high-frequency librations and internal rotations with extremely low energies or even tunneling (the well known case of fast-spinning CH_3 groups in solids) to slower reorientations of whole small molecules and segmental motions in polymers and the slowest tumbling bulky molecules. Some of the motions can become cooperative, providing additional contributions to the frequency spectrum of motions and their energetics. In this context, dynamically heterogeneous solids (e.g., porous materials, glasses) represent a special class. These systems (e.g., biological tissues, gels, plasticized polymers, zeolites) can be formulated as solids with an open matrix that is filled with a liquid. The small sizes of the liquid spaces, increasing from tens of angstroms to a few microns, can create the conditions for very complex liquid dynamics, dependent on the structural parameters of the matrix. Finally, translational diffusion, particularly in immobilized catalysts in the presence of solvents, is also of great interest and requires quantitative characterizations.

In general, the frequency spectrum of molecular mobility can be represented conditionally by three regions: *slow* dynamics within the frequency diapason between 1 Hz and 1 kHz, *moderate* dynamics characterized by kHz frequencies, and *very fast* dynamics, when molecular or group motions can reach frequencies up to ~100 MHz.[1] The quantitative analysis of these motions in solutions and solids is the subject of studies performed with dynamic NMR (DNMR). Experimental methods and techniques used in DNMR depend on the time scales of the motions being investigated. Figure 4.1 illustrates motions and processes that are associated with

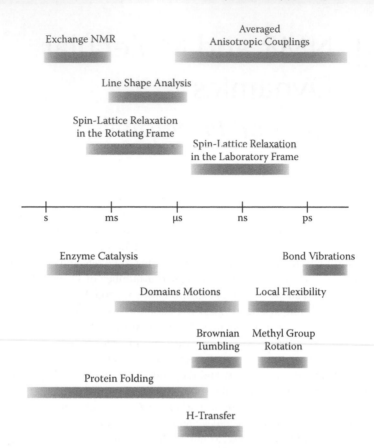

FIGURE 4.1 Dynamic NMR experiments and representative time scales of processes associated with protein molecules. (Adapted from Bakhmutov, V.I., *Solid-State NMR in Materials Science: Principles and Applications*, CRC Press, Boca Raton, FL, 2011.)

the dynamics of protein molecules. As can be seen, the time scales of the motions cover the diapason between seconds and picoseconds, thus requiring the application of specific experimental methods. Relaxation time measurements are directed toward determining the correlation times (τ_C) that characterize the fast molecular reorientations. Slower molecular motions associated with reorientations of chemical shift anisotropy tensors or quadrupolar coupling tensors occur on the same time scale (or slightly faster) as the free induction decay (FID) time is collected. Because these reorientations correspond to the condition $\tau_C \leq T_2$ and strongly affect line shapes, line-shape analysis can be applied for their characterization. Slower dynamics (e.g., conformational transitions or segmental reorientations of bulky polymer molecules in solids) can be detected due to changes of orientation-dependent NMR frequencies and probed by exchange NMR spectroscopy. Finally, very slow processes on the NMR time scale (e.g., chemical reactions) can be investigated by recording NMR spectra over time.

4.1 KINETICS OF CHEMICAL REACTIONS BY NMR SPECTROSCOPY

When a chemical reaction, considered as a special type of chemical movement, occurs significantly more slowly than collection of the FID data and when the time to mix reagents and transfer a sample into the magnet is negligible, then the kinetics of this reaction can be quantitatively studied by regular NMR. In some sense, this work can be considered as being both spectroscopic and kinetic. The spectroscopic aspect is aimed at accurate determinations of time-dependent integral intensities in the NMR spectra which reduce or increase for initial compounds or products, respectively. These measurements should obviously be performed at delay times corresponding to full nuclear relaxation in each cycle, and NMR spectra should be recorded with careful phase/baseline corrections to provide an accuracy in these measurements with minimal errors (<3 to 5%). It should be noted that generally the time-dependent NMR spectra can be recorded automatically by standard software of NMR spectrometers.

The kinetic aspect is directed toward interpretations of the obtained data on the basis of a well-known arsenal of formal chemical kinetics. For example, when compound A reacts in accordance with the mechanism of Equation 4.1, then the kinetic equation (Equation 4.2) can be written as

$$nA + mB \rightarrow C \tag{4.1}$$

$$dA/dt = k[A]^n[B]^m \tag{4.2}$$

where n and m represent kinetic orders of the reaction. In turn, the rate of the reaction (k) is determined as follows:

$$k = (k_B T/h) \exp(-\Delta G^*/RT) \tag{4.3}$$

$$k = (k_B T/h) \exp(-\Delta H^*/RT) \exp(\Delta S^*/R) \tag{4.4}$$

where ΔG, ΔH, and ΔS are the free activation energy, activation enthalpy, and activation entropy, respectively. Thus, the experimental dependencies of intensity vs. time can be recalculated to obtain the k values, while the kinetic orders and energies of the reaction can be found by NMR kinetic experiments performed with variations in concentration and temperature.

For very fast and irreversible reactions, special NMR techniques have been developed to obtain information about reagents with half-lives as fast as 0.1 s. The techniques for rapid injecting and stirring reactant solutions into an NMR tube, as well as stopped-flow NMR tube tools, can be found in the recommended literature. In addition to the stopped-flow technique, treatment of NMR data, including numerical simulations of NMR spectra, should take into account the effects of chemical reactions on the collected FIDs. In fact, reactants are mixed and detected simultaneously. For example, the NMR spectrum obtained after a 90° pulse for a fast irreversible, first-order reaction demonstrates an important feature: The NMR signal of a reactant strongly broadens as a function of the reaction rate. This effect is due to a decrease in the coherence lifetime caused by the chemical reaction. Under these conditions, the line width of the signal is given by Equation 4.5:

$$\Delta v = k/\pi + 1/\pi T_2 \tag{4.5}$$

where the second term is the natural line width in the absence of the reaction.

It is clear that at a reaction rate of >1 s^{-1}, the line width is dominated by the rate constant, k, which can be determined experimentally. Currently, several methods are available that provide simulations of the results for any arbitrary kinetic model, including second-order kinetics or even complex multistep reactions.

4.2   CHEMICAL EXCHANGE

Chemical exchange, such as between two non-equivalent spin states of nuclei, noted as A and X and shown below:

$$\text{Spin}(A) \Leftrightarrow \text{Spin}(X) \tag{4.6}$$

$$v^A;\tau^A \qquad\qquad v^X;\tau^X$$

will affect the line shapes of A and X resonances when the exchange frequencies, v_{EXCH}, and chemical shift differences, $v^A - v^X$ (expressed in Hz), are commensurable. It should be emphasized that these spins can belong to one molecule or different molecules, corresponding to an intramolecular or intermolecular exchange, respectively. Under these conditions, the NMR spectra show the temperature evolution typical of *two-center exchange*, which is shown in Figure 4.2A, where the A and X resonances are sharp at low temperatures, thus showing the natural line widths dictated by T_2 times. Then, upon heating, the resonances broaden, coalesce, and finally transform to a singlet when the exchange becomes very fast. At equal populations of the spin states, this singlet is observed exactly at the frequency of $v = (v^A + v^X)/2$. Similar but more complex behavior can be observed in the case of multicenter exchanges. Figure 4.2B illustrates a four-center exchange observed for the olefin protons C=CH–NH in ^{1}H NMR spectra of nitroenamines that experience E/Z isomerization in solutions.[2] In the presence of the exchange, all of the lines are strongly broadened; however, in contrast to a simple fast two-center exchange, this fast four-center exchange can give a doublet (Figure 4.2B) or a singlet resonance if, due to the exchange, the spin–spin coupling ^{1}HN–C^{1}H is remaining or lost, respectively. In the absence of the highest temperature NMR spectra, the choice of the exchange model is difficult, but it can be done on the basis of theoretical calculations performed within the framework of a line-shape analysis. As follows from the data in Figure 4.2B, the theoretical line shape is much better fitted to the experimental shape when equal rates are obtained for the change of the isomer states and the loss of the $J(^1$H–N–C–^{1}H) constant. In other words, the exchange is initiated by dissociation of N–H bonds.[2]

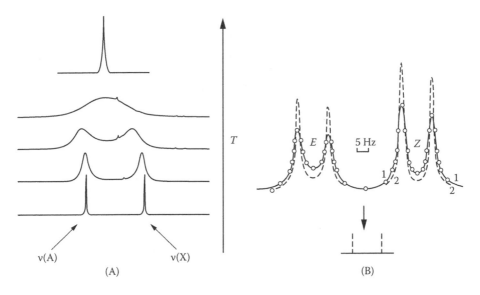

FIGURE 4.2 (A) Temperature evolution of NMR signals typical of two-center exchange. (B1) Experimental spectrum of olefin protons =CH–NH in ^{1}H NMR spectra of nitroenamines undergoing E/Z isomerization in solutions. (B2) Theoretical ^{1}H NMR spectrum calculated to account for two cases of the exchange: when the J(HN–CH) coupling constant is lost (points) and when it remains (dashed line).

4.2.1 LINE-SHAPE ANALYSIS

As noted in Chapter 1, the line shape of a resonance is well predicted on the basis of solving the Bloch equations. In the case of chemical exchange between A and X, the lifetimes τ^A and τ^X characterizing this exchange should be added to the Bloch equations to give the McConnell expressions that describe the line shapes as a function of τ^A and τ^X values. Thus, the line shapes documented experimentally can be calculated to obtain the lifetimes of the exchanging spin states.

In practice, this full line-shape analysis can be realized by calculation of the theoretical shapes by conventional computer programs, which should be accurately fitted to the experimental line shapes. These programs use various algorithms to calculate the lifetimes, including variations in chemical shift differences, which are often strongly temperature dependent (this is particularly important for relatively small $\Delta\delta$ values); populations of spin states, which are also temperature dependent in many cases; and line widths in the absence of an exchange (i.e., controlled by T_2 times). Some of these programs use all of the experimental data collected at different temperature to determine the lifetimes and activation energies characterizing a particular exchange. However, it should be emphasized that the complex mechanisms of exchanges can result in a surprising phenomenon where the rate of exchange varies *inversely* with the temperature, corresponding to negative activation energies.[3]

In general, full line-shape analysis performed, for example, for two-center A,X exchange gives the lifetimes of spin states A and X belonging to the same molecule (an intramolecular exchange) or different molecules (an intermolecular exchange). In these terms, the lifetime of spin state A, for example, can be expressed via common Equation 4.7, where n and m are again kinetic orders of the exchange:

$$1/\tau^A = k[A]^{n-1}[B]^m \tag{4.7}$$

Thus, for any monomolecular process or any intermolecular exchange characterized by the first kinetic order, Equation 4.7 transforms to Equation 4.8:

$$1/\tau = k \tag{4.8}$$

In addition to the computer line-shape analysis, the conditions of the slow or fast exchange also allow simple determination of the k constants via Equations 4.9 and 4.10, respectively:

$$k_{slow} = \pi(\Delta\nu_{obs} - \Delta\nu_0) \tag{4.9}$$

$$k_{fast} = \pi\Delta\delta_0^2 \, (\text{Hz})/2(\Delta\nu_{obs} - \Delta\nu_0) \tag{4.10}$$

Here, δ_0 is the chemical shift difference in the absence of the exchange (measured in Hz), and $\Delta\nu_{obs}$ and $\Delta\nu_0$ are the line widths in the presence and absence of the exchange, respectively.

When the exchanging signals coalesce at temperature T_C, then rate constant k_C can be calculated via Equation 4.11. Under these conditions, the free activation energy can be found by using Equation 4.12:

$$k_C = 2.22\delta_0 \, (\text{Hz}) \tag{4.11}$$

$$\Delta G_C^{\neq} = 2.3RT_C(10.32 + \log (T_C/k_C)) \tag{4.12}$$

As in the case of regular NMR studies directed toward characterization of kinetics, spectroscopic DNMR consists of accurate collection of variable-temperature NMR data applied for full (or approximate) line-shape analysis, while interpretation of the data is again based on formal chemical kinetics. Figure 4.3A shows the ^1H signals of the $COOCH_3$ group experiencing two-center exchange. The line shapes unexpectedly demonstrate the presence of a primary kinetic isotropic effect, when bond N–H is replaced for bond N–D. Thus, even rotation around the partial double bond C=C can occur via the complex mechanism (Figure 4.3B) where dissociation of the N–H bond provides a significant contribution to the total rate of the process. In reality, this process is described by the complex kinetic equation (Figure 4.3C).[4]

Full line-shape analysis can be applied for different nuclei for which the chemical shift differences $\nu^A - \nu^X$ (expressed in Hz) differ, thus corresponding to different lifetimes. In the case of protons, the lifetimes τ measured by line-shape analysis are

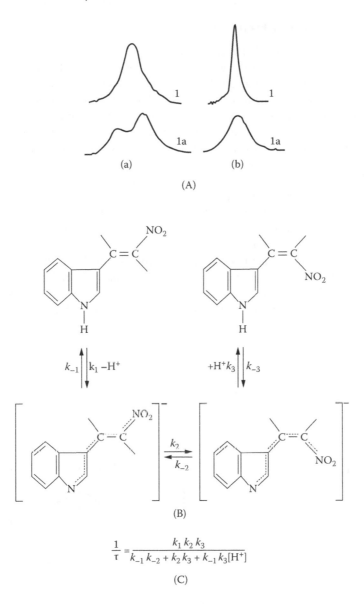

FIGURE 4.3 (A) The ^{1}H signals of COOCH$_3$ groups neighboring NO$_2$ groups in the *E/Z* enamine with (1) bond N–H and (1a) N–D in (a) pyridine and (b) acetonitrile. (B) Mechanism of *E/Z* isomerization (rotation around bond C=C). (C) Kinetics of the isomerization process.

typically between 10^{-4} and 10^{-1} s, corresponding to free energies $\Delta G^{\neq}$ of 11 to 18 kcal/mol. This time scale can be enlarged by probing nuclei with larger chemical shift differences at the exchanging sites, including paramagnetic species (zone of faster exchanges) or saturation transfer techniques (zone of slower exchanges).

4.2.2 SLOW CHEMICAL EXCHANGE

A slow chemical exchange between two sites A and X corresponds to the relationship $v_{EXCH} \ll (v^A - v^X)$. In this case, resonance lines in the NMR spectra are well separated and remain quite narrow. In spite of this effect, T_1 relaxation times of the signals can be partially averaged. Then, the FIDs collected by inversion–recovery experiments can give relaxation times $T_1(A)$ and $T_1(X)$, which are dependent on lifetimes τ^A and τ^X. Generally speaking, this situation can lead to additional errors in T_1 determinations, because T_1 calculations based on inversion–recovery curves will require separations of two superimposed exponentials and independent determinations of magnitudes $1/\tau^A$ and $1/\tau^X$. In fact, standard treatments will result in effective magnitudes $T_1(eff)$. On the other hand, this feature can be used for determining the lifetimes of exchanging spins by another approach, based on the above effect. If the resonance lines are sharp, as shown schematically in Figure 4.4A, a slow exchange can be detected by *saturation transfer experiments*.

These experiments, generally applied for protons, involve a double resonance technique. In Figure 4.4A, nuclei A are observed and nuclei X are irradiated. As follows from NMR principles, the irradiation leads to transitions of X spins between the Zeeman levels, equalizing their populations. As a result, the X resonance will

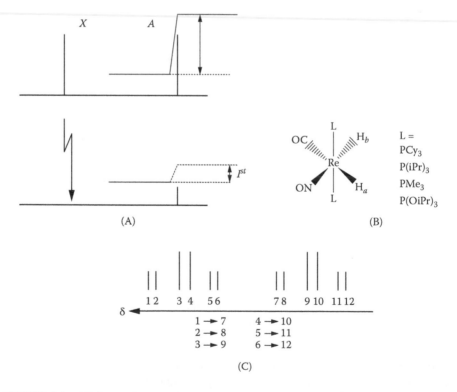

(A)

(B)

(C)

FIGURE 4.4   (A) Saturation transfer experiment shown schematically. (B) Transition metal hydrides with the H_a and H_b ligands, which experience a slow chemical exchange in solutions. (C) The 12-center exchange of the H_a/H_b ligands.

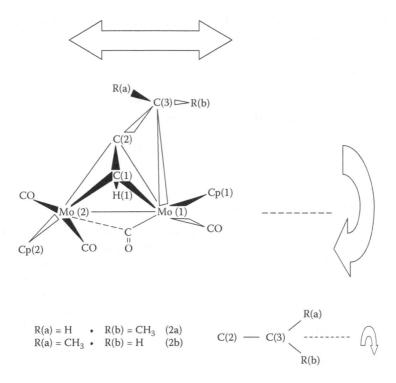

FIGURE 4.5   The bimetallic cluster system showing non-rigidity in solutions.

be *saturated*. In the presence of a slow A,X exchange, the saturating effect can be transferred from X resonance to A resonance to reduce its integral intensity, noted as I^{ST}. In turn, measurements of the I^{ST} values give the exchange rate constants k via Equation 4.13 if the T_1 times are known:

$$I^{ST}/I^0 = 1/(1 + kT_1) \qquad (4.13)$$

This saturation transfer technique offers two advantages. First, its application enlarges the NMR time scale to characterize the lifetimes τ up to several seconds, corresponding to $\Delta G^{\neq}$ values of 20 to 21 kcal/mol, as demonstrated in Figure 4.4B for the H_a/H_b exchange in solutions of Re complexes.[5] Second, as shown in Figure 4.4C, this process is formally a 12-center exchange due to the spin–spin coupling constants $J(^1H–^1H)$ and $J(^1H–^{31}P)$. Such an exchange is difficult to analyze even by computer programs used for line-shape analysis, but the saturation transfer technique simplifies such an analysis. In addition, the saturation transfer effect in the presence of a slow exchange can be very useful as a criterion in interpretations of variable-temperature ^1H NMR data. This is illustrated in Figure 4.5 by a bimetallic cluster system that is not rigid in solutions. The cluster shows two types of non-rigidity: migration of the carbon–metal bond from Mo(1) to Mo(2) and rotation around bond C(2)–C(3). The ^1H NMR spectra recorded in the temperature range of 244 to 229 K correspond to the situation when interconversions are frozen on the NMR time scale. In fact, two lines of the Cp ligands are observed for each conformer (2a and 2b). However, upon further cooling to 192 K, these Cp signals show consecutive broadening and narrowing, a finding that is difficult to interpret.

Theoretically, on the basis of DNMR principles, this broadening and narrowing effect could be attributed to a chemical exchange with participation of a new poorly populated state. In accordance with this assumption, the ^{1}H NMR spectrum recorded at 187 K actually exhibits two new resonances in the Cp region with very low intensities. These resonances could be explained by a rotation around bond Mo–Mo; however, this hypothesis can be confirmed only by saturation transfer upon irradiation of these low-populated signals.[6]

4.2.3   EXCHANGE NMR SPECTROSCOPY

Whereas the saturation transfer technique can be successfully applied for chemical exchanges probed by protons in ^{1}H NMR, other nuclei are generally studied by two-dimensional (2D) exchange spectroscopy (EXSY) NMR[1,8,9] to identify and characterize slow chemical exchanges. This approach is obviously suitable for protons. EXSY experiments carried out in solutions are based on 2D NMR techniques, where the operating 90° radiofrequency pulses create the 2D pattern after double Fourier transformation due to a mixing time in the pulse sequence as shown in Figure 4.6A. It is obvious that for $\tau_{mix} = 0$ the data collected during times t_1 and t_2 will be identical and no cross peaks will appear in the 2D spectrum. Thus, the mixing time plays a critical role in EXSY experiments, allowing nuclei to detect the presence of a chemical exchange during time τ_{mix}. It should be noted that the same pulse sequence is applied for nuclear Overhauser effect spectroscopy (NOESY) experiments, when mixing times are specially adjusted to register NOEs. As in the case of the saturation transfer technique, EXSY enlarges the time scale for the study of slow chemical exchanges, providing quantitative characterization of rate constants between 0.2 and 100 s^{-1}.

One of the simplest practical 2D EXSY applications is detection of a slow chemical exchange in systems under investigation. This case is represented in Figure 4.6B, where protons H_a and H_b in transition metal hydrides (see Figure 4.4B) are coupled by the slow H_a/H_b exchange. In fact, the corresponding cross peaks can clearly be seen in the 2D NMR spectrum.[5] Because the cross peaks appear in 2D EXSY NMR spectra due to the mixing time, their intensities will depend on exchange rates. A quantitative analysis of these intensities is not simple and requires considerable experience. Nevertheless, standard computer procedures can provide analysis of cross-peak intensities taken from experiments to calculate rate constants. Generally, two EXSY experiments are needed: during a mixing time and in its absence. Then, the calculations are carried out for systems with an arbitrary number of exchange sites, spins, populations, and T_1 times. Because the final results of computer calculations depend strongly on the quality of the collected data, the experiments should be performed within parameters that have been properly adjusted and are correct. It is obvious that 2D EXSY NMR can be applied for both solutions and solids.

The rate of chemical exchange can be also determined with a one-dimensional (1D) EXSY NMR technique based on relaxation-type experiments performed with selective inversion.[7] The simplest case, shown in Figure 4.7A, is an exchange between

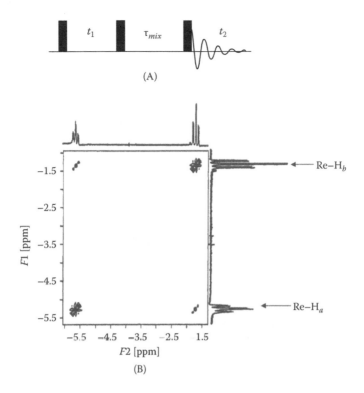

FIGURE 4.6 (A) Pulse sequence applied for EXSY (or NOESY) 2D NMR experiments. (B) The EXSY 2D NMR spectrum of complex $Re(CO)(H)_2(NO)(PPri)_3)_2$ in toluene-d_8 at 65°C.[5]

two spins A and B. As can be seen, the pulse sequence shown in Figure 4.7A (top) selectively inverts spin A, which experiences an exchange due to the presence of mixing time. Under this condition, non-inverted spin B will produce a resonance upon reducing the intensity, again as a function of mixing time due to this chemical exchange. A non-selective inversion–recovery experiment (bottom) is used to show how the exchange rates can be separated from the usual T_1 properties.

In the case of multicenter exchanges for selectively inverted spin i, the relaxation rate (R_i) follows Equation 4.14, where T_{1i} is the relaxation time and k_{ij} is the rate constant associated with each exchange process:

$$R_i = 1/T_{1i} + \Sigma k_{ij} \text{ at } i \neq j \qquad (4.14)$$

Accurate quantitative determinations of exchange rates can be achieved by modeling the experimental data.

4.2.4 CARR–PURCELL–MEIBOOM–GILL RELAXATION DISPERSION

The time scale between 3×10^{-1} and 10 ms is of a great interest for researchers investigating motions in complex polymer systems and biomolecules. These motions represent chain reorientations, secondary structure changes, and hinged domain

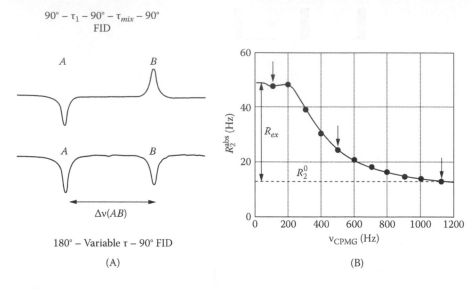

FIGURE 4.7 (A) The 1D EXSY NMR relaxation-type experiment performed with selective inversion (top) and the regular inversion–recovery pulse sequence (bottom). (B) Typical dispersion curve showing molecular dynamics in the coordinates relaxation rate R_2(OBS) vs. refocusing frequency ν_{CPMG}. (From Kletchner, I.R. and Foster, M.P., *Biochim. Biophys. Acta*, 1814, 942, 2011. With permission.)

movements. Different NMR approaches can be applied to characterize these motions. Among these approaches described in the recommended literature, the method based on Carr–Purcell–Meiboom–Gill relaxation dispersion (CPMG RD) plays a very important role. As was shown in Chapter 3, the CPMG pulse sequence is generally applied for measurements of T_2 times. The same sequence can be used for characterization of motions corresponding to the condition of a slow exchange; the constant rate k is $\sim\Delta\delta_0$ (expressed in Hz). Such an exchange causes additional broadening of detected resonances according to Equation 4.15:

$$1/T_2(obs) = 1/T_2^0 + R(ex) \tag{4.15}$$

where $R(ex)$ is the exchange rate. In fact, the CPMG experiments are aimed at observation of echo signals appearing after the action of a 180° pulse (see Equation 3.43 in Chapter 3). In turn, this spin echo completely refocuses the magnetization vectors when each individual magnetization vector is characterized by the same chemical shift during the first and second time period τ (i.e., echo delay). However, in the presence of an exchange, spins will experience different chemical shifts during one τ period. As a result, the magnetization cannot be refocused after time 2τ and the detected signal will be additionally broadened. This broadening effect obviously decreases for $\tau < 1/k(ex)$.

The main goal of CPMG RD experiments is to refocus this broadening due to the exchange by the addition of a set of spin echo pulses to transverse magnetization during a special relaxation delay. Because the degree of refocusing depends on the difference between the average shifts in the first and second τ periods (larger differences lead to larger broadenings), the CPMG relaxation dispersion reveals the quantitative relationship between echo delay τ and signal broadenings. Generally, a CPMG RD experiment consists of the following steps:

1. The spin echo pulses are applied during the fixed relaxation time in order to find the CPMG frequency as $\nu = 1/4\tau$.
2. The CPMG RD pulses are used at different relaxation delays to refocus the exchange broadening.
3. The remaining signal intensity is detected and used for the calculation of effective relaxation rate $R_2(obs)$, shown in Figure 4.7B.
4. The dispersion curves demonstrate the molecular dynamics by the patterns of relaxation rate vs. refocusing frequency.

4.3 MOLECULAR MOBILITY FROM RELAXATION TIMES

Figure 4.1 shows that spin–lattice relaxation corresponds to a region of fast molecular motions described in terms of the common NMR time scale. However, within the relaxation time scale, the motions can again be classified as fast and slow relative to the Larmor resonance frequency ν_0 (or ω_0) in Figure 4.8. According to the theory of nuclear relaxation occurring via dipolar, quadrupolar, and chemical shift anisotropy (CSA) mechanisms, the plots of T_1 vs. $1/T$ are symmetrical and V-shaped to show T_1 minima in semilogarithmic coordinates (Chapter 3) because molecular

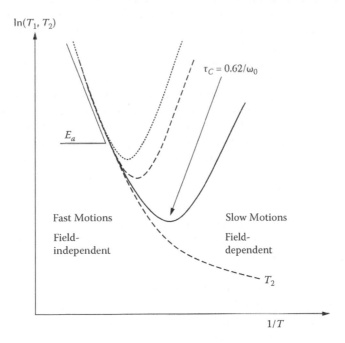

FIGURE 4.8   The plots T_1 vs. $1/T$ in semilogarithmic coordinates for spin–lattice nuclear relaxation occurring via dipolar and quadrupolar mechanisms.

motion correlation times are temperature dependent. As seen in Figure 4.8, the T_1 times are field independent or field dependent in areas of fast or slow molecular motions, respectively, while T_1 minima correspond to $\tau_C = 0.62/\omega_0$. This equation is valid for dipolar and quadrupolar relaxation, but a similar expression can be easily obtained for chemical shift anisotropy relaxation. Thus, the T_{1min} times allow calculation of the relaxation force constant (internuclear distances, quadrupolar coupling, or chemical shift anisotropy). The τ_C values can then be calculated at each temperature. Because the slopes of T_1 curves in the area of fast and slow motions correspond to activation energies E_a, the T_1 time measurements in the large temperature diapason can characterize molecular mobility in terms of correlation times and energies of molecular reorientations. Here, the best approach is based on fitting treatments of the experimental curves to determine the necessary parameters. However, the experimental observation of full temperature T_1 curves (as shown in Figure 4.8) is not trivial because the real experimental conditions are dictated by the nature of molecular systems, their solubility in solvents, properties of the solutions, etc. In fact, the left part of a curve is generally typical of solutions, whereas the right part is typical of solids, which require cooling and heating, respectively, to observe the T_{1min} values. Nevertheless, the τ_C/E_a calculations are still possible when the relaxation force constants are determined independently.

In the context of quantitative interpretations of relaxation studies, the relaxation mechanism should be well established (or the corresponding relaxation contribution should be well evaluated), all of the undesirable factors affecting relaxation times should be eliminated, and molecular motions would be isotropic.

Some conclusions can be made on the basis of T_1 and T_2 measurements. As seen in Figure 4.8, the left (high-temperature) wing of the T_1 curve is field independent and $T_1 = T_2$, whereas the right (low-temperature) wing depends on the field and $T_1 > T_2$. Thus, even in the absence of a large temperature diapason, experiments performed with different magnetic fields and the comparison of T_1 and T_2 times can identify what kind of motional regime (fast or slow) is taking place. The temperature T_1 dependence itself can discriminate the mechanism that controls relaxation. For example, in contrast to dipolar and quadrupolar relaxation, the spin–rotational mechanism corresponds to decreasing T_1 values with increasing temperature. The CSA relaxation mechanism shows T_1 times that decrease when the B_0^2 factor increases, whereas dipolar/quadrupolar relaxation times are field independent in the fast motional regime. However, in reality, the situation can be more complex. In fact, the relaxation mechanism can be mixed, thus requiring the application of special approaches to identify the corresponding contributions.

Among the undesirable factors affecting nuclear relaxation, the possible presence of paramagnetic impurities plays a very important role (see discussion relating to paramagnetic relaxation in Chapter 3). Because relaxation rates are additive magnitudes, the effect of the paramagnetic contribution to total relaxation rates will depend remarkably on natural relaxation T_1 times of the nuclei being investigated. The longer natural T_1 times will be more strongly distorted. In general, any paramagnetic impurities (paramagnetic metal ions and organic/organometallic radicals or molecular oxygen in solvents or in solids) should be carefully removed by conventional methods because the relaxation mechanism via unpaired electrons is extremely effective. Finally, it is obvious that the measurement of molecular motion correlation times from T_1 values requires minimizing disparities between the temperature settings and the real temperature in NMR samples.

4.3.1 DIPOLE–DIPOLE RELAXATION, NUCLEAR OVERHAUSER EFFECT, AND MOLECULAR MOBILITY

Dipolar relaxation is often used for the determination of correlation times and activation energies of molecular motions in solutions and in the solid state when this mechanism is completely dominant. However, determining this domination requires an independent test to measure the nuclear Overhauser effect (NOE). When nuclei A and X are coupled by dipole–dipole interactions and the distance between them is relatively small, they can show the NOE. This phenomenon, important in the context of structural and relaxation studies, consists of perturbation of the equilibrium X magnetization, measured experimentally as the integral intensity of signal X, in the presence of a radiofrequency field applied for nuclei A (Figure 4.9A). At positive γ magnitudes of nuclei A and X, the NOE corresponds to an *enhancement* of the X integral intensity.

At the same time, the NOE can be negative if the γ constant for one of the coupled nuclei is negative. It is valid, for example, in pairs $\{^1H\}/^{15}N$ or $\{^1H\}/^{29}Si$. The effect can be more complex, such as in a structural fragment containing three protons $H_a \cdots H_b \cdots H_c$. Here, the H_a irradiation generally results in positive and negative NOE enhancement being detected for H_b and H_c protons, respectively. An important element in NOE experiments is the fact that any comparison of integral intensities should be carried out in two experiments, with the irradiation frequency being strongly shifted in one of the experiments.

Because the NOE is based on dipole–dipole internuclear interactions and the correlation times of molecular tumbling depend on the temperature, the NOE is also temperature dependent. Figure 4.9B illustrates the dependence expected for protons (NOE enhancement vs. $\omega_0 \tau_C$). As can be seen, increasing the τ_C value on cooling reduces the positive NOE enhancement to zero. The enhancement then becomes negative. In practical terms, small-size molecules reorienting in non-viscous media with τ_C correlation times of 10^{-10} to 10^{-11} s should show the positive NOE enhancements at moderate temperatures. In contrast, bulky molecules, such as organic polymers, complex inorganic aggregates, or proteins, tumble with τ_C values of 10^{-7} s; therefore, their NOE enhancements are generally negative. All of these features can be seen in NOE measurements performed on solutions using 1D or 2D NMR (NOESY) experiments. For 2D NMR, the cross pikes will be responsible for the NOE.

Theoretically, full domination of the dipole–dipole mechanism in the relaxation of nuclei A and X causes the NOE enhancement detected for nuclei X, expressed by Equation 4.16:

$$M^X\{A\}/M_0^X = 1 + \gamma_A/2\gamma_X \qquad (4.16)$$

where $M^X\{A\}$ and M_0^X are the X integral intensities measured in the presence and in the absence of the A irradiation, respectively. Thus, in the case of two protons ($\gamma_A = \gamma_X = \gamma_H$), a NOE enhancement of 50% will be maximal. At the same time, dipole–dipole

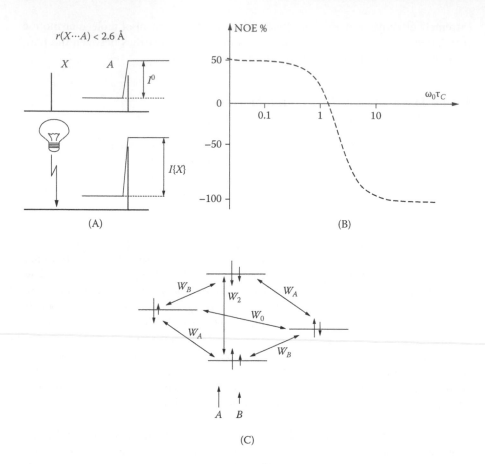

FIGURE 4.9 (A) The NOE enhancement for protons A and X. (B) The NOE enhancement expected for a pair of protons as a function of correlation times τ_C. (C) Energy levels for a two-spin AB system, where W is the probability of the corresponding nuclear transitions.

coupling is reduced with increasing internuclear distances. Therefore, the NOE can readily be detected experimentally at proton–proton distances less than 2.6 Å to prove the presence of dipolar relaxation.

The NOE measurements are particularly important when target nuclei can relax by several relaxation mechanisms simultaneously. In fact, measurements of $T_1(total)$ times and the NOE enhancements for nuclei X and A result in an accurate evaluation of the dipole–dipole relaxation contribution T_1DD, as shown below:

$$M^X\{A\}/M_0^X = 1 + (\gamma_A/2\gamma_X) * (T_1DD(X-A))^{-1}/T_1(total)^{-1} \qquad (4.17)$$

This evaluation can also be carried out by measuring the selective relaxation times (T_{1sel}) to determine their contributions into non-selective (regular) T_1 times. The physical principle of the evaluation can be well rationalized by consideration of the two spin systems where protons A and B are spatially close and coupled. Figure 4.9C shows the energy levels corresponding to the spin states. These states are combined and responsible for cross-relaxation. When the inverting 180° pulse in the inversion–recovery experiment excites only proton B, then its T_{1sel} time can be written as in Equation 4.18:

$$1/T_{1sel}(B) = 2W_B + W_2 + W_0 \qquad (4.18)$$

where W_B, W_2, and W_0 are probabilities of the corresponding spin transitions. In turn, Equation 4.19 shows the $T_1(B)$ time determined for the non-selective 180° pulse:

$$1/T_1(B) = 2W_B + 2W_2 \qquad (4.19)$$

Thus, a 100% dipolar relaxation can be proven if the T_{1sel}/T_1 ratio determined experimentally is equal to 1.5 found theoretically. It follows that smaller T_{1sel}/T_1 magnitudes will show the presence of additional relaxation mechanisms, the contributions of which can be well evaluated.

4.3.2 Effects of Molecular Motional Anisotropy on Nuclear Relaxation

As was noted earlier, for isotropic molecular motions quantitative interpretations of relaxation studies are relatively simple. However, even small organic or inorganic molecules are not spherical, and they generally experience anisotropic motions in solutions and in solids. For example, in solids, a typical case is the extremely fast rotation of methyl groups while a molecule remains immovable on the regular NMR time scale. These groups can reorient in molecular crystals either by thermally activated jump rotation over the very low energy barrier or by tunneling through the barrier.[10] Such a rotation is observed in a sample of solid toluene CH_3–C_6D_5 mixed with CD_3–C_6D_5 to remove dipole–dipole interactions of methyl protons with protons outside the CH_3 group. According to the measurements, two types of activation processes take place: random jump reorientation over barrier $\Delta E_a = 1.3$ kJ/mol and torsional excitation at $\Delta E_a = 0.55$ kJ/mol, associated with tunneling.

Because nuclear relaxation is directly connected with molecular reorientations, the anisotropy of motions can strongly affect the relaxation times. This effect can be rationalized quantitatively on the basis of the simple model, where a symmetrical ellipsoid moves in a continuous medium. This ellipsoid obviously has two principal rotational orientations (Figure 4.10A), which correspond to two different inertia moments. Thus, as was noted in Chapter 3, the motions of the ellipsoid are characterized by two correlation times noted as $\tau_\perp$ and $\tau_\parallel$. They will obviously correspond to two different activation energies.

Considering an idealized ^{1}H dipole–dipole, spin–lattice relaxation for two protons fixed in this symmetric ellipsoid leads to the important conclusion in the presence of a thermally activated anisotropic rotational diffusion of the ellipsoid: The Bloembergen–Purcell–Pound (BPP) spectral density function will adequately describe the proton relaxation if the isotropic motional correlation times τ_C (see

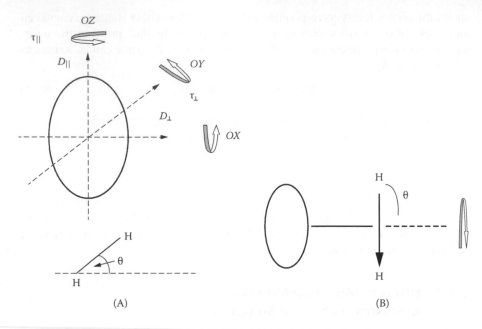

FIGURE 4.10 (A) Idealized ^{1}H dipole–dipole spin–lattice relaxation for two protons fixed in a symmetric ellipsoid and experiencing thermally activated anisotropic rotational diffusion. (B) Fast internal rotation (rotational diffusion) of a H_2 ligand in transition metal dihydrogen complexes, leading to a fourfold elongation of ^{1}H T_1 relaxation times.

Chapter 3) are replaced by the effective correlation times τ_{eff}.[11] These effective correlation times are given below:

$$\tau_{eff} = A(\theta) * \tau_A + B(\theta) * \tau_B + C(\theta) * \tau_C \qquad (4.20)$$

In this equation, coefficients A, B, and C, depending on the angle θ formed by the H–H vector and the rotational axis (see angle θ in Figure 4.10A) are determined via Equation 4.21:

$$A(\theta) = (1/4)(3\cos^2\theta - 1)^2$$

$$B(\theta) = 3\sin^2\theta\cos^2\theta \qquad (4.21)$$

$$C(\theta) = (3/4)\sin^4\theta$$

In turn, the correlation times τ_A, τ_B, and τ_C noted in Equation 4.20 are given by Equation 4.22:

$$\tau_A = \tau_\perp$$

$$(\tau_B)^{-1} = (5/6)\tau_\perp^{-1} + (1/6)\tau_\parallel^{-1} \qquad (4.22)$$

$$(\tau_C)^{-1} = (1/3)\tau_\perp^{-1} + (3/2)\tau_\parallel^{-1}$$

In practice, this theory can be successfully applied for ellipsoidal molecules when their structure is known and T_1 measurements are carried out for several nuclei differently located in the ellipsoids to find the $\tau_\perp$ and $\tau_\parallel$ values.

This model can be represented more obviously to show clearly how the motional anisotropy can have an effect on the ^{1}H dipole–dipole, spin–lattice relaxation by changing the character of the temperature dependence of ^{1}H T_1 times. Again, two protons are located in the same ellipsoid, and the anisotropy of motions can be described by parameter ρ as shown below:

$$\rho = D_\parallel / D_\perp \qquad (4.23)$$

where $D_\parallel$ and $D_\perp$ are the rotational diffusion constants corresponding to rotation around the axes $\parallel$ and $\perp$, respectively. For simplicity, the θ angle can be taken as $90°$. It should be noted that such a situation is often encountered in, for example, transition metal complexes containing dihydrogen ligands (see Figure 4.10B). Here, the protons participate in whole molecular tumbling and, simultaneously, in intramolecular rotation. The rate of the ^{1}H dipolar relaxation, $1/T_1$, on these conditions can be expressed as follows:

$$1/T_1\,(\mathrm{H}\cdots\mathrm{H}) = (3/40)\left(\mu_0/4\pi\right)^2 \gamma_H^4 \hbar^2 r(\mathrm{H–H})^{-6}\,\tau_C \qquad (4.24)$$

$$\times \left[\begin{array}{l} 1/\left(1+\omega_H^2\tau_C^2\right)+4/\left(1+4\omega_H^2\tau_C^2\right) \\ +3a/\left(a^2+\omega_H^2\tau_C^2\right)+12a/\left(a^2+4\omega_H^2\tau_C^2\right) \end{array} \right]$$

where $a = (2\rho +1)/3$ and $\tau_C = 1/6D_\perp$.[12] It is remarkable that at $\rho = 1$ Equation 4.24 converts to the BPP function (i.e., written for the case of isotropic molecular reorientations; see Chapter 3). On the other hand, at $\rho \to \infty$ and with the fast internal rotation occurring on the time scale of whole molecular tumbling, the function $J(\omega_H,\tau_C)$ in Equation 4.24 transforms to Woessner's form in Equation 4.25:

$$J(\omega_H,\tau_C) = [(3\cos^2\theta - 1)^2/4](\tau_C/(1 + \omega_H^2\tau_C^2) + 4\tau_C/(1 + \omega_H^2\tau_C^2) \qquad (4.25)$$

because factor $(3\cos^2\theta - 1)^2$ becomes equal to 1 at $\theta = 90°$.

Finally, for visualization of the effect, the distance $r(\mathrm{H–H})$ can be taken as 2 Å to compute the ^{1}H T_1 time as a function of the correlation time τ_C. Figure 4.11 illustrates these results, as the plots of the $\ln(T_1)$ vs. τ_C, obtained for variations in the ρ value from 1 to 2, 5, 10, 50, and ∞ and at the proton resonance frequency of 200 MHz. Because the τ_C time is temperature dependent, the T_1 curves reproduce the temperature dependencies of relaxation times for increasing τ_C. This computation leads to several important conclusions:

1. Increasing the anisotropy ρ causes remarkable elongation of the observed ^{1}H T_1 times. The elongating T_1 effect is particularly strong in high- and low-temperature regions corresponding to left and right wings of the curves, respectively.

$R(\mathrm{H-H}) = 2\ \text{Å}$

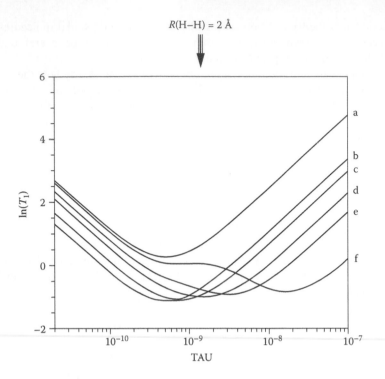

FIGURE 4.11 Plots of $\ln(T_1)$ vs. τ_C (equivalent to the T_1 temperature dependencies) computed via Equation 4.24 at ρ values from (a) ∞ to (f) 50, (e) 10, (d) 5, (c) 2, and (b) 1. (From Gusev, D.G. et al., *Inorg. Chem.*, 32, 3270, 1993.)

2. Anisotropic reorientations described by intermediate ρ values (e.g., $\rho = 10$ and 50) can lead to non-symmetric plots relative to the τ_C values at T_1 minima. Moreover, at $\rho = 50$, the $\ln(T_1)$ plot shows two ^{1}H T_1 minima demonstrating directly the presence of two different molecular motions. At the same time, these features remain invisible at low anisotropy (e.g., $\rho = 2$ to 5) or at very high anisotropy ($\rho = \infty$). Thus, under these conditions, the anisotropic and isotropic molecular reorientations cannot be discriminated experimentally.

3. When the ρ value changes from 1 to 50, disposition of T_{1min} times in the T_1 curves is visibly shifted toward lower temperatures, complicating their experimental observation.

4. In spite of the motional anisotropy, the $T_1(\tau_C)$ curves can show equal slopes (i.e., the equal E_a energies determined by the left and right wings).

5. When the variable-temperature T_1 relaxation curves are treated by an isotropic motional model (i.e., by the BPP function) while molecular systems are ellipsoidal, then the calculated parameters will be strongly distorted. In fact, if one of the motions is very fast and the dipolar H–H vector is perpendicular to the internal motional axis, then $T_1(aniso) = 4\ T_1(iso)$. In other words, the measured T_1 time is strongly elongated.

6. It is important that the isotropic model applied for anisotropic motions gives activation energies (E_a) that are meaningful, while the correlation time constants τ_0 (see Equation 3.1 in Chapter 3) are fictitious.

The common conclusion is that for accurate quantitative characterizations of molecular mobility by relaxation time measurements, the type of motions should be established independently. In fact, if the correlation function is unknown, quantitative interpretations of relaxation times will be connected with uncertainty in the form of this function.[13]

The unknown correlation function can be approximated by a simple model containing a small number of dynamic parameters. These parameters can be determined from the relaxation times, such as by the so-called model-free approximation shown below, where $C(t)$ is the correlation function of any anisotropic motion:

$$C(t) = S^2 + (1 - S^2)f(t) \tag{4.26}$$

Here, S^2 is taken as an order parameter and $f(t)$ is a decay function that satisfies the condition $f(0) = 1$ and $f(\infty) = 0$. The simplest form of the $f(t)$ function is given by Equation 4.27:

$$f(t) = \exp(-t/\tau_C) \tag{4.27}$$

It should be noted that the S^2 parameter characterizes averaging the dipolar coupling or quadrupolar interactions related to the motional amplitude. Finally, the order parameter S^2 can be expressed by Equation 4.28:

$$S^2 = <P_2(\cos\theta \ (0,\infty))> \tag{4.28}$$

thus showing the dependence of the S^2 parameter on the type of molecular motions.

For many motions, such as jumps between two equally populated states, rotational diffusion, or wobbling with a cone, the S^2 factor is already known as a function of angular amplitudes. This factor is found theoretically or experimentally and can be successfully applied.[13] It should be noted that within the framework of the model-free approach, molecular motions can be formally described by combining slow isotropic global molecular reorientations, corresponding to correlation times τ_M, and additional fast internal motions characterized by the correlation times τ. When $\tau \ll \tau_M$, the spectral density function takes the following form:

$$J(\omega) = (2/5)[S^2\tau_M/(1+ \omega^2\tau_M^2)] \tag{4.29}$$

When the correlation times of the global and internal motions are comparable, then these motions can be characterized by the effective correlation times, τ_{EFF}, used in the spectral density function given by Equation 4.30:

$$J(\omega) = (2/5)[\tau_{EFF}/(1+ \omega^2\tau_{EFF}^2)] \tag{4.30}$$

Finally, this model-free approximation can be applied for studies of molecular mobility in both solutions and solids.

4.3.3   MOLECULAR DYNAMICS IN THE PRESENCE
OF CORRELATION TIME DISTRIBUTIONS

In contrast to dilute solutions, molecular motions (or reorientations of molecular fragments) in viscous media or amorphous solids (e.g., glasses) can be very complex. They cannot be described by a single molecular correlation time τ_C at a given temperature. Such molecular dynamics can be represented by an ensemble of reorienting units, where there are sub-ensembles, each of which is characterized by a single correlation time τ_C. However, this time changes from one sub-ensemble to another. Generally speaking, a basic reason for this phenomenon is molecular motions that are correlated and cooperative. As a result, the BPP theory poorly describes the variable-temperature relaxation data collected in such systems even when accounting for motional anisotropy.

In order to simplify quantitative treatments of the relaxation data obtained for such systems, the concept of the correlation time distribution can be applied. Here, the regular correlation time τ_C will be at the center of the distribution, while the width can be represented by a distribution parameter. In some sense, this correlation time distribution approach can be considered as a special mathematical treatment of relaxation experimental data that deviate from the BPP theory, where the τ_C time is replaced by its distribution.

Several correlation time distributions are widely used in practice, including Davidson–Cole, Cole–Cole,[14] Fouss–Kirkwood,[15] and Havriliak–Negami.[16] For quantitative interpretations of relaxation data, these distributions can be incorporated into spectral density functions; for example, incorporating the Fouss–Kirkwood distribution gives the spin–lattice relaxation rate for a pair of protons as follows:

$$1/T_1\,(\mathrm{H}\cdots\mathrm{H}) = \left(DC_{\mathrm{H-H}}\tau_C\beta/\omega_{\mathrm{H}}\tau_C\right)\times\left[\begin{array}{c}\left(\omega_{\mathrm{H}}\tau_C\right)^{\beta}\Big/\left(1+\left(\omega_{\mathrm{H}}\tau_C\right)^{2\beta}\right)\\[4pt]+2\left(2\omega_{\mathrm{H}}\tau_C\right)^{\beta}\Big/\left(1+\left(2\omega_{\mathrm{H}}\tau_C\right)^{2\beta}\right)\end{array}\right] \tag{4.31}$$

where $DC_{\mathrm{H-H}}$ is the proton–proton dipolar coupling and β is the width of the distribution taking the value of $0 < \beta \le 1$. Equation 4.31 converts to the BPP spectral density function at $\beta = 1$. Similarly, in the presence of the Cole–Davidson correlation time distribution, the ^{1}H T_1 relaxation time for the same pair of protons is given by Equation 4.32:

$$1/T_1\,(\mathrm{H}\cdots\mathrm{H}) = DC_{\mathrm{H-H}}\tau_C \times\left[\begin{array}{c}\sin\left(\beta\arctan\tau_C\omega_{\mathrm{H}}\right)\Big/\omega_{\mathrm{H}}\tau_C\left(1+\left(\omega_{\mathrm{H}}\tau_C\right)^2\right)^{\beta/2}\\[4pt]+4\sin\left(\beta\arctan 2\tau_C\omega_{\mathrm{H}}\right)\Big/2\omega_{\mathrm{H}}\tau_C\left(1+\left(2\omega_{\mathrm{H}}^2\tau_C^2\right)^2\right)^{\beta/2}\end{array}\right] \tag{4.32}$$

In the context of variable-temperature T_1 measurements and their interpretations, the correlation time distributions can be represented by two important classes: symmetric distributions (e.g., Fuoss–Kirkwood) and asymmetric distributions (e.g., Cole–Davidson). They are shown schematically in Figures 4.12 and 4.13, respectively. As can be seen, the influence of the distributions on the T_1 data differs quite a bit.

For example, the asymmetric Cole–Davidson distribution affects the T_{1min} values only insignificantly. In contrast, the symmetric distribution can result in remarkable underestimations of activation energies E_a and overestimations of T_{1min} values. This is valid when the temperature-dependent T_1 data are analyzed in terms of isotropic models as is shown in Figure 4.12A. It should be noted that, under these conditions, the width of the β distribution plays a major role. The above effect is particularly important if the presence of a correlation time distribution is not recognized experimentally but actually exists. In this case, there is a significant chance of overestimating, for example, internuclear distances calculated from relaxation data.

In addition, in the presence of the symmetric correlation time distribution, the slopes of the $\ln(T_1)/(1/T)$ curves corresponding to magnitudes $E_a(obs)/R$ are proportional to the distribution parameters β in Equation 4.33:

$$E_a(obs) = \beta E_a \qquad (4.33)$$

where E_a and $E_a(obs)$ are the real and experimentally observed (effective) activation energies, respectively, and $0 < \beta \leq 1$.

In general, this feature makes it possible to recognize experimentally the presence of a correlation time distribution. In fact, the real activation energy of molecular motions can be accurately measured by localization of the T_{1min} values in the variable-temperature relaxation curves observed for different magnetic fields. Then, these T_{1min} values obviously give correlation times τ_{Cmin}, being at the center of the distribution and thus determined at different temperatures. Finally, the dependence $\ln(\tau_{Cmin})$ vs. $1/T$ results in energy E_a. Now, if this value does not correspond to an E_a value obtained from the slopes of the $\ln(T_1)$ vs. $1/T$ curves, then a correlation time distribution is present. This test is particularly useful in studies of solids, where

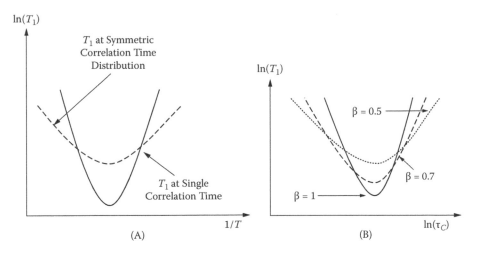

FIGURE 4.12 (A) Schematic illustration of how a symmetric correlation time distribution affects the variable temperature relaxation times where a motion is described by a single correlation time at a given temperature. (B) The T_1 times in the presence of a symmetric correlation time distribution at different β parameters.

correlation time distributions often characterize their complex dynamics. Potentially, the variable temperature data collected experimentally for such systems can be fitted to curves with incorporated distributions; however, the reliability of such fittings is not always obvious.

Because the Cole–Davidson distribution is asymmetric relatively to T_{1min} (Figure 4.13A), its presence can be found experimentally via asymmetric plots of $\ln(T_1)$ vs. $\ln(\tau_C)$. The latter can be the basis for application of the Cole–Davidson model to analyze T_1 data in solids or even in viscous solutions and liquids. For example, ^{1}H spin–lattice relaxation in 1,2,3,4-tetrahydro-5,6-dimethyl-1,4-methanonaphthalene,

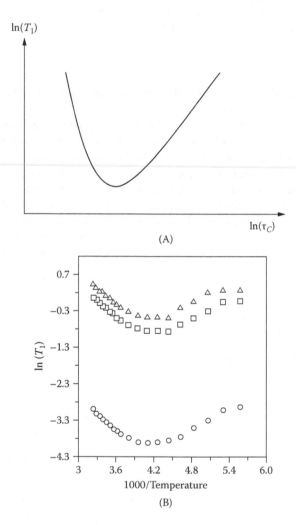

(A)

(B)

FIGURE 4.13   (A) Plot of $\ln(T_1)$ vs. $\ln(\tau_C)$ in the presence of an asymmetric Cole–Davidson correlation time distribution. (B) Variable-temperature ^{1}H T_1 times measured for $RuH_2(H_2)$ $(PPh_3)_2$ in toluene-d_8: (O) hydride ligands; ($\triangle$) *para* and *meta* protons of PPh_3; ($\square$) *ortho* protons of PPh_3.

classified as an organic glass-forming liquid, remarkably deviates from the isotropic motional model. The corresponding calculation gives the correlation time distribution characterized by the β values between 0.9 and 0.8. In contrast, glycerol shows a narrower correlation time distribution with β = 0.97.

Theoretically, the existence of the reorienting ensembles and sub-ensembles can be assumed in regular liquids at the lowest temperatures that increase their viscosity. Figure 4.13B shows the variable-temperature ^{1}H T_1 times measured for all the protons in a toluene-d_8 solution of $RuH_2(H_2)(PPh_3)_2$. The data cannot be treated with the BPP model because the slopes of the curves in high- and low-temperature zones differ remarkably. Such data obviously require application of a correlation time distribution, which can be a useful mathematical tool for estimating possible errors in the calculation of molecular motion correlation times or internuclear distances when the data are treated with the help of an isotropic model.[17]

4.3.4 DISTRIBUTION OF ACTIVATION ENERGIES

The concept of activation energy distributions gives a similar (but more detailed) understanding of the nuclear relaxation that occurs in the presence of correlation time distributions. This concept is widely applied for NMR studies of dynamics in heterogeneous systems, such as structurally disordered glasses[18] or zeolites with pores containing small guest molecules.[19] The idea behind this concept can be explored by using zeolites with ^{2}H-labeled guest molecules to measure ^{2}H T_1 times completely dominated by quadrupolar interactions. Generally speaking, distributions of activation energies characterizing molecular motions can originate (1) from the random distribution in the system of aluminum atoms bonding guest molecules, (2) from different surroundings of the guest molecules, and (3) from their different binding energies. As a result, ^{2}H relaxation observed in these systems is strongly non-exponential. Such a relaxation can be treated by a stretched exponential, $\exp(-(t/T_1)^\beta)$, to obtain a T_1 time that is only a representative value. In fact, even in spite of the good quality of the fittings it is unclear how the dynamic parameters, activation energies, and correlation times will be connected with this T_1 time and the β parameter in the stretched function. In contrast, the concept of an activation energy distribution explains the non-exponentiality of the relaxation behavior and results in numerical values that characterize the distribution width and molecular motion correlation times.

This approach is based on the fact that ^{2}H nuclei, for example, belonging to the CD_3 groups of a guest methanol molecule in a zeolite are assumed to participate in motions that correspond to activation energy E_0. Around this value, the possible activation energies describing the motion of other groups can be distributed as shown in Equation 4.34:

$$E_k = E_0 + \Delta E_k \qquad (4.34)$$

The probability of the E_k energy is W_k, given, for example, by the Gaussian distribution in Equation 4.35:

$$W_k = C \exp(-\Delta E_k^2/2\sigma^2) \qquad (4.35)$$

where σ is the width of the distribution. Under these terms, the correlation times τ_k are determined via Equation 4.36:

$$\tau_k = \tau_0 \exp[(E_0 + \Delta E_k)/RT] \tag{4.36}$$

and the relaxation rates $(1/T_{1k})$ can now be written by Equation 4.37:

$$1/T_{1k} = A[J(\tau_k,\omega_0) + 4J(\tau_k,2\omega_0)] \tag{4.37}$$

where the BPP spectral density function is written as shown in Equation 4.38:

$$J(\tau_k,\omega) = \tau_k/[1 + \tau_k^2\omega^2] \tag{4.38}$$

Coefficient A depends on a motional model.

It should be added that the non-exponential magnetization recovery curves observed experimentally (for example, in the presence of several CD_3 groups) can be treated with several exponentials, where each subsystem will again be characterized by mean activation energy E_0, distribution width σ, and the τ_0 value. Other models addressing energy activation distribution (e.g., Vogel–Fulcher model) are also available and can be found in the recommended literature.

4.4   NMR RELAXOMETRY: DIFFUSION COEFFICIENTS

An important element in the dynamics of liquids (bulk liquids or liquids in porous solids) is their self-diffusion, which can be characterized by diffusion NMR spectroscopy. The theoretical basis of diffusion magnetic resonance can be found in the recommended literature. In general, diffusion NMR experiments are based on the use of static or pulsed field gradients (PFGs) to label the spatial position of nuclear spins to determine the molecular diffusion coefficient due to translational displacement over a given time period. It is of interest that the influence of an inhomogeneous magnetic field on the formation and decay of the spin echo signal was noted by Hahn in 1950. When a static field gradient G is present, then a coherent transverse magnetization created by the first 90° pulse in the Hahn echo pulse sequence and expressed as follows,

$$90° \rightarrow G(\tau) \rightarrow 180° \rightarrow G(2\tau) - \text{FID} \tag{4.39}$$

experiences an evolution in gradient G. Due to a dephasing time τ, the second 180° pulse inverts the sign of the net accumulated phase and the spins reduce their accumulated phases due to the Larmor precession. When the rephasing and dephasing rates are equal, then the magnetization reaches its coherent initial state after time 2τ, thus creating a spin echo. It is obvious that the spin echo will decay because of T_1/T_2 relaxation, as well as factors directly related to molecular displacement on the time scale of the spin echo experiments. Finally, if the field gradients are well defined, the time correlation functions describing the spin echo decays can be applied to probe translational molecular diffusion.

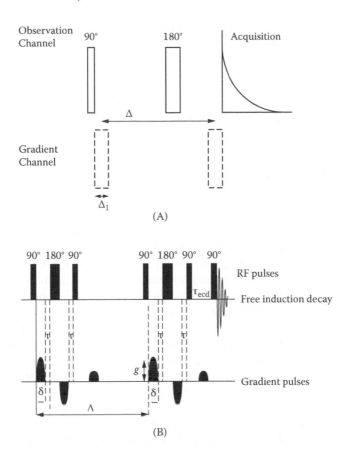

FIGURE 4.14 (A) Simplest diffusion NMR experiment using a channel for detection and a gradient channel. (B) A MAS PFG NMR experiment with radiofrequency pulse sequence and gradient pulse scheme, where Δ is diffusion time, δ is gradient pulse width, g is gradient pulse strength, τ_{ecd} is eddy current delay, and τ is intergradient delay. (From Romanova, E.A. et al., *J. Magn. Reson.*, 196, 110, 2009. With permission.)

Technically, diffusion NMR experiments can be realized in the presence of static field gradients (SFGs) or pulsed field gradients. Figure 4.14A illustrates the simplest experiment using a channel for detection and a gradient channel. Application of the pulse sequence leads to an echo intensity proportional to $\exp[-(\gamma g \Delta_1)^2 D(\Delta - \Delta_1)/3]$, where g is the gradient strength, Δ_1 and Δ are the time delays necessary for action of the gradient and collection of the echo, and D is the diffusion coefficient. Thus, an analysis of the echo intensity as a function of time results in determination of the D coefficients in solutions. It should be noted that a combination of large static field gradients with short radiofrequency pulse spacing provides measurements of very small diffusion coefficients (10^{-15} m² s⁻¹) at high spatial resolution (10 nm).

More complex pulse sequences are applied in solids containing a liquid phase.[21,22] For example, Figure 4.14B illustrates a pulsed field gradient experiment adjusted for solids spinning at high rates (MAS NMR experiment) that can be used for studies of self-diffusion in porous glasses. Due to application of these pulses, the diffusion coefficient D can be determined via the echo signal intensity:

$$I = I_0 \exp[-D(4\gamma\delta g/\pi)^2(\Delta - \tau/2 - 2\delta/3 - p_\pi)] = I_0 \exp(-D_k) \qquad (4.40)$$

where δ is the gradient pulse width, Δ is the observation time, τ is the inter-gradient delay, and p_π is the duration of the pulse. Some practical examples are considered below.

4.5   MOLECULAR DYNAMICS FROM LOW-FIELD NMR

Commonly, low-field NMR is based on the application of magnets with strengths less than $2T$. This is a disadvantage of this technique because of the low sensitivity coming from small differences in the equilibrium populations of spin states; however, it is clear that when strongly heterogeneous systems are probed a high and homogeneous magnetic field is not necessary. For this reason, low-field NMR has great potential for use in the food and oil industries, as well as many other fields, particularly for characterizations of highly heterogeneous samples by measurements of relaxation times as parameters of molecular dynamics.

The low field is the main advantage of this technique because the relaxation curves can be obtained with the help of simple NMR devices having inhomogeneous magnetic fields, where NMR signals are detected as echoes by applying the Hahn echo or multiple echo (CPMG) pulse sequences. T_1 time measurements can also be performed by inversion–recovery or saturation–recovery pulse sequences that precede the multi-echo detection sequence echo trains. Details regarding such experiments can be found in the recommended literature.

The principles of low-field NMR and its applications can be demonstrated by the study of soils, which is important in the context of petroleum-induced water repellency in soils. The relaxation measurements are performed by the CPMG pulse sequence, and the collected data are plotted as patterns of echo intensity vs. T_2 time. In other words, complex inhomogeneous mixtures can be quantitatively characterized to show the relative amount of each component and its mobility.

Suitable software program packages can be used to obtain final T_2 distributions. These distributions are normalized to show a spectrum on a scale of percent, where 100% represents the integral under the total T_2 distribution as is shown in Figure 4.15. Similar approaches are generally applied in food chemistry to characterize, for example, the internal water change in sweet corn.[22]

REFERENCES AND RECOMMENDED LITERATURE

1. Azevedoa, E.R., Bonagambaa, T.J., and Reichertb, D. (2005). *Prog. Nucl. Magn. Reson. Spectrosc.*, 47: 137.
2. Bakhmutov, V.I., Kochetkov, K.A., and Fedin, E.I. (1980). *Izv. Akad. Nauk. Ser. Khim.* 1295.
3. Korniets, E.D., Fedorov, L.A., Kravtsov, D.N., and Fedin, E.I. (1979). *Izv. Akad. Nauk. SSSR Khim.*, 3: 525.
4. Bakhmutov, V.I. and Fedin, E.I. (1984). *Bull. Magn. Reson.*, 6: 142.

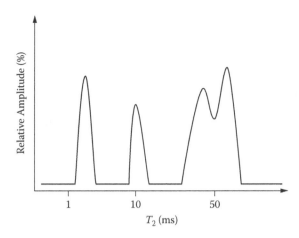

FIGURE 4.15 T_2 time distribution for an inhomogeneous sample: T_2 time vs. normalized echo intensity

5. Bakhmutov, V.I., Burgi, T., Burger, P., Ruppli, U., and Berke, H. (1994). *Organometallics*, 13: 4203.
6. Galakhov, M.V., Bakhmutov, V.I., and Gruselle, M. (1991). *J. Organomet. Chem.*, 321: 65.
7. Davis, L.J.M., Ellis, B.L., Ramesh, T.N., Nazar, L.F., Bain, A.D., and Goward, G.R. (2011). *J. Phys. Chem. C*, 115: 22603.
8. Hagemeyer, A., Schmidt, K., and Spiess, H.W. (1989). *Adv. Magn. Reson.*, 13: 85.
9. Dejong, A.F., Kentgens, A.P.M., and Veeman, W.S. (1984). *Chem. Phys. Lett*, 109: 337.
10. Horsewill, A.J. (1999). *Prog. Nucl. Magn. Reson. Spectrosc.*, 35: 359.
11. Woessner, D.E. (1962). *J. Chem. Phys.*, 36: 1.
12. Gusev, D.G., Nietlispach, D., Vymenits, A.B., Bakhmutov, V.I., and Berke, H. (1993). *Inorg. Chem.*, 32: 3270.
13. Krushelnitsky, A. and Reichert, D. (2005). *Prog. Nucl. Magn. Reson. Spectrosc.*, 47: 1.
14. Cole, K. and Cole, R.J. (1941). *Chem. Phys.*, 9: 341.
15. Fuoss, R.M. and Kirkwood, J.G. (1941). *J. Am. Chem. Soc.*, 63: 385.
16. Geppi, F.C., Malvaldi, M., and Mattoli, V. (2004). *J. Phys. Chem. B*, 108: 10832.
17. Gusev, D.G., Vymenits, A.B., and Bakhmutov, V.I. (1991). *Inorg. Chim. Acta*, 179: 195.
18. Svare, I., Borsa, F., Torgeson, D.R., and Martin, S.W. (1994). *J. Non-Crystal. Solids*, 172–174: 1300.
19. Stoch, G., Ylinen, E.E., Birczynski, A., Lalowicz, Z.T., Gora-Marek, K., and Punkkinen, M. (2013). *Solid-State Nucl. Magn. Reson.*, 49–50: 33.
20. Romanova, E.A., Grinberg, F., Pampel, A., Kärger, J., and Freude, D. (2009). *J. Magn. Reson.*, 196: 110.
21. Viel, S., Ziarelli, F., Pagès, G., Carrara, C., and Caldarelli, S. (2008). *J. Magn. Reson.*, 190: 113.
22. Shao, X. and Li, Y. (2013). *Food Bioprocess. Technol.*, 6: 1593.

RECOMMENDED LITERATURE

Bakhmutov, V.I. (2005). *Practical NMR Relaxation for Chemists*. Chichester: Wiley.
Blumich, B., Casanova, F., and Appelt, S. (2009). NMR at low magnetic fields. *Chem. Phys. Lett.*, 477: 231.

Brand, T., Cabrita, E.J., and Berger, S. (2005). Intermolecular interaction as investigated by NOE and diffusion studies. *Prog. Nucl. Magn. Reson. Spectrosc.*, 46: 159.

Canet, D. (2005). Introduction: general theory of relaxation. *Adv. Inorg. Chem.*, 57: 3.

Case, D.A. (2002). Molecular dynamics and NMR relaxation in proteins. *Acc. Chem. Res.*, 35: 325.

Christianson, M.D. and Landis, C.R. (2007). Generalized treatment of NMR spectra for rapid chemical reactions. *Concepts Magn. Reson. Part A*, 30A: 165.

Griesinger, C. (2001). NMR spectroscopy as a tool for the determination of structure and dynamics of molecules. In *Essays in Contemporary Chemistry: From Molecular Structure towards Biology* (Quinkert, G. and Kisakuerek, M.V., Eds.), pp. 35–106. Zurich: Verlag Helvetica Chimica Acta.

Hoffman, R.A. (1970). Line shapes in high-resolution NMR. *Adv. Magn. Reson.*, 4: 87.

Koay, C.G. and Özarslan, E. (2013). Conceptual foundations of diffusion in magnetic resonance. *Concepts Magn. Reson. Part A*, 42A: 116.

Levitt, M. (2008). *Spin Dynamics: Basics of Nuclear Magnetic Resonance*, 2nd ed. Chichester: John Wiley & Sons.

Martin, M.L., Delpuech, J.-J., and Martin, G.J. (1980). *Practical NMR Spectroscopy*. Philadelphia, PA: Heyden & Son.

Wasylishen, R.E. (1987). NMR relaxation and dynamics. In *NMR Spectroscopy Techniques* (Dybowski, C. and Lichter, R.L., Eds.), Chap. 2. New York: Marcel Dekker.

5 NMR Spectroscopy in Solutions

Practice and Strategies of Structural Studies

Since the publication of pioneering works on nuclear magnetic resonance, two domains—NMR spectrometry and NMR relaxation—have come to represent its practical applications in physics, chemistry (fundamental chemistry, food chemistry, biochemistry, geochemistry), biology, geology, archeology, pharmaceuticals, and materials science. The final result of any spectroscopic experiment is an NMR spectrum recorded as a pattern of signal intensity vs. frequency, where the lines and integral intensities indicate the number of magnetically and chemically non-equivalent nuclei in proportion to their content. This is key information for structural analyses based on the principles of NMR and the accurate assignment of signals using well-established spectral–structural relationships. In turn, the reliability of the data should be ensured by correct NMR adjustments and parameter setting to minimize instrumental errors and artifacts potentially affecting interpretation of NMR spectra. This is particularly important for two-dimensional (2D) or three-dimensional (3D) NMR experiments, for which the appearance of artifacts is a regular phenomenon. In this context, the preparation of samples, adjustment of spectral resolution, and choice of a reference line for further calculations of chemical shifts are major factors.

5.1 PREPARATION OF NMR SAMPLES: MINIMAL REQUIREMENTS

A general requirement for the preparation of samples is that a compound must be sufficiently soluble in an appropriate ^2H–solvent, the signal of which is used for field-frequency lock. Also, the solution must be homogeneous, and the volume must be larger than the size of the radiofrequency (RF) coil in order to avoid a vortex effect during spinning. One should remember that solvents with smaller viscosities provide the better spectral resolution required to determine the number of lines. For compounds with low solubility, the undesirable intense signal coming from the solvent (e.g., HOD) can be suppressed by conventional techniques. Various NMR tubes for NMR studies in solutions are available, including regular tubes (in different sizes), coaxial tubes, tubes for microsamples to achieve maximal resolution, and high-pressure NMR tubes. Tubes for microsamples should be located at the center of the RF coil, where the radiofrequency field is maximally homogeneous. Air-sensitive samples can be prepared under an argon or a nitrogen atmosphere or in a vacuum in commercially available NMR tubes designed to seal samples. Finally, in order to achieve the best resolution and to measure nuclear Overhauser effect (NOE) enhancements or relaxation times, solvents and solutions must be oxygen free. Deoxygenation can be carried out with three or four freeze–pump–thaw cycles. The samples can then be sealed under vacuum or in an inert atmosphere, if necessary.

5.1.1 ADJUSTMENT OF SPECTRAL RESOLUTION AND SPECTRAL MANIPULATIONS

Adjustment of the spectral resolution is a simple but very valuable step in preparation for NMR experiments. This is particularly important for beginners. Generally speaking, spectral resolution is the ability of a NMR spectrometer to resolve lines located close together in the NMR spectrum. It can be defined as $R = v_0/\Delta v$, where Δv is the smallest difference in frequencies that can be distinguished for resonance frequency v_0. The resolution is directly connected with the natural width of a Lorenz-shaped signal, as shown in Figure 5.1A, from uncoupled spins in an ideal magnetic field, dictated by spin–spin relaxation time T_2. In room-temperature solutions, $T_2 = T_1 \approx 10$ s (e.g., protons in small organic compounds), and the expected line width would be 0.032 Hz. However, in reality, the line width depends on the inhomogeneity of the external magnetic field defined via the effective relaxation constant T_2^*. In practice, this constant can be increased in order to reduce the line width by standard shimming-coil manipulations. In general, the goal of these manipulations is observation of the Lorentz shape (Figure 5.1A) at a line width of 0.1 to 0.2 Hz.

Figure 5.1B illustrates typical results of resolution adjustments for a sample placed into an inhomogeneous magnetic field in the presence of Z1, Z2, Z3, and Z4 gradients reduced with the help of corresponding shimming coils. These signal shapes can be used as a guide for manual adjustments of spectral resolution.[1] It should be

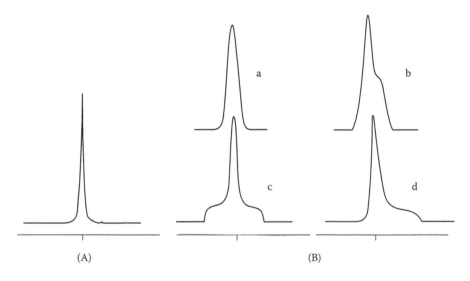

FIGURE 5.1   (A) Lorentz shape of NMR signals typical of solutions. (B) Line shape of the resonance in the presence of magnetic field gradients (a) Z1, (b) Z2, (c) Z3, and (d) Z4.

emphasized that, for example, in the presence of significant gradient Z3 (see shape c), the spinning sample will show sidebands. In turn, these sidebands can mask lower intensity signals that can be observed in the NMR spectrum. In the highest magnetic fields, the number of shimming coils is generally added and the shimming manipulations can become more sophisticated. In such cases, the pulsed field gradient technique can be very useful for mapping the shims to achieve good resolution. To avoid misunderstanding, it should be noted again that (1) the spectral resolution measured by line width cannot be any better than that determined by a natural T_2 time, and (2) recording the test NMR spectra should be supported by computer software and hardware to provide sufficient digital resolution.

Because the reliability of NMR interpretations based on the determination of integral intensities and chemical shift values depends on the quality of the final NMR spectra, they should be recorded with accurate phasing and baseline correction procedures after Fourier transformation. This is particularly important for detection and integration of the low-intensity signals or relatively broad lines in NMR spectra in solutions that appear due to scalar relaxation of the second kind or chemical exchange, among other reasons, and can be overlapped. Currently, several methods have been developed specifically for the automatic calculation of integrals and their allocations to multiplets, even when they are overlapping. Generally, in the context of integration, such methods give overall errors of 7%, whereas manual procedures can reduce such errors to 3 to 5%.[2]

Multiple-pulse NMR experiments are often phase sensitive; therefore, they require the application of special phase cycling during collection of the NMR data. For example, double quantum ^{13}C–^{13}C NMR experiments require 256 cycles in the phase programs applied. Such phase cycling procedures simplify phase corrections in the final NMR spectra, and they eliminate or at least minimize artifact signals appearing in multiple-pulse experiments. In addition, special phase cycling can even suppress undesirable signals by selection of the corresponding coherence pathways; for example, two spectra obtained with a phase shift of 180° can be summed to cancel certain lines. This approach is very successful for ^{13}C–^{13}C INADEQUATE (incredible natural abundance double quantum transfer experiment) NMR, where the resonances of uncoupled ^{13}C nuclei can be removed to observe the antiphase ^{13}C–^{13}C satellites.[3]

5.1.2 REFERENCE LINES

In general, any resonance lines observed in NMR spectra and characterized by known chemical shifts can be used as references to obtain internal δ values in a compound under investigation. However, to identify this compound, the measured shifts are compared with NMR data available in the literature that have been obtained with the use of various spectrometers and sample manipulations and under various experimental conditions. Therefore, it is generally accepted that the best approach is to use standard compounds for which the chemical shifts are 0 ppm (e.g., TMS for ^1H and ^{13}C NMR spectra). Such standard compounds should be stable enough in the chemical sense to be added to a sample, and all of the signals detected in a compound would be observed in the low- or high-field parts of the NMR spectra.

In the absence of such standard compounds, signals coming from solvents can be also applied as secondary references. It should be noted, however, that in spite of even good digital accuracy, the chemical shifts identified depend on the experimental conditions, such as TMS chemical shifts that depend on temperature (0.0005 ppm K^{-1}) and solvents that cause changes between −0.8 and 0.2 ppm.

In many cases, an external standard can be used to report the chemical shift values observed. However, because magnetization in a sample depends on its magnetic susceptibility ($M = \chi_{BO}$), the external standard cannot be entirely accurate. Nevertheless, it is good enough, particularly for nuclei having large chemical shift diapasons.

All of the above requirements are easy to realize when probing solutions of regular organic and inorganic compounds, but they become problematic in NMR studies of biomolecules generally soluble only in water. In fact, the ^1HOD chemical shift is remarkably temperature (and pH) dependent. In addition, biomolecules show chemical shifts due to the formation of secondary structures. Because such shifts are small, the line dispositions should be assigned within ~0.010 ppm to be meaningful in this sense.

According to the protocol, $\delta(^1H)$ values of biomolecular systems are measured in ^1H NMR spectra relative to the signal of sodium 2,2-dimethyl-2-silapentane-5-sulfonate (DSS) in a buffer/DSS water solution. For the nuclei ^{13}C and ^{15}N, another approach is generally acceptable. In this case, δ measurements are carried out by the frequency method based on the gyromagnetic ratios of nuclei that are known with very good accuracy. For example, the $\gamma/2\pi$ values of ^1H and ^{15}N nuclei have been measured as 42.577481337 and −4.314338631 MHz/T, respectively. Thus, their ratio is calculated as 0.101329118. Then, multiplication of the ^1H frequency corresponding to 0.00 ppm in the ^1H NMR spectrum by this factor gives the ^{15}N frequency, which corresponds to 0.00 ppm in ^{15}N NMR. A similar approach can also be applied for NMR spectra of other heavy nuclei.

5.2   STRUCTURAL STUDIES BY SOLUTION NMR: GENERAL STRATEGIES AND 1D AND 2D NMR EXPERIMENTS

One-dimensional (1D) NMR spectra in solutions obtained with direct excitation of nuclei do not suffer from artifacts; thus, they exhibit a number of non-equivalent nuclei in accordance with structure. For the structural analysis of a compound, for example, 1D ^1H NMR is always the preferred way to begin. In fact, the results of such an experiment will dictate later strategy (e.g., multiple pulse sequences leading to one- or two-dimensional NMR patterns, double quantum experiments, or multiple quantum filtration). Because these approaches vary and are constantly in development, only some of them are considered here as classic methods often applied for structural analysis in solutions. A deep understanding of the actions of the radiofrequency pulses applied in such experiments is desirable and can be attained with experience.

5.2.1   FROM 1D NMR TO 2D CORRELATION SPECTROSCOPY

Because $\delta(^1H)$ and $\delta(^{13}C)$ values and the spin–spin coupling constants $^nJ(^1H-^1H)$ and $^nJ(^1H-^{13}C)$ have been well established for various molecular fragments and can be easily found in the literature, the structures of regular organic or organometallic compounds with high solubility can be deduced through a combination of 1D ^{1}H NMR with 1D ^{13}C{^{1}H} and 1D ^{13}C NMR, resulting in direct measurements of δ and J values for both nuclei. It should be emphasized that ^{13}C NMR data, collected for long relaxation delays and the {^{1}H} field switched on for the acquisition time only (to avoid NOE enhancements), can provide integration of ^{13}C resonances important for signal assignments. In addition, the naturally long relaxation times of ^{13}C nuclei can be reduced in the presence of paramagnetic relaxation reagents, such as Cr(acac)$_3$.

For compounds with limited solubility, the problem of low sensitivity in ^{13}C experiments can be partially solved by distortionless enhancement by polarization transfer (DEPT) based on the pulse sequence, as shown in Figure 5.2A. DEPT is a technique that combines radiofrequency pulses in first sections of the pulse sequence that are acting at the proton and carbon frequency to provide polarization transfer from protons to carbons. Due to this transfer, an experiment performed, for example, for ^{1}H–^{13}C pairs will give the proton-decoupled ^{13}C NMR spectrum at better NMR sensitivity. In addition, DEPT experiments are capable of distinguishing CH, CH$_2$, and CH$_3$ groups. Generally, they are carried out to give three NMR spectra recorded at θ pulses of 45°, 90°, and 135° operating at the ^{1}H frequency (decoupler). The first spectrum will show the proton-attached carbon signals, the phase of which is positive. The second spectrum will exhibit only CH groups. Finally, the third experiment will result in CH and CH$_3$ signals during the positive phase, while the phase of CH$_2$ signals will be negative. As follows from the pulse sequence, the major advantage of DEPT experiments is that relaxation delays can be reduced to provide ^{1}H relaxation only. The disadvantage is obvious in that carbons without protons are not detected.

The pulse sequence known as INEPT (insensitive nuclei enhanced by polarization transfer) is shown in Figure 5.2B. The general idea of this approach is to invert selectively one of the proton transitions in a doublet originating from heteronuclear spin–spin coupling in the AX spin system, where $A = {}^1$H and $X = {}^{13}$C, ^{15}N, ^{29}Si, etc. This excitation leads to a non-equilibrium population in the spin system and then, due to selective inversion, the X nuclei can be observed with a greater intensity,

(A) DEPT

$90^H_{\phi1} \rightarrow \tau \rightarrow 90^C_{\phi2}, \quad 180^H \rightarrow \tau \rightarrow 180^C_{\phi2}, \quad \theta^H_1 \rightarrow \tau \rightarrow$ decouple ^{1}H, acquire ^{13}C

(B) INEPT

$90^H \rightarrow \tau \rightarrow 180^H, \quad 180^C \rightarrow \tau \rightarrow 90^H_{\phi1}, \quad 90^C_{\phi2} \rightarrow \Delta \rightarrow 180^H, \quad 180^C_{\phi2} \rightarrow \Delta \rightarrow$ decouple ^{1}H, acquire ^{13}C

(C) NOESY

$90_{\phi1} \rightarrow t_1 \rightarrow 90_{\phi2} \rightarrow \tau \rightarrow 90_0 \rightarrow$ acquire

(D) HETCOR

$90^H \rightarrow t_1/2 \rightarrow 180^C \rightarrow t_1/2 \rightarrow \Delta_1 \rightarrow 90^H_{\phi1}, \quad 90^C \rightarrow \Delta_2 \rightarrow$ decouple protons, acquire ^{13}C

FIGURE 5.2 Pulse sequences applied for (A) DEPT, (B) INEPT, (C) NOESY, and (D) HETCOR NMR experiments.

where the enhancement factor depends on $\pm\gamma_H/\gamma_X$. For this reason, the INEPT technique is more often applied for structural studies of molecular systems containing nitrogen atoms. As seen in Figure 5.2B, the INEPT pulse sequences include sections for observation of spin echo.

One-dimensional NMR techniques are less productive for complex molecular systems, so 2D NMR takes on a dominant role. The simplest version of 2D NMR experiments is homonuclear correlation spectroscopy (COSY). For proton–proton correlation, such an experiment produces a 2D NMR spectrum where the ^{1}H chemical shift range is plotted on both axes (1D projections); the diagonal represents the 1D ^{1}H NMR spectrum, and cross peaks show spin–spin coupling between the coupled protons. Time $t(1)$ in the pulse sequence shown in Figure 1.10A (see Chapter 1) is an incrementable delay for mapping chemical shifts. In some sense, this experiment corresponds to double resonance experiments performed consistently for all of the protons.

As it follows from the COSY pulse sequence, digital resolution in the final 2D NMR spectrum depends strongly on the number of increments used in $t(1)$ section to create indirect dimension. Higher digital resolution obviously requires larger numbers of increments; however, the latter increases the total time needed for the experiment, with twice as many points in this dimension.

Addition of a constant delay time before each 90° pulse in the COSY sequence allows observation of cross peaks that correspond to small spin–spin coupling constants (long-range COSY, or COLOC). More complex pulse sequences can be found in Reference 4. COLOC experiments are directed, for example, toward the observation of correlations between quaternary carbons and neighboring protons; however, they do not work properly for long-range correlations involving carbons with attached protons.

Manipulations with pulses and time delays in the HETCOR (heteronuclear correlation) sequence (Figure 5.2D) give a 2D NMR pattern where the chemical shift range of the proton spectrum is plotted on one axis and the chemical shift range of the ^{13}C spectrum is plotted on the second axis. Again, due to the mixing time, ^{1}H–^{13}C spin–spin coupling will be responsible for the appearance of cross peaks. This technique is applied for H–X pairs, where $X = {}^{15}$N, ^{29}Si, ^{31}P etc., and the pulse delays are generally equal to $2(^1HT_1)$. Details regarding heteronuclear multidimensional NMR experiments on solutions of proteins using pulsed field gradients can be found in Reference 5.

An analog of the COSY experiment is the pulse sequence shown in Figure 5.2C, which allows observation of the NOE in 2D ^{1}H NMR spectra. Here, mixing time τ before the third (registering) 90° pulse develops the magnetization by dipolar coupling. This mixing section leads to the appearance of cross peaks in the 2D pattern representative of protons that are spatially close (generally, the proton–proton distance should be less than 3 Å). One should remember that positive NOE enhancement reduces to zero and then takes negative values with an increase in the motional correlation time (τ_C) of bulky molecules. This problem can be avoided in rotating-frame Overhauser effect spectroscopy (ROESY) experiments designed to measure the NOE for molecules with a molecular weight of 1000 to 2000. The technique includes a spin-lock phase to transform the laboratory coordinate system to a rotating frame. ROESY enhancements are always positive (i.e., the NOE cross peaks show a positive phase), whereas the phase of the diagonal peaks is negative.

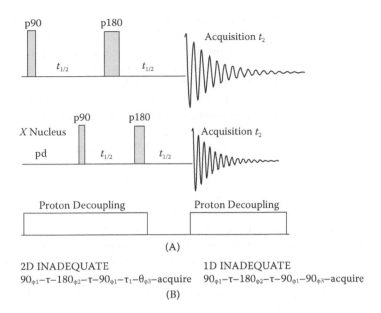

FIGURE 5.3 (A) Homonuclear (top) and heteronuclear (bottom) versions of 2D *J*-resolved NMR experiments. (B) The 2D and 1D INADEQUATE pulse sequences.

Total correlation spectroscopy (TOCSY) is an important experiment that allows determination of connectivity between scalar-coupled nuclei. ^{1}H–^{1}H TOCSY is particularly useful when proton signals in the ^{1}H NMR spectra overlap and spin systems are complex. The pulse sequence includes a section for homonuclear cross-polarization under spin-lock conditions and the application of magnetic field gradients. Generally, this technique is used for large molecules such as peptides, proteins, and polysaccharides.

Structural analysis by NMR often requires accurate measurement of spin–spin coupling constants. The measurements can be carried out by 2D *J*-resolved NMR spectroscopy, available in both homonuclear and heteronuclear versions, as shown in Figure 5.3A. The experiments are particularly useful for determining spin–spin coupling constants in overlapped multiplets. As seen, the *J*-resolved pulse sequence is based on a spin echo principle where the magnetization vectors are associated with their chemical shifts due to the 180° refocusing pulse. Then, the 2D *J*-resolved NMR pattern gives two coordinates. The well-resolved spin–spin coupling constants can be seen along one coordinate, while chemical shifts are seen along the other one. It should be noted that *J*-resolved experiments are particularly important for identifying the conformational states of organic molecules (see below).

5.2.2 Multi-Quantum NMR

As shown above, the NMR spectra obtained with excitation of single-quantum transitions are powerful tools for structural studies in solutions, but they are not sufficient to determine the structure of complex organic molecules where knowledge of carbon–carbon connectivity is necessary. Generally speaking, the carbon–carbon connectivity or, in other words, the molecular skeleton can be determined by observations of the ^{13}C–^{13}C couplings. In practice, they can be seen directly in ^{13}C NMR spectra only for ^{13}C-enriched molecules. At the same time, in molecules with natural abundance these couplings can be detected by stimulation of double and multiple quantum transitions. In fact, homonuclear coupling creates a double quantum coherence that can be discriminated by exploiting its unique phase behavior.

One-dimensional ^{13}C INADEQUATE NMR is based on the pulse sequence shown in Figure 5.3B which provides detection of ^{13}C–^{13}C couplings for neighboring carbons at their natural abundance. The time delay between the last two pulses should be short enough for the phase shift, thus working as a double quantum filter. It should be noted that the ^{1}H decoupler is turned on during the experiment. Because the sensitivity of these 1D experiments is low, a reasonable experimental time is needed to collect the NMR data. Generally, the experiments require samples with neat liquids. The sensitivity can be improved by the application of modern NMR spectrometers equipped with cryogenic probes. Typically, tens of milligrams quantities are sufficient to collect NMR data in an overnight experiment. Unfortunately, 1D ^{13}C INADEQUATE experiments suffer from artifact signals, which can be removed only by very complex phase cycling.

The 2D version of INADEQUATE NMR is also shown in Figure 5.3B, where again the ^{13}C–^{13}C constants through one chemical bond (between 30 and 50 Hz) can be well determined. As in the case of the 1D version, a spin echo (followed by a 90° pulse) converts an initial polarization into double quantum coherence. Typically, time delays τ should be adjusted to the $1/4J(^{13}C$–$^{13}C)$ value (around 8 ms).

The heteronuclear multiple quantum coherence (HMQC) experiment, or reverse polarization transfer (Figure 5.4A), reveals heteronuclear correlations between X nuclei (e.g., ^{13}C or ^{15}N with increased sensitivity) and protons. Protons are detected directly with indirect detection of X nuclei. In order to eliminate undesirable signals belonging to protons that are not coupled by the heteronuclei, the echo-difference method can be applied. This approach requires very accurate calibrations of 90° ^{1}H pulses and calibration of the decoupler, working at a carbon frequency with well-adjusted power and pulse length. Both 1D and 2D versions of the HMQC NMR are available.[6]

Basic heteronuclear multiple bond correlation (HMBC) experiments (Figure 5.4B) are similar to HMQC NMR.[7] For organic compounds, it allows detection of ^{13}C–^{1}H chemical shift correlations based on long-range ^{13}C–^{1}H spin–spin couplings in combination with an increased sensitivity for ^{13}C nuclei. This is particularly important for samples with small amounts of compounds. Typically, these C–H correlations can be visible in HMBC NMR when evolution delays are 0.06 s, corresponding to $J(^{1}H$–$^{13}C)$ of 4 to 10 Hz, or 0.1 s for $J(^{1}H$–$^{13}C)$ of 2 to 7 Hz. In other words, this technique allows detection of quaternary carbons. In some sense, the HMBC method can be considered as an alternative to 2D INADEQUATE NMR experiments, which are known to be insensitive. The HMBC pulse sequence does not include the ^{13}C decoupling section; therefore, cross peaks that appear due to coupling through one, two, and three chemical bonds can be discriminated. Finally, other NMR methods applied for structural analysis in solutions can be found in the recommended literature.

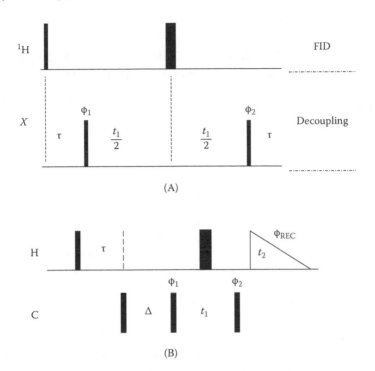

FIGURE 5.4 (A) Pulse sequence applied for standard HMQC NMR experiments. (B) Pulse sequence used for a classical HMBC NMR experiment.

5.2.3 EXAMPLES OF STRUCTURAL STUDIES

A summary of the most popular NMR approaches is presented in Table 5.1,[8] which shows how the total experimental time (TET) and the amount of a compound in a solution depend on the type of experiment performed with a magnetic field of 7 to 9 T with a NMR spectrometer equipped with a regular NMR probe and a 5-mm NMR tube. It should be emphasized that the times shown in Table 5.1 (estimated by assuming relaxation times as ^1H $T_1 = 1$ s and ^{13}C $T_1 = 3$ s) are the times required to obtain interpretable results (i.e., NMR spectra with good signal-to-noise ratios). It is obvious that the data in Table 5.1 will change with applications of cryo-probes and highest magnetic fields, although the tendency will be similar.

When there is limited NMR sensitivity, the total experimental time and the concentration of solutions increase significantly going from nuclei with a high natural abundance to rare nuclei, particularly for applications of advanced NMR techniques. This is the primary factor in choosing the type of NMR experiment and the appropriate strategy for a NMR study with regard to the nature of the problem to be solved. In many cases, the significance of the problem is inversely proportional to the amount of a compound available. Between the two criteria that are important in the selection of NMR experiments—minimization of the experimental time (spectrometer time) and minimization of operator time—the first criterion usually dominates, and many 2D NMR approaches can be ruled out as a result.[8]

TABLE 5.1
Total Experimental Time Required to Collect NMR Data Regarding Unlimited Solubility of a Compound as a Function of NMR Experiments

NMR Experiment	Total Experimental Time (min)	Minimal Amount of Compound (mg)
¹H (1D NMR)	—	5×10^{-3}
¹⁹F (1D NMR)	—	5×10^{-3}
¹³C{¹H} 1D NMR	0.1	5
DEPT (or INEPT)	0.1 (single θ (Δ) value)	5 (single θ (Δ) value)
	0.5 (complete edit)	10 (complete edit)
INADEQUATE (1D)	1.5	30
Homonuclear COSY	4, 30, 60 (low, normal, and high digital resolution)	2 (at low digital resolution)
		5 (at regular digital resolution)
Heteronuclear COSY	4	10 (at high digital resolution)
INADEQUATE (2D)	60	30 (long-range correlation)
HMQC	16	100
HMBC	32	1 to 2
Long-range COSY or COLOC	4	1 to 2
		30 to 40

Note: Minimal amount of a compound is the amount required for NMR data collection in an overnight experiment based on a molecular weight of 500 g/mol.

Sometimes even the simplest approach, such as ¹H 1D or ¹⁹F 1D NMR, showing multiplets with directly measured spin–spin coupling constants, in combination with a ¹H or ¹⁹F homonuclear decoupling experiment is sufficient to elucidate the final structure. The simple version of the COSY experiment is also appropriate in such cases. Thus, a reasonable strategy in a NMR study is to begin with the simplest experiment and then move to more sophisticated approaches as necessary until either sufficient information is obtained or the limit of sensitivity is reached.[8] In other words, a researcher should find a reasonable compromise between the complexity and duration of an NMR experiment and the amount of a compound. In fact, as follows from Table 5.1, a 1D proton spectrum can be obtained for micrograms of a compound in solutions, whereas double quantum 2D NMR experiments exploiting ¹³C–¹³C coupling at a natural abundance will require >1 g.

An additional factor is artifact signaling or missing the signals possible in 2D NMR. In some cases, phase cycling removes artifacts due to poorly calibrated pulses. In fact, because potential artifacts follow a different coherence pathway, canceling them by repeating the same pulse sequence with different phases can be quite effective.[9] However, in carbon-detected heteronuclear correlation experiments, for example, phase cycling does not help. Even carefully calibrated pulses (or the application of composite pulses) only minimize artifact signals. Also, missing signals can often occur in long-range correlation NMR experiments.

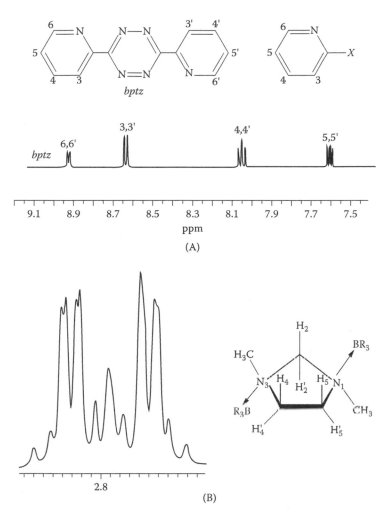

FIGURE 5.5 (A) Aromatic heterocycle system *bpzt* and its one-pulse ¹H NMR spectrum, recorded in CD₃CN. (B) The *AA'BB'* pattern observed for protons 5(4') and 5'(4) in the RT ¹H NMR spectrum of a cyclic compound (R=H) in C₆D₆.

Several examples can help to illustrate the strategy and logic behind structural studies of small organic molecules in solutions. The compound *bpzt* shown in Figure 5.5A represents an aromatic heterocycle system. Its ¹H NMR spectrum—recorded in CD₃CN with relaxation delays providing complete ¹H relaxation—shows four 1/1/1/1 well-resolved resonances, and their splitting is in accordance with the ¹H–¹H spin–spin couplings expected for this simple spin system. Thus, the number of the signals and their chemical shift values correspond well to the molecular symmetry. It is obvious, however, that the structure has not been definitely determined, because the fragment in Figure 5.5A could also show a similar ¹H NMR spectrum. In the presence of an elemental analysis, the observation of six signals in the ¹³C{¹H} NMR spectrum could finally confirm the structure because the carbons neighboring with nitrogen atoms can be well assigned. In the absence of the elemental analysis, the final conclusion requires ^{15}N NMR to observe two ^{15}N resonances. Because a one-pulse ^{15}N NMR experiment is insensitive due to the very low natural abundance of ^{15}N nuclei, advanced NMR techniques, such as $^{1}H-^{15}N$ HMBC cross-polarization experiments or $^{1}H-^{15}N$ HSQC techniques, can be applied.[10]

The structure of the five-membered cycle shown in Figure 5.5B can be deduced by one-pulse ^{1}H, ^{13}C, $^{10,11}B$, and ^{19}F NMR (R=F) spectra,[1] even in the absence of elemental analysis. At the same time, the ^{1}H NMR spectrum is difficult to predict due to its strong dependence on molecular symmetry. The experimental spectrum exhibits the signals corresponding to equivalent protons 2 and 2', while non-equivalent protons 5(4') and 5'(4) form the complex pattern shown in Figure 5.5B. The appearance of such a multiplet does not necessarily identify the structure or even correct signal assignments until the pattern is recognized as belonging to the group of signals forming a spin system with reasonable spin–spin coupling constants.[11] The computations indicated that the group of lines in Figure 5.5B represent spin system $AA'BB'$, where the geminal $^{1}H-^{1}H$ coupling constants are negative, whereas the $J(5-4')$ and $J(5'-4)$ constants take positive (but different) values of 9.5 and 2.5 Hz, respectively. Only this step leads to reaching the correct final conclusion. In this context, an illustrative example was considered in Chapter 2, where the action of SbF_5 on the olefin $CF_2=C(F)-CF_2-CF=CF_2$ in solutions results in a ^{19}F NMR spectrum exhibiting more than 140 lines. Again, only full analysis of the ^{19}F seven-spin system expected for the allyl cation proves its structure.

Because the resonance shape is mathematically known, line widths observed in the spectra should also be a focus of researchers. For example, some lines in NMR spectra can be further broadened by, for example, the unresolved long-range spin–spin coupling constants or the presence of scalar relaxation of the second kind. Observation of such broadenings can be very useful for signal assignments and elucidation of structures. Figure 5.6A shows two tri-substituted nitro-olefins that can be recognized by ^{1}H NMR. In general, the configuration of tri-substituted nitro-olefins in solutions can be found by predicting the chemical shifts of olefinic protons via Equation 5.1, where increments Z are known for many groups:

$$\delta(CH) = 5.28 + Z(gem) + Z(cis) + Z(trans) \qquad (5.1)$$

However, the reliability of this approach is not always obvious. As seen in Figure 5.6A, the ^{1}HC signals of the Z and E isomers show different temperature-dependent line widths. This effect is caused by scalar relaxation of the second kind due to the unresolved constant $^{3}J(^{1}H-^{14}N)$. This constant is larger for translocations of the coupled nuclei, thus providing reliable signal assignments. Similar selective broadenings can be seen in $^{13}C\{^{1}H\}$ NMR spectra in the presence of carbons neighboring with quadrupolar nuclei such as ^{14}N or $^{10,11}B$. Signal assignments are important elements of structural studies, but they can often be simplified by experiments on different solvents. Generally, polar molecules show substantially different chemical shifts in aromatic solvents, such as benzene, pyridine, or C_6F_6, compared to less magnetically interactive solvents, such as CCl_4, $CDCl_3$, acetone-d_6, or CD_3CN. This tendency,

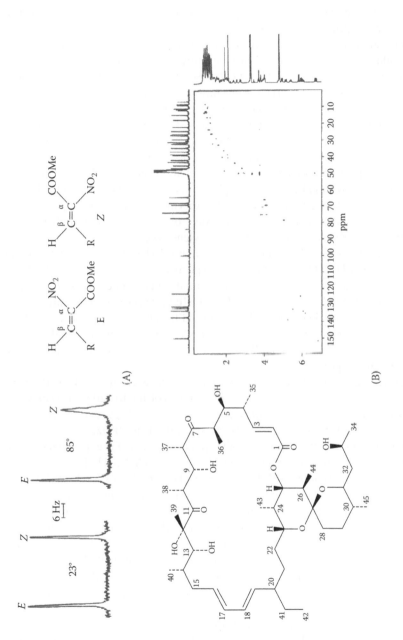

FIGURE 5.6 (A) The β-CH proton signals in the ¹H NMR spectrum of nitro-olefin in DMSO. (B) The 2D ¹H–¹³C correlation NMR spectrum of a natural compound. (From Morris, G.A., *Magn. Reson. Chem.*, 24, 371, 1986. With permission.)

known as aromatic solvent-induced shift (ASIS),[12] can be explained by magnetic anisotropy of the solvent molecules. The effect is very similar to magnetic anisotropy contributions to total chemical shifts, as discussed in Chapter 2. In practice, experiments in different solvents can help with the assignment of signals or can even simplify complex spin–spin systems.

The optimal NMR strategy changes remarkably from simple molecules to complex molecular systems. The natural product oligomycin is shown in Figure 5.6B. The single-pulse ^1H or $^{13}C\{^1H\}$ NMR spectra (see the corresponding projections in 2D patterns) are too rich[8] to carry out accurate signal assignments and structural identification, respectively, even on the basis of well-known spectrum–structure relationships. In contrast, assignments for both ^1H and ^{13}C nuclei can be simplified by the use of 2D heteronuclear shift correlation experiments; however, the 2D NMR spectrum shown in Figure 5.6B requires a 14-hr acquisition time. Nevertheless, the structure is still uncertain because the carbon skeleton is not clear. Determination of the molecular skeleton and hence the complete ^{13}C assignments can be reliable only within the framework of an INADEQUATE experiment. This point is particularly important for molecules containing a wide range of carbons—from aliphatic, aromatic, and olefin carbons to carbons of the acetylene type.

Figure 5.7 shows a steroid compound and regular INADEQUATE ^{13}C NMR spectrum recorded in a DMSO-d_6 solution.[13] The spectrum was obtained at τ delays of 0.005 s to account for $^1J(^{13}C-^{13}C)$ constants around 50 Hz and exhibits cross peaks corresponding to the $^{13}C-^{13}C$ correlations for the aliphatic carbons. This carbon–carbon connectivity is sufficient to identify the steroid framework. At the same time, no cross peaks are observed for carbons of the acetylene group with chemical shifts of 88.98 and 74.96 ppm or for carbons 17 and 19. In the absence of such information the structure remains unproven.

The physical reason for the missed cross peaks is the large $^1J(^{13}C-^{13}C)$ difference. The $^1J(^{13}C-^{13}C)$ coupling constant reaches magnitudes of 160 to 220 Hz for acetylene carbons (vs. 50 Hz used as the parameter in regular INADEQUATE experiments), whereas the $^1J(^{13}C-^{13}C)$ value between sp and sp^3 carbons is 70 Hz.

Following the logic of the INADEQUATE pulse sequence, the experiment should be modified and performed as long-range $^{13}C-^{13}C$ INADEQUATE at different delays: 0.015 s, corresponding to $3/4J(C-C) = 50$ Hz, and 0.05 s for $1/4J(C-C) = 5$ Hz. This combination results in observation of long-range carbon–carbon correlations through two and three chemical bonds. It should be noted that at the first τ delay prior to the second 90° pulse (which is longer than that in the regular experiment) an anti-phase magnetization is created via larger constant $^1J(^{13}C-^{13}C)$ that is equal to 164 Hz (carbons 19 and 20), as well as a smaller constant $^1J(^{13}C-^{13}C)$ of 72 Hz for carbons 17 and 19. Thus, even $^{13}C-^{13}C$ INADEQUATE NMR experiments in solutions are not single valued and require modifications to provide the final proof of structure.

Finally, it should be emphasized again that the lack of pulse perfection, in addition to phase alternation and so-called zero-frequency effects, can lead to the observation of confusing artifacts in multidimensional NMR spectra. This is particularly important when the structure is unknown. For example, HMBC NMR spectra often show so-called $^1J(C-H)$ artifacts, and differentiation between the artifacts and the

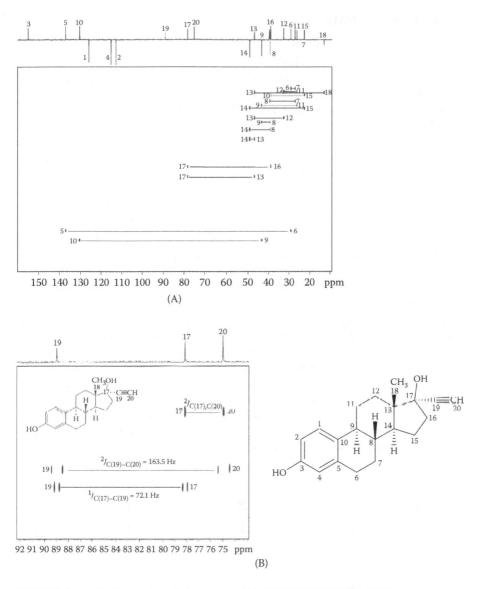

FIGURE 5.7 (A) Expansion of the regular $^{13}C-^{13}C$ INADEQUATE NMR spectrum for observation of correlations corresponding to $^1J(C-C) = 50$ Hz. (B) Expansion of the long-range INADEQUATE NMR spectrum demonstrating the one-bond correlation between acetylene carbons 19 and 20 and 19 and 17, as well as the two-bonded cross signals between carbons 17 and 20. (From Bain, A.D. et al., *Magn. Reson. Chem.*, 48, 630, 2010. With permission.)

long-range cross peaks is difficult, particularly in the absence of spectroscopic experience. Moreover, computer programs aimed at automatic signal assignments can treat the artifacts as real peaks, leading to incorrect structural conclusions. Discussions regarding the comprehensive analysis of artifacts and their recognition can be found in the recommended literature.

5.2.4   ULTRAFAST NMR

As noted above, multi-dimensional and multi-quantum NMR generally requires hundreds of scans to collect the valuable NMR data in solutions; therefore, in practice, long-term experiments are necessary even for highly concentrated solutions, where the duration of an experiment can be several hours or longer. The duration can be reduced by applying so-called ultrafast NMR (UF-NMR). The UF-NMR technique provides the unique opportunity to perform multi-dimensional and multi-quantum NMR experiments (e.g., TOCSY, HSQC, HMBC) with the use of a single scan. It is obvious that such a time scale allows probing molecular systems that experience fast rearrangement, chemical exchange, or even fast chemical reactions. The central idea of this approach is spatial encoding of NMR interactions that can be realized by special NMR probes equipped with high-amplitude Z-axis gradients.[14] In addition, special software is used to analyze the final results.

Figure 5.8 shows, as an example, the complex pulse sequence used to perform UF-HSQC NMR experiments that allow monitoring of fast chemical reactions. Here, the INEPT block, including a delay time of 1.72 ms to account for the target average $^1J(^{13}C-^1H)$ coupling constant of 145 Hz, is directed toward enhancement of signal intensities. Two 15-ms pulses are applied in the encoding block to act simultaneously with the ±15 G/cm encoding gradients covering a frequency dispersion much larger than the ^{13}C frequency range. The amplitude of gradients G_1 and G_2 should be optimized to fold resonances in the spatially encoded dimension (e.g., –35 G/cm and –14.3 G/cm). It should be noted that this pulse sequence optimizes the spatial encoding parameters in order to obtain the best compromise between sensitivity and resolution in NMR spectra. Finally, the NMR data are treated with special computing procedures.[14] A simple estimation shows that UF-HSQC NMR spectra can be recorded at times ranging from 0 to 40 min to observe how proton–carbon correlations change during a fast chemical reaction.

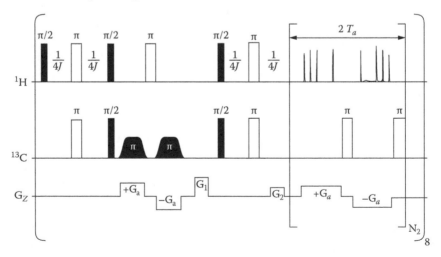

FIGURE 5.8   Ultrafast 2D HSQC pulse sequence applied to monitor fast chemical reactions. (From Queiroz, L.H.K. et al., *Magn. Reson. Chem.*, 50, 496, 2012. With permission.)

5.3 CONFORMATIONAL (ISOMERIC) ANALYSIS BY NMR IN SOLUTIONS

An important aspect of structural analysis of organic molecules is determination of their isomeric and conformational states, which is where NMR plays a major role. Traditionally, these states are identified by determination of chemical shifts or spin–spin coupling constants in variable-temperature NMR experiments. In addition, NOE and residual dipolar coupling measurements can be performed for so-called frozen conformations or for systems experiencing fast interconversions.

5.3.1 COMMON PRINCIPLES

By definition, conformational analysis, as part of stereochemistry, involves the study of molecular conformations and the energies necessary for their transformations. This is important because each conformation has a unique property and their totality affects the physical and chemical properties of the systems being investigated. The simplest case is the ethane molecule, for which rotation around a C–C bond leads to the molecular states shown in Figure 5.9A. As follows from the energy diagram, four conformations are possible, but one of them is energetically preferable. This energy preference is governed by proton–proton space interactions. Geometrically, the conformations and space interactions are characterized by the dihedral angles H–C–C–H. At the angle close to zero, the interactions and energy are maximal. By analogy, the relative stability of isomeric forms (e.g., *cis* or *trans*, *E* or *Z*) is also controlled by intramolecular interactions and again by the corresponding dihedral angles.

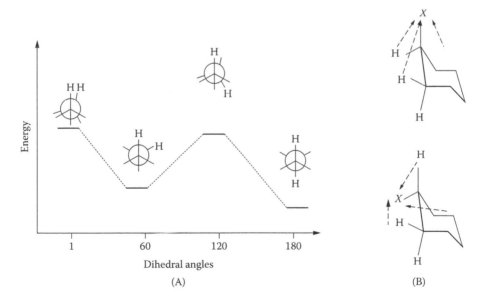

FIGURE 5.9 (A) Ethane conformations and their energies as a function of dihedral angles (only two of the hydrogens are shown for clarity). (B) Two conformations of a six-member cycle (frozen on the NMR time scale) showing different magnetic environments of nuclei *X*.

The principal idea behind NMR recognition of the configurations (*cis* or *trans*, *E* or *Z*) or conformations (e.g., chair, half-chair, boat, or twist-boat in cyclic organic compounds) of simple rotamers is illustrated in Figure 5.9B. Here, a six-member cycle shows two conformations. When the conformation transitions are frozen on the NMR time scale, different spatial dispositions of *X* nuclei give rise to their different magnetic environments. In turn, these environments lead to chemical shift differences in the conformational states due to the magnetic anisotropy effects of neighboring groups or chemical bonds, different spin–spin coupling constants through three chemical bonds and/or through space, and different dipolar coupling that cannot be averaged by molecular motions. All of these factors can be used for a conformational analysis in solutions.

5.3.2 DETERMINATION OF MOLECULAR GEOMETRY (CONFORMATIONS, ISOMERS) VIA CHEMICAL SHIFTS AND SPIN–SPIN COUPLING CONSTANTS THROUGH CHEMICAL BONDS

Because a chemical shift is due to various contributions (see Chapter 2), chemical shift differences in molecular conformations depend on the nature of the target nuclei. Figure 5.10 illustrates this point, where the ^1H chemical shift difference is not large. For example, in the case of the axial and equatorial methylene protons in cyclohexane, this difference is only 0.5 ppm or slightly more in compound (C). Four-member cycle (A) and cycle (D) show $\Delta\delta(^1H)$ of 1.03 and 1.65 ppm, respectively. Thus, typically, the $\delta(^1H)$ difference takes values between 0.1 and 1 ppm. Even in the case of compound (B), the *cis/trans* orientation of the three-member cycle does not significantly perturb the $\Delta\delta(^1H)$ value.

The ^{13}C NMR chemical shifts are obviously more sensitive to conformational factors, especially in heterocyclic compounds, where the so-called angle-dependent γ effects play a very important role. For example, large chemical shift differences are observed for ^{13}C nuclei in the CH_3 axial and equatorial groups in 1,4-dimethylcyclohexane (see Figure 5.10E) or for carbons in the cycle. The very large δ difference is observed for ^{31}P nuclei in *cis*- and *trans*-4-methylcyclohexyldimethylphosphine (Figure 5.11A).[15] Finally, ^{19}F nuclei represent the most sensitive labels for conformational studies, where the $\Delta\delta(^{19}F)$ values can reach several ppm. For example, difluorohexane in Figure 5.11B shows a low-temperature ^{19}F NMR spectrum with signals at δ of −196.5 ppm for the diaxial form and −182.9 ppm for the diequatorial form (referred to as fluorobenzene).[16]

The free activation energy for the conversion of cycles in solutions is generally between 9 and 12 kcal/mol; therefore, conformations (or configurations) can be determined by NMR experiments performed at low temperatures but requiring the corresponding solvents or their mixtures. Signals in well-resolved NMR spectra thus should be completely assigned. This important step is relatively simple in the case of single-labeled compounds, such as 4-methylcyclohexyldimethylphosphine, as shown in Figure 5.11A, where the ^{31}P NMR spectrum exhibits singlet resonances. However, signal assignments in mixtures of conformers are usually not simple, particularly for small chemical shift differences.

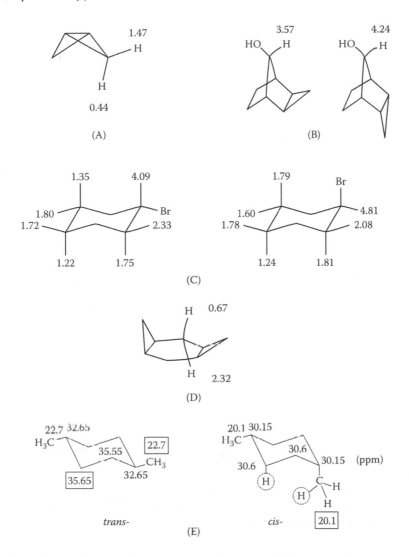

FIGURE 5.10 The ¹H (A, B, C, D) and ¹³C (E) chemical shift difference typical of conformations in cyclic organic molecules.

Several examples illustrate the methodology of conformational studies by NMR in solutions. The conformational states of bromocyclohexane (see Figure 5.10C) can be found by ¹H NMR spectra because the number of resonances in the δ(¹H) aria between 4.0 and 5.0 ppm observed for protons at bonds C–Br will correspond to the number of conformations, and their relative populations can be found by integration. To stop fast ring inversion on the ¹H NMR time scale, ¹H NMR experiments have been performed in a $CDCl_3$–$CFCl_3$ solvent mixture at −80°C.[17] Multiplets in the low-temperature ¹H NMR spectrum are overlapped, so full assignment of the signals and determinations of their chemical shifts have been carried out by COSY and by heteronuclear correlation

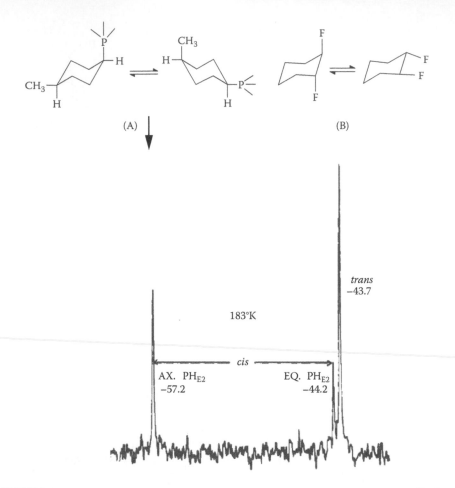

FIGURE 5.11 (A) *Cis-* and *trans-*4-methylcyclohexyldimethylphosphine and its $^{31}P\{^1H\}$ NMR spectrum at 183°C. (From Gordon, M.D. and Quin, L.D., *J. Am. Chem. Soc.*, 98, 15, 1976. With permission.) (B) Two conformations in difluorohexane.

(HECTOR) spectroscopy NMR. Based on the assignments and $\delta(^1H)$ values in Figure 5.10, the major conformer is equatorial (82%), while the minor conformer is axial (18%). Finally, the origin of the substituent effect on the $\delta(^1H)$ shifts has been discussed in terms of the steric influence of the Br atom and the electric field of the C–Br bond.

Signal assignments in the ^1H NMR spectra of conformers are not trivial when the chemical shift differences are small and solvent dependent. The spectral behavior of *trans-*1,4-dibromo-1,4-dicyanocyclohexane in solutions (Figure 5.12A) prompts the use of conformational studies in such cases.[18] The room-temperature ^1H NMR spectrum of a compound depends strongly on the solvents; for example, a poorly resolved signal is observed in magnetically anisotropic toluene (Figure 5.12B), while the complex pattern of a strong coupled spin system is found in DMSO (Figure 5.12C). Generally speaking, if the spin system in this compound is intrinsically complex, then an inverse solvent effect should be expected at room temperature.

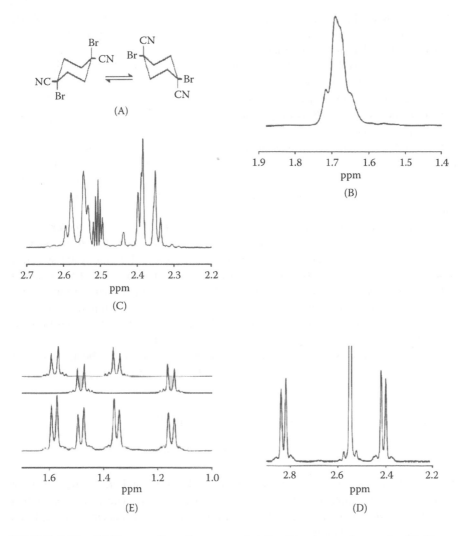

FIGURE 5.12 (A) Two conformations expected for this six-member cycle. (B) Room-temperature ¹H NMR spectrum in toluene-d_8. (C) Room-temperature ¹H NMR spectrum in DMSO-d_6. (D) The ¹H NMR spectrum in acetone-d_6 (230 K). (E) Experimental (bottom) and simulated (top) ¹H NMR spectra in toluene-d_8 (245 K). (From Bain, A.D. et al., *J. Phys. Chem. A*, 115, 9207, 2011. With permission.)

According to molecular modeling calculations, two conformations with two equatorial Br atoms and two axial Br atoms, respectively, can be stable for this cycle. Because energy barriers of transformations are low, averaged chemical shifts will be observed at room temperature. However, it is remarkable that, given the unequal populations, the ¹H NMR spectrum should show separately axial and equatorial protons, forming an *AA'BB'* spin system even in the presence of fast exchange. The ¹H NMR pattern in DMSO is actually well simulated by this spin system. In turn, equally populated conformations will correspond to the fully averaged δ(¹H) and

J(H–H) values in the presence of fast exchange to give a singlet in the ^{1}H NMR spectrum. In some sense, this situation corresponds to "flat" molecules. When conformational populations are similar (but not exactly equal), the ^{1}H NMR pattern can be poorly resolved. All of these statements can be verified by low-temperature experiments only (e.g., in acetone-d_6 or toluene-d_8). As seen in Figure 5.12D,E, the low-temperature ^{1}H NMR spectra, recorded at 230 K in acetone-d_6 and at 245 K in toluene-d_8 show two $AA'BB'$ spin systems for each conformer, one of which is obviously less populated.

Spin–spin or J coupling (homonuclear and heteronuclear) via chemical bonds is also a powerful tool for conformational analysis. Because the Fermi mechanism dominates in this coupling and s-electron density plays the most important role, the dependence of J on mutual spatial orientations of chemical bonds is theoretically justified.

The vicinal coupling constants, such as $^3J(^1H–^1H)$ in $^1H–X–Y–^1H$ structural fragments or $^3J(^{13}C–^1H)$ in $^1H–X–Y–^{13}C$ fragments are actually sensitive to the stereochemistry of organic compounds, particularly olefin molecules and cyclic systems. In addition, they change only slightly in different solvents in contrast to chemical shifts. Both of these factors suggest their applications for the determination of conformations (or configurations) in solutions, particularly at low temperatures, when conformational transitions are frozen on the NMR time scale. Under these conditions, ^3J values characterize dihedral angles in individual states.

Initially, Karplus used the valence bond theory to formulate the influence of molecular geometry on vicinal coupling constants in derivatives of ethane via Equation 5.2:

$$^3J(^1H–^1H) = K_1 + K_2 \cos^2\phi \text{ (for } \phi = 0 – 90°) \tag{5.2}$$

$$^3J(^1H–^1H) = K_1 + K_3 \cos^2\phi \text{ (for } \phi = 90 – 180°)$$

Here, K_1, K_2, and K_3 are constants dependent on the steric and electronic effects of substituents in the molecular fragment under investigation. The character of this dependence is shown in Figure 5.13A,[19] and Equation 5.3 represents the valence bond (VB) σ-electron calculations:

$$^3J(^1H–^1H) = A + B \cos\phi + C \cos2\phi \tag{5.3}$$

where constants A, B, and C take the values of 4.22, –0.5, and 4.5 Hz, respectively. Another useful expression can also be found in the literature:

$$^3J(^1H–^1H) = A \cos^2\phi + B \cos\phi + C \tag{5.4}$$

where A, B, and C are equal to 9.0, –0.5, and –0.3 Hz, respectively. In accordance with these equations, the *trans* ^3J(H–H) constant in olefin molecules is always larger than the *cis* ^3J(H–H) constant, with their relative magnitudes depending on the size of the cycles in cyclic organic molecules. In cyclopropanes, for example, the *cis* ^3J(H–H)

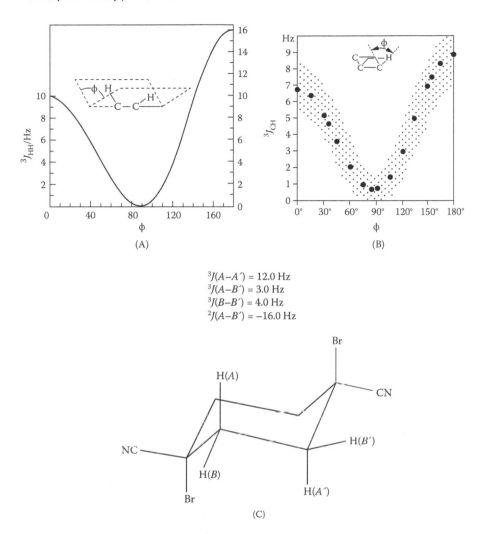

$$^3J(A-A') = 12.0 \text{ Hz}$$
$$^3J(A-B') = 3.0 \text{ Hz}$$
$$^3J(B-B') = 4.0 \text{ Hz}$$
$$^2J(A-B') = -16.0 \text{ Hz}$$

FIGURE 5.13   (A) The Karplus curve for $^3J(^1H-^1H)$ coupling constant vs. the dihedral angle. (B) The Karplus-type dependence of $^3J(^{13}C-^1H)$ on the dihedral angle. (C) The $J(^1H-^1H)$ spin–spin coupling constants in *trans*-1,4-dibromo-1,4-dicyanocyclohexane. (From Feeney, J., *Proc. R. Soc. Lond. A*, 345, 61, 1975. With permission.)

constants are larger than the *trans* $^3J(H-H)$ constants, which, in turn, are larger than the geminal $^2J(H-H)$ constants: $^3J(cis)$, $^3J(trans)$, and $^2J(gem)$, equal to 6 to 12 Hz, $-4 \div +8$ Hz, and $-3 \div -1$ Hz, respectively. Similarly, $^3J(cis)$, $^3J(trans)$, and $^2J(gem)$ are equal to $5 - 9$ Hz, $2 - 7$ Hz, and $0.9 - 4$ Hz, respectively, in aziridine cycles.

Thus, $^3J(H-H)$ constants measured for conformations (configurations) frozen on the NMR time scale lead to molecular geometry via Karplus-type dependencies. The classical example is shown in Figure 5.13C, where one of two conformations of *trans*-1,4-dibromo-1,4-dicyanocyclohexane is characterized by ^1H NMR. Because

the low-temperature ^{1}H NMR spectrum exhibits resonances forming strongly coupled spin systems, classified as $AA'BB'$, the J(H–H) values can be obtained with
computer simulations only. The results agree well with the shown molecular geometry, where again the *trans* ^3J(H–H) constant is much larger than the *cis* ^3J(H–H)
constant.[19]

Because Fermi-contact interactions dominate spin–spin couplings of nuclei other
than protons ^{13}C, ^{19}F, ^{31}P, or ^{15}N, their ^3J constants are also sensitive to molecular
geometry. Many Karplus-like expressions describing the ^3J values as functions of
dihedral angles are currently known and can be used for conformational analysis.
For example, Equations 5.5 and 5.6 have been deduced for ^3J(^{1}H–C–C–^{13}C) constants by an analysis of norbornane molecules with a known geometry:[20]

$$^3J(HCCC) = 4.50 - 0.87\cos\phi + 4.03\cos2\phi \tag{5.5}$$

$$^3J(HCCC) = 8.06\cos^2\phi - 0.87\cos\phi + 0.47 \tag{5.6}$$

Dihedral angles ^{1}HCC^{15}N and the ^3J values in amino sugars and peptides have
resulted in Equations 5.7 and 5.8, respectively:

$$^3J(HCCN) = 3.1\cos^2\phi - 0.6\cos\phi + 0.4 \tag{5.7}$$

$$^3J(HCCN) = -4.6\cos^2\phi + 3.0\cos\phi + 0.8 \tag{5.8}$$

The ^3J(^{1}H–C–N–^{19}F) coupling constants in *N*-fluoro-amide derivatives follow a
Karplus-type equation:

$$^3J(HCCF) = 70.8\cos^2\phi - 44.1\cos\phi - 7.2 \tag{5.9}$$

Being based on the same physical principle, all of the above equations are similar, and the first coefficient depends on the absolute magnitudes and signs of the J.
Expressions involving other nuclei in different structural fragments are also available
in the literature. For example, a discussion regarding the stereochemical behavior of
J(^{13}C–^{13}C) constants in *N*-vinylpyrroles can be found in Reference 21. One remark
is important in the context of J analysis: When the measurements give abnormal
spin–spin coupling constants that are unexpected i n terms of the Karplus curves,
then conformers still rapidly convert on the NMR time scale.

5.3.3 Molecular Geometry and Spin–Spin Coupling through Space

Through-space (TS) spin–spin coupling[22] occurs via non-bonding interactions (see
Chapter 2) and supports the following important statement: The observation of a
spin–spin coupling constant does not necessarily indicate the presence of a chemical bond. Despite this, long-range coupling constants can be successfully used in
conformational (or configuration) analysis. Through-space coupling is particularly
important for F-containing molecules that show strong ^4J(^{19}F–^{19}F)$_{TS}$ constants.

According to the theory, the TS mechanism consists of overlapping two lone-pair orbitals belonging to spatially proximate fluorine atoms. The overlapping is obviously significant when the atoms are located closer than the sum of their van der Waals radii (≤ 2.7 Å). Similarly, other nuclei can also show TS spin–spin coupling, such as $J(^{13}C-^{77}Se)_{TS}$, $J(^{13}C-^{31}P)_{TS}$, and $J(^{199}Hg-^{31}P)_{TS}$,[23] again due to a specific molecular geometry. In general, TS coupling can be observed when (1) the coupled nuclei are not magnetically equivalent, (2) the nuclei are in close proximity even for fast motions in solutions, or (3) the nuclei are surrounded by at least one electron lone pair.[24]

As expected, the J_{TS} constant, based on the idea of electron overlapping, decreases with increasing internuclear distance. Various theoretical approaches to identifying the transmission mechanisms of TS coupling confirm this statement. However, the theoretical dependencies of J_{TS}–internuclear distance are different. Therefore, in practice, the empirical relationships are more valuable and can be written as follows:

$$J_{TS} = A \exp^{-Br} \tag{5.10}$$

where coefficients A and B change with the nature of the coupled nuclei and compounds. For example, constants $J(^{19}F-^{19}F)_{TS}$ in 1-pentafluorophenyl-1-trifluoromethylethylenes can be predicted at $A = 6800$ and $B = -1.99$. For a series of 1,8-difluoronaphthalene, A and B take the values of 1.1×10^7 and -4.96, respectively. The $J(^{13}C-^{19}F)$ coupling constants in polycyclic aromatic hydrocarbons follow Equation 5.10 for $A = 5541$ and $B = -2.44$, where parameter r is the non-bonded H–F distance. Finally, $J(^1H-^{19}F)_{TS}$ constants are also significant in alkylfluorobenzenes, where hydrogens and fluorine atoms are separated by five chemical bonds.

Figure 5.14A illustrates some of the dependences for spin–spin coupling constants through four or five chemical bonds which can be applied to determine internuclear distances in solutions via the J_{TS} values. As seen, for short F–F distances, the J_{TS} constants can be exclusively large to reach magnitudes of 100 to 200 Hz. In this context, it is obvious that the unusual long-range constants observed in NMR spectra are the key factor in identifying conformations, configurations, or rotamers of molecules frozen on the NMR time scale.[25] Even for small J_{TS} values, structural assignments are still possible with accurate measurements. For example, the long-range $J(^1H-^{19}F)$ constants in the compounds shown in Figure 5.14B can be used to determine the structure. In fact, the ^1H NMR spectra show very similar $^5J(^1H_X-^{19}F)$ and $^6J(^1H_B-^{19}F)$ values of 0.35 and 0.55 Hz, respectively. The through-space constants $^5J(^1H-^{19}F)$ can help with assignments of the s-trans and s-cis conformations (Figure 5.14C).[26] Finally, the low-temperature ^{19}F NMR spectrum of 1,2,3,4-tetrafluorocyclohexane shown in Figure 5.14D has four ^{19}F signals in CD_2Cl_2 at 200 K.[27] It is obvious that the F atoms can be non-equivalent in the chair conformation. As it follows from the data,[27] the two 1,3-diaxial fluorine nuclei in the chair conformation show the $^4J(^{19}F-^{19}F)$ constant, which is very large and measured as 29 Hz. This value is twice as large as that for the vicinal gauche $^{19}F-^{19}F$ coupling constants, which generally take values around 14 Hz. Thus, it is obvious that this coupling occurs through space due to the close proximity of the axial fluorine atoms.

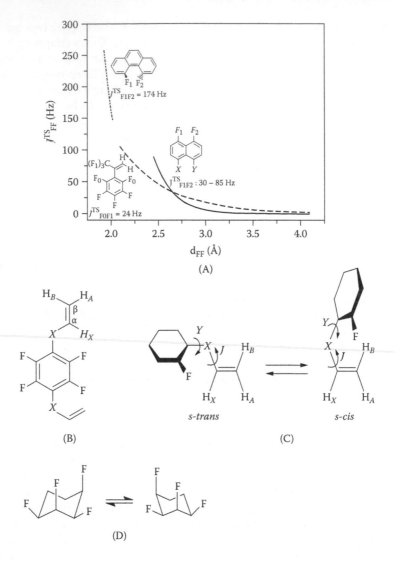

(A)

(B)   (C)

(D)

FIGURE 5.14   (A) Dependencies of the $J_{TS}(^{19}F–^{19}F)$ constants on the F–F distances for three types of F-containing compounds. (From Xie, X. et al., *Magn. Reson. Chem.*, 47, 1024, 2009. With permission.) (B, C, D) The F-containing compounds with conformational-dependent long-range $^1H–^{19}F$ and $^{19}F–^{19}F$ spin–spin coupling constants.

5.3.4   Conformational Analysis in the Presence of Fast Interconversions

As shown earlier, the determination of conformations, configurations, or molecular rotamers on the basis of spectral parameters δ and J is not complex if molecular transformations are stopped on the NMR time scale. In fact, under these conditions, the measured parameters characterize individual forms. However, in practice, the observation of frozen states in solutions is not always technically possible for a variety of reasons: limited solubility at low temperatures, nature of the solvents, or very high transformation rates. Also, conformations (configurations) that are characterized at low temperatures are energetically different; therefore, low-temperature data showing the relative populations of states can be extrapolated for the area of moderate (or room) temperatures only at known (or determined independently) energies of these states. Because the energies are often unknown and interconversions are fast, conformational analysis of such systems requires a quite different approach.

In the presence of fast exchanges, experimentally measured conformation-dependent parameters are averaged, and their values depend on the contributions of the conformations. Under these conditions, the experimental data are generally treated by the so-called discrete model. Within the limits of this model, for example, the spin–spin coupling constants are considered to be geometrically dependent in accordance with equations of the Karplus type. The discrete model assumes that only a limited number of low-energy conformers (configurations or rotamers) cover the whole conformational space of a flexible molecule. Figure 5.15A shows three rotamers, a, b, and c, which are associated with low-energy rotation around the single C–C bond where A and B are protons. The dihedral angles in these rotamers correspond to the $^1H-^1H$ spin–spin coupling constants through three chemical bonds. In addition, two coupling pathways are possible for each proton. In the presence of fast exchange between the rotamers, the measured vicinal spin–spin coupling constants are averaged and written in Equation 5.11:

$$^3J^{(1)} = P^{(1)}(a)^3 J^{(1)}(a) + P^{(1)}(b)^3 J^{(1)}(b) + P^{(1)}(c)^3 J^{(1)}(c)$$

$$^3J^{(2)} = P^{(2)}(a)^3 J^{(2)}(a) + P^{(2)}(b)^3 J^{(2)}(b) + P^{(2)}(c)^3 J^{(2)}(c) \qquad (5.11)$$

$$P(a) + P(b) + P(c) = 1$$

where P represents fractions of rotamers or their populations, and the J values are the so-called limiting spin–spin coupling constants. They should be accurately determined experimentally or calculated theoretically. Finally, superscripts (1) and (2) in Equation 5.11 represent two coupling pathways corresponding to the same chemical bond.

Analysis of the equation written for three rotamers shows that only two experimentally measured coupling constants are needed in order to determine populations of the rotamers. It is also obvious that increasing the number of the molecular states increases the number of the measured coupling constants.

As noted above, in order to solve equations such as Equation 5.11 the limiting spin–spin coupling constants should be well determined. Because these constants (e.g., in cycle molecules) depend on the molecular geometry (via Karplus-type expressions) and properties of substituents (their space properties and electronegativity), the solution is not simple, particularly for new compounds. In fact, NMR data obtained for flexible molecules in such cases will represent only the average molecular structures over time. Therefore, the data for compounds should be supplemented by modeling (i.e., molecular dynamics or Monte Carlo simulations) to identify all of

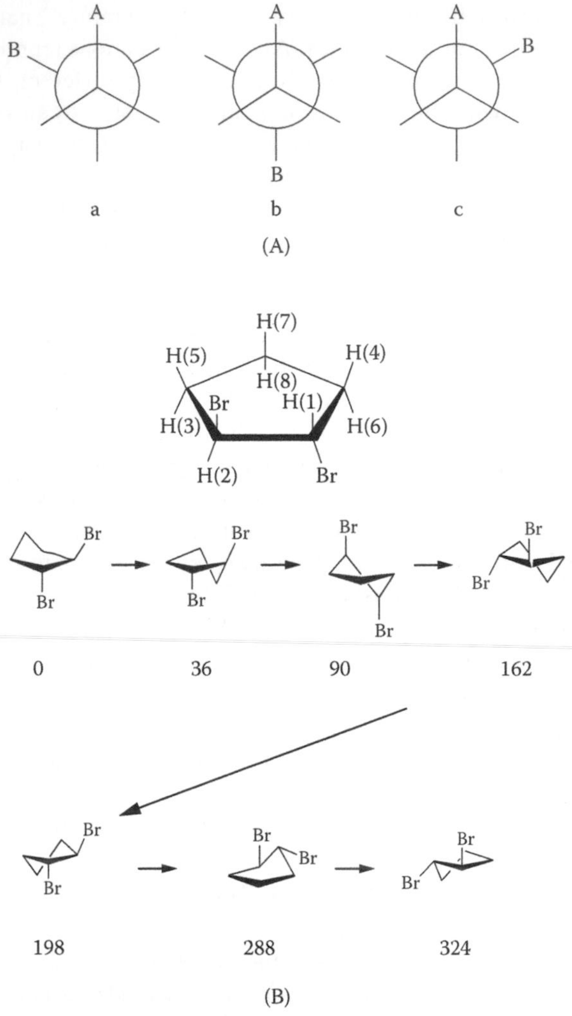

FIGURE 5.15   (A) Three rotamers associated with low-energy rotation around the C–C bond. (B) Some of the canonical conformations of cyclopentanes, shown as a function of the phase angles at the pseudo-rotation process.[29]

the energy minima. Then, optimization of the molecular geometry and calculating the spin–spin coupling constants via density functional theory (DFT) and Karplus-type equations can give the limiting J constants used for conformational analysis.[28]

Currently, several strategies are available for the conformational analysis of flexible molecular systems, including NMR experiments and computer automation procedures. One of them is represented by the study of *trans*-1,2-dibromocyclopentane, as shown in Figure 5.15B. The ^{1}H room-temperature NMR spectrum of this molecule in solutions shows the complex strongly coupled spin systems of protons 1 to 8 due to the commensurable values $\Delta\delta(^1H)$ and $J(^1H-^1H)$.[29] This spectrum consists of around 300 lines, which should be accurately analyzed to obtain the $^3J(^1H–^1H)$ vicinal constants related to the dihedral angles in the molecule. This step, necessary for characterization of the average conformational state of the molecule, can be achieved with any conventional computer program for total line-shape analysis. Then, the experimental $^3J(^1H–^1H)$ coupling constants can be compared with the ^3J values obtained for all of the possible conformations on the basis of Karplus-type Equation 5.12:

$$^3J(^1H–^1H) = P_1 \cos^2\psi + P_2 \cos\psi + P_3 + \Sigma\Delta\chi i[P_4 + P_5 \cos^2(\xi_i\psi + |\Delta\chi i|)] \quad (5.12)$$

where $\Delta\chi i$ represents the difference in electronegativities of the bromine and hydrogen atoms (to account for the presence of the Br substituents), ξ_i is the sign characterizing the relative disposition of the Br substituents, and P_i ($i = 1, 2, ..., 5$) are the empirical constants dependent on the number of substituents. As it follows from the data,[29] the *trans*-1,2-dibromocyclopentane molecule can form 20 possible conformations (some of them are shown in Figure 5.15B) experiencing the fast pseudo-rotation transformation. The geometry of each conformer can be optimized by, for example, the *ab initio* HF/6–31G* method to obtain the dihedral angles noted as ψ in Equation 5.12. Finally, this cycle of theoretical and experimental procedures leads to the conclusion that only the diequatorial conformation of the molecule is highly populated at room temperature.

5.4 NOE AND RESIDUAL DIPOLAR COUPLING MEASUREMENTS IN STRUCTURAL STUDIES

Except for chemical shifts and spin–spin coupling constants through bonding and non-bonding electrons, the direct dipolar interactions between nuclear magnetic moments can be useful quantities to define the structural features of compounds in solutions more precisely. Some of the experiments require an anisotropic medium. Due to fast molecular motions in isotropic solutions, the dipolar coupling is reduced to zero (Chapter 2). Nevertheless, even in solutions, dipolar interactions are still present intrinsically as cross-relaxation rates contributing to total dipole–dipole relaxation, and they can be observed experimentally via NOE detected qualitatively or measured quantitatively by 1D or 2D NOE NMR experiments in solutions of small molecules and complex biological macromolecules. Figure 5.16A shows a $[Ir_2(\mu–H_a)$ $(\mu–Pz)_2(H_b)(H_c)(H_d)(NCCH_3)(PPr^i_3)_2]$ complex containing four hydride ligands. The ligands appear in the ^1H NMR spectrum as four resonances forming a simple spin system. It is clear that the structure of the complex can be deduced from the ^1H, ^{31}P, and ^{13}C NMR spectra by analysis of the chemical shifts and spin–spin coupling constants; however, the NOE experiment is sufficient to confirm the structural motif of H ligands. The experiment can be carried out with selective irradiation of the resonances (double resonance); however, 2D NOESY NMR is more effective. As seen in Figure 5.16A, the 2D ^1H NOESY NMR spectrum shows the cross peaks $H_c\cdots H_a$, $H_d\cdots H_a$, and $H_d\cdots H_b$, which support their *cis* orientations.

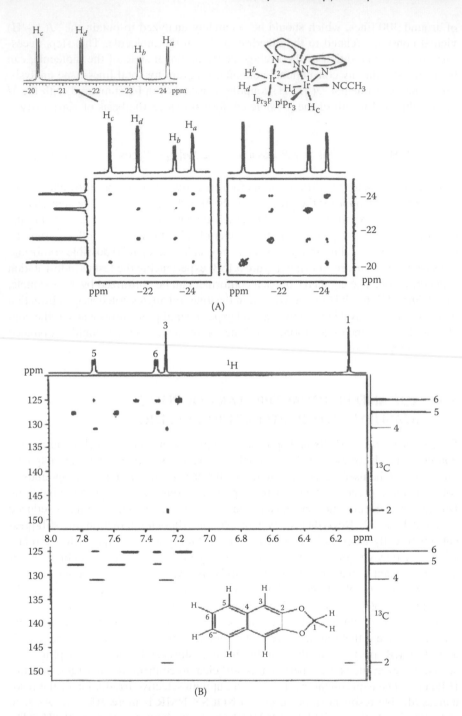

FIGURE 5.16 (A) The ^1H and ^1H 2D NOESY NMR spectra of the Ir complex in C_6D_6. (B) The inverse (top) and direct (bottom) HOESY NMR spectra of the compound with the structure shown. (From Walker, O. et al., *Magn. Reson. Chem.*, 41, 776, 2003. With permission.)

Such data show that homonuclear (proton–proton) NOE experiments are simple technically and generally lead to results that are easy to interpret. In the case of 2D ^1H–X heteronuclear Overhauser effect spectroscopy (HOESY), the situation is less trivial, particularly for X nuclei with low natural abundance, such as ^{13}C. Here, the major problem is low ^{13}C NMR sensitivity, which requires a long experimental duration. The detection of ^{13}C nuclei, direct or inverse, can be used to obtain 2D ^1H–^{13}C HOESY patterns. Details regarding HOESY pulse sequences and procedures can be found in Reference 30.

Figure 5.16B shows the 2D HOESY spectra obtained by experiments on a symmetric organic compound performed with direct and inverse detection, where the duration of the experiments was ~24 hours.[30] The NMR patterns exhibit the NOE interactions necessary for signal assignments and determination of structure. Both of the spectra reveal the same cross peaks corresponding to correlations between carbons and directly bonded protons, remote protons (see the long-distance correlation C_2 and H_1, ~3 Å), and protons arising from chemically equivalent sites. The latter is particularly important for an analysis of symmetric compounds. One feature of inverse detection should be noted. A doublet for the carbon atoms C_5 and C_6 is not symmetrical due to the effect of chemical shift anisotropy (CSA) (^{13}C)–dipolar (CH) cross-correlation developed during the mixing time.

According to the theory discussed in Chapter 2, the force of dipolar coupling depends on the angle term ($3\cos^2\theta - 1$), which is not zero in the absence of high-amplitude molecular motions. This is the case for solids, where dipolar interactions are typically measured as tenths of kilohertz. In solutions, where molecular reorientations are fast and often isotropic, this term goes to zero and thus the structural information based on dipolar coupling is lost. However, for weakly oriented molecules placed in an alignment medium, for example, the molecular tumbling becomes anisotropic. When this occurs, favorable orientations of dipolar internuclear vectors relative to the external magnetic field can appear, and the term ($3\cos^2\theta - 1$) will not be zero. Due to this effect, one can observe residual dipolar coupling, estimated as tenths of a hertz. In other words, at known angle θ, residual dipolar coupling can be easily interpreted in structural terms.

In practice, residual dipolar coupling measurements are carried out for molecules oriented in a liquid crystal or in a stretched gel.[31] The choice of alignment medium depends on the nature of the compounds under investigation. In the case of carbohydrates, the best choice is a liquid crystalline medium in an aqueous solvent. Liquid crystals form mesophases strongly oriented in the external magnetic field, resulting in weak alignments of molecules. The appropriate temperatures and concentrations should be optimized experimentally. Under these conditions, signals observed in NMR spectra show additional splitting, which can be well discriminated by comparison of the data obtained in isotropic solutions and liquid crystals. Figure 5.17A shows the ^1H NMR spectra recorded for methyl-D-xylopyranoside in isotropic and aligned media, where the additional splitting or line broadening is well observed for each ^1H signal. This additional splitting is required for determination of the residual dipolar coupling constant and its sign. The latter is particularly important when the residual dipolar coupling is smaller than the regular J coupling.

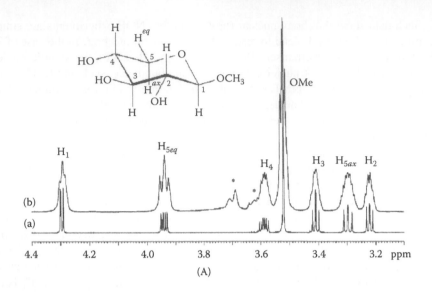

(A)

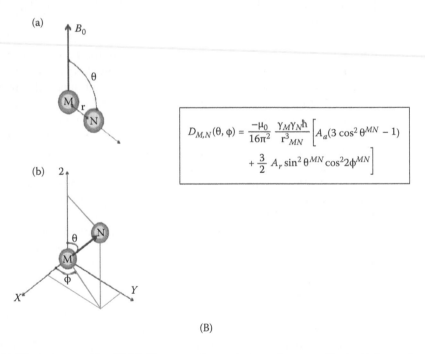

(B)

FIGURE 5.17 (A) The ^{1}H NMR spectra for methyl-D-xylopyranoside in (a) isotropic and (b) aligned media. (From Hinchley, S.L. et al., *J. Am. Chem. Soc.*, 126, 13100, 2004. With permission.) (B) Part (a) shows the dipolar interaction for nuclei M and N, which depends on internuclear distance r and the angle formed by vector r and external magnetic field B_0; part (b) shows the angles θ and ϕ used to obtain dipolar coupling $D_{M,N}$ in the equation shown. (From Canales, A. et al., *Magn. Reson. Chem.*, 50, S80, 2012. With permission.)

Interpretation of the data obtained to further refine structure is based on calculations of molecular geometry in terms of the equation in Figure 5.17B, written for spins of 1/2. The so-called alignment tensor is defined by five key parameters representing the magnitude and rombicity of the alignment and three principal axes of the molecular alignment relative to the direction of the external magnetic field. For rigid molecules or rigid molecular fragments, this tensor can be accurately determined by using five residual dipolar coupling values pointing in non-parallel directions. Then, the axial and rhombic components, marked in the equation as A_a and A_r, respectively, can be used to calculate dipolar coupling D_{MN}; θ^{MN} is the angle formed by vector M–N and the Z-axis in the molecular frame, and ϕ^{MN} is the azimuth angle shown in Figure 5.17B.[32]

Different pulse sequences can be applied for the measurement of residual dipolar coupling to obtain at least five independent parameters providing accurate refinement of structures. The sequences and methodologies for oriented carbohydrates are discussed in Reference 32. Because dipolar proton–carbon and carbon–carbon coupling increases the accuracy of structural determinations, ^{13}C labeling is applied.

Typical experiments performed on carbohydrates, free or binding to proteins, illustrate the basic methodology of such studies. Dipolar interactions of ^{13}C nuclei and ^1H via one chemical bond, $^1D(CH)$, can be measured by 2D heteronuclear single quantum coherence (HSQC) NMR spectra, where regular or improved pulse sequences are applied. The dipolar coupling ^1H–^1H, important in a structural context, is difficult to measure in simple ^1H NMR spectra because the target protons interact with all of the neighboring protons, resulting in unresolved multiplets, as can be seen in Figure 5.17A. Therefore, $^nD(HH)$ values are generally measured by J-modulated (or modified COSY) NMR experiments. ^{13}C labels allow determination of long-range heteronuclear dipolar coupling $^nD(CH)$ and $^nD(CC)$, where n is the number of chemical bonds. The $^{2,3}D(CH)$ couplings are generally measured with the J-modulated HMBC pulse sequence or 2D selective J-scaled HSQC method. The $^1D(CC)$ values can be extracted from 1D gate-decoupled ^{13}C NMR (or 2D ^{13}C–^{13}C COSY) experiments. Finally, with the ultrafast HSQC NMR technique, single-scan 2D NMR can be applied for structural refinements in oriented media.[33]

REFERENCES AND RECOMMENDED LITERATURE

1. Castolo, A.A., Bakhmutov, V., Theurel, R.C., et al. (2006). *Ejemplos practicos del uso de la resonancia magmnetica nuclear en la quimica*. Gustavo A. Madero, Mexico: Cinvestav.
2. Griffiths, L. (2001). *Magn. Reson. Chem.*, 39: 194.
3. Buddrus, J. and Lambert, J. (2002). *Magn. Reson. Chem.*, 40: 3.
4. Kempgens, P. (2011). *Concepts Magn. Reson. A*, 38A: 74.
5. Sattlera, M., Schleucherb, J., and Griesinger, C. (1999). *Prog. Nucl. Magn. Reson. Spectrosc.*, 34: 93.
6. Mandal, P.K. and Majumdar, A. (2004). *Concepts Magn. Reson. A*, 20A: 1.
7. Furrer, J. (2012). *Concepts Magn. Reson. A*, 40A: 101.
8. Morris, G.A. (1986). *Magn. Reson. Chem.*, 24: 371.
9. Bain, A.D., Burton, I.W., and Reynolds, W.F. (1994). *Prog. Nucl. Magn. Reson. Spectrosc.*, 26: 59.

10. Pinto, D., Santos, C.M.M., and Silva, A.M.S. (2007). *Recent Res. Dev. Heterocycl. Chem.*, 397.
11. Guizado-Rodriguez, M., Ariza-Castolo, A., Merino, G., Vela, A., Noth, H., Bakhmutov, V.I., and Contreras, R. (2001). *J. Am. Chem. Soc.*, 123: 9144.
12. Laszlo, P. (1967). *Prog. Nucl. Magn. Reson. Spectroscosc.*, 3: 231.
13. Bain, A.D., Hughes, D.W., Anand, C.K., Niec, Z., and Robertsond, V.J. (2010). *Magn. Reson. Chem.*, 48: 630.
14. Queiroz, L.H.K., Giraudeau, P., dos Santos, F.A.B., de Oliveira, K.T., and Ferreira, A.G. (2012). *Magn. Reson. Chem.*, 50: 496.
15. Gordon, M.D. and Quin, L.D. (1976). *J. Am. Chem. Soc.*, 98: 1.
16. Wiberg, K.B., Hinz, W., Jarret, R.M., and Aubrecht, K.B. (2005). *J. Org. Chem.*, 70: 8381.
17. Abraham, R.J., Warne, M.A., and Griffiths, L. (1997). *J. Chem. Soc. Perkin Trans.*, 2: 2151.
18. Bain, A.D., Baron, M., Burger, S.K., Kowalewski, V.J., and Rodríguez, M.B. (2011). *J. Phys. Chem. A*, 115: 9207.
19. Feeney, J. (1975). *Proc. R. Soc. Lond. Ser. A,* 345: 61–72.
20. Coxon, B. (2009). *Adv. Carbohydr. Chem. Biochem.*, 62: 17.
21. Rusakov, Y.Y., Krivdin, L.B., Shmidt, E.Y., Vasil'tsov, A.M., Mikhaleva, A.I., and Trofimov, B.A. (2007). *Russ. J. Org. Chem.*, 43: 880.
22. Bakhmutov, V.I., Galakhov, M.V., and Fedin, E.I. (1985). *Magn. Reson. Chem.*, 23: 11.
23. Pecksen, G.O. and White, R.F. (1989). *Can. J. Chem.*, 67: 1847.
24. Hierso, J.C. (2014). *Chem. Rev.*, 114: 4838.
25. Xie, X., Yuan, Y., Kruger, R., and Broring, M. (2009). *Magn. Reson. Chem.*, 47: 1024.
26. Afonin, A.V. (2010). *Russ. J. Org. Chem.*, 46: 1313.
27. Durie, A.J., Slawin, A.M.Z., Lebl, T., Kirschb, P., and O'Hagan, D. (2011). *Chem. Commun.*, 47: 8265.
28. Napolitano, J.G., Gav, J.A., Garcia, C., Norte, M., Fernandez, J.J., and Daranas A.H. (2011). *Chem. Eur. J.*, 17: 6338.
29. Zubkov, S.V., Golotvin, S.S., and Chertkov, V.A. (2002). *Izv. Akad. Nauk. Ser. Khim.*, 7: 1129.
30. Walker, O., Mutzenhardt, P., and Canet, D. (2003). *Magn. Reson. Chem.*, 41: 776.
31. Hinchley, S.L., Rankin, D.W.H., Liptaj, T., and Uhrn, D. (2004). *J. Am. Chem. Soc.*, 126: 13100.
32. Canales, A., Jiménez-Barbero, J., and Martín-Pastor, M. (2012). *Magn. Reson. Chem.*, 50: S80.
33. Giraudeau, P., Montag, T., Charrier, B., and Thiele, C.M. (2012). *Magn. Reson. Chem.*, 50: S53.

RECOMMENDED LITERATURE

Friebolin, H. (1991). *Basic One- and Two-Dimensional NMR Spectroscopy*. Weinheim: Wiley-VCH.
Günther, H. (2013). *NMR Spectroscopy: Basic Principles, Concepts and Applications in Chemistry*. Weinheim: Wiley–VCH.
Harris, R.K. (1986). *Nuclear Magnetic Resonance Spectroscopy: A Physicochemical View*. Avon: Bath Press.
Ludwig, C. and Viant, M.R. (2010). Two-dimensional *J*-resolved NMR spectroscopy: review of a key methodology in the metabolomics toolbox. *Phytochem. Anal.*, 21: 22.
Sanders, J.K.M. and Hunter, B.K. (1994). *Modern NMR Spectroscopy: A Guide for Chemists*. Oxford: Oxford University Press.

6 NMR Relaxation in Solutions
Applications

The theory of nuclear relaxation is the basis for various relaxation applications that provide accurate measurements of quadrupolar and dipolar coupling constants or chemical shift anisotropy, in connection with the symmetry of nuclear environments, structural features of compounds, and characterization of molecular mobility, including the type of motions, identification of weak intermolecular interactions, and even determination of interatomic distances in solutions. In addition to sophisticated relaxation experiments directed toward solving very specific problems, the measurement of relaxation times T_1 and T_2, utilizing standard hardware and software (or knowledge of their values) to assist in recording one-dimensional (1D) and two-dimensional (2D) NMR spectra, can reduce potential errors, such as perturbation of integral intensities, missing signals, or the appearance of artifacts. They can also help with the assignment of signals based on their unusual widths. In addition, the modification of pulse sequences to account for relaxation behavior can help to overcome difficulties in the accumulation of signals with long relaxation times. The driven equilibrium Fourier transform (DEFT) pulse sequence, for example, is effective for ^{29}Si NMR in solutions of biological systems.[1] Finally, there are pulse sequences that allow detection of small amounts of compounds in the presence of large amounts of other compounds when their T_2 relaxation times are strongly different. Some aspects of such applications are considered in this chapter to demonstrate the methods and strategies of studies for which the choice of nuclei (high or low sensitivity, quadrupolar or not) and the experimental techniques applied play major roles. The techniques should be adequate and correspond well to the specified goals, as relaxation measurements generally represent long-term experiments.

6.1 PARTIALLY RELAXED NMR SPECTRA: RESOLVING THE UNRESOLVED SIGNALS AND ASSIGNING NMR SIGNALS

As noted in Chapter 3, standard inversion–recovery experiments for T_1 relaxation time measurements produce partially relaxed NMR spectra when delay times (τ) between 180° and 90° pulses are less than T_1 times. An analysis of such spectra can be very useful for the assignment of signals. Figure 6.1 shows the products of room-temperature ionic hydrogenation between the Re dihydride and acetone. Both of the products, the Re monohydride complex and the alcohol, are easily identified by regular ^1H NMR spectra; however, at low temperatures, new ^1H resonances are observed for the reaction mixture which can be attributed to an intermediate located on the

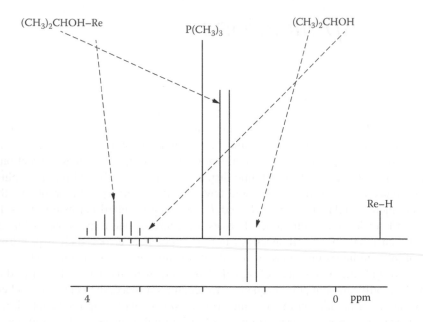

FIGURE 6.1 (Top) Two-step ionic hydrogenation mechanism for the reaction between the Re dihydride and acetone, resulting in the intermediate complex at –70°C in a toluene solution. (Bottom) Partially relaxed ^{1}H NMR spectrum of the intermediate complex (–70°C) obtained by inversion–recovery at $\tau < 0.5$ s. (The signals with positive and negative phases belong to the alcohol complex [ReH(CO)(NO)(P((CH$_3$)$_3$)$_2${(CH$_3$)$_2$CH–OH}]$^+$[CF$_3$COO]$^-$ and free (CH$_3$)$_2$CH–OH, respectively.)

coordinate of this reaction. Following chemical logic, the alcohol molecule in the intermediate could be binding to the metal center; however, in the absence of reliable assignments, this statement remains a hypothesis only. Proof can be obtained through the use of low-temperature, partially relaxed ^{1}H NMR spectra. As seen in these spectra, the resonances of molecular fragments corresponding to the transition metal hydride and alcohol are already relaxed, while the negatively phased lines of the free alcohol molecule are far from full relaxation. In terms of NMR relaxation, this observation suggests that molecular reorientations with participation of the hydride complex and alcohol moieties are correlated to chemical bonding in the intermediate. Now the assignment of signals can be considered to be reliable.[2]

Similarly, the assignment of signals can be simplified in ^{13}C{^{1}H} NMR spectra in solutions of complex organic molecules, such as vitamin B12 or coenzyme B12.[3] In fact, because the signals of carbons neighboring with protons relax significantly faster than proton-free carbons due to the stronger dipolar interactions occurring in the first case, different intensities of the signals and their evolution in partially relaxed ^{13}C spectra help with assignments. Moreover, the careful choice of delay times (τ) in inversion–recovery sequences leads to partially relaxed ^{13}C NMR spectra where non-protonated carbons are not visible but all of the resonances of the protonated carbons show high (positive) intensities. The approach may be archaic but it is still effective.[3]

In many cases, partially relaxed NMR spectra can resolve resonances that remain unresolved in regular NMR spectra. In fact, the signals in regular NMR spectra are often broad due to a superposition of several lines, the chemical shift differences of which are small. For example, treatment of the hydride complex $RuH_4(PPh_3)_3$ in a toluene-d_8 solution with D_2 (the hydride ligands are capable of a D_2/H_2 exchange) leads to the ^1H NMR spectrum where the initial integral intensity of the hydride resonance (–7.2 ppm) is reduced due to the above exchange. However, the isotopomers expected after the D_2 treatment are not resolved because the residual hydride resonance is broadened to 30 Hz (Figure 6.2A). In contrast, an analysis of the partially relaxed ^1H NMR spectra shows clearly that the broad hydride signal consists of two lines with $\delta(^1H)$ of –7.21 and –7.11 ppm. The former resonance relaxes more slowly due to weak dipole–dipole proton–deuterium interactions; therefore, it can be assigned to the complex $RuHD_3(PPh_3)_3$ with some confidence. The second line relaxes faster and hence belongs to the isotopomer $RuH_2D_2(PPh_3)_3$.[4]

The same methodology obviously can be applied for any nuclei in solutions, even for nuclei with pronounced quadrupolar properties. For example, the regular ^7Li NMR spectra of human red blood suspensions prepared in the presence of LiCl show only one line (Figure 6.2B). In contrast, two components, relaxing differently, can clearly be seen in the partially relaxed ^7Li NMR spectra.[5] This is important for interpretation of the NMR data. Similarly, the resolution can be enhanced in ^{11}B NMR spectra of boron compounds, as is shown in Reference 6 for experiments with partial relaxation.

In addition to improvements in spectral resolution, partial relaxation experiments are useful for the detection of signals that might be unobservable in regular NMR spectra. For example, because of the low natural abundance of ^{15}N nuclei, the ^{31}P NMR spectra of phosphorus compounds in Figure 6.3 (bottom) do not show the satellites ^{31}P–^{15}N, which are needed to measure $J(^{15}N$–$^{31}P)$ spin–spin coupling constants. At the same time, these satellites can be detected in Carr–Purcell–Meiboom–Gill (CPMG) experiments aimed at measurements of T_2 times. Because ^{14}N nuclei relax much faster than ^{15}N nuclei and this relaxation difference affects formation of the echo signal, variations in the τ values optimize conditions for observing satellites ^{31}P–^{15}N at the natural abundance of ^{15}N nuclei.[7] The same effect can be used for determination of $J(^{15}N$–$^{31}P)$ values by the Hahn echo extended pulse sequence when $T_2 < T_1$.

6.2 RELAXATION TIMES IN SOLUTIONS: QUADRUPOLAR COUPLING CONSTANTS AND CHEMICAL SHIFT ANISOTROPY

Because quadrupolar coupling constants (QCCs) and chemical shift anisotropies (CSAs) allow chemical interpretation and can be useful in this context, their measurement is of great interest. In solids, QCC and CSA values can be directly extracted from the line shapes of resonances in NMR spectra. In contrast, this information is lost in NMR spectra of solutions due to fast molecular reorientations averaging these

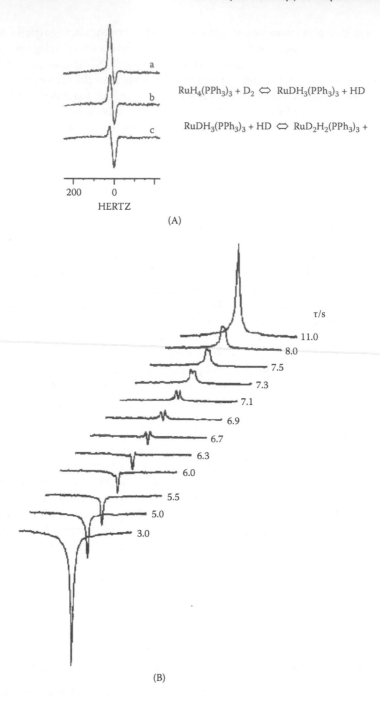

RuH$_4$(PPh$_3$)$_3$ + D$_2$ $\Leftrightarrow$ RuDH$_3$(PPh$_3$)$_3$ + HD

RuDH$_3$(PPh$_3$)$_3$ + HD $\Leftrightarrow$ RuD$_2$H$_2$(PPh$_3$)$_3$ +

FIGURE 6.2 (A) The ^{1}H NMR signals of hydride ligands in partially relaxed spectra collected for a deuterated complex, RuH$_4$(PPh$_3$)$_3$, at 280 K in toluene-d_8. (B) Partially relaxed ^{7}Li NMR spectra recorded for a human red blood suspension in the presence of LiCl. (From Rong, Q. et al., *Biochemistry*, 32, 13490, 1993. With permission.)

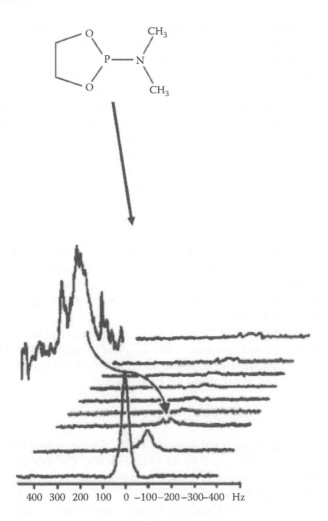

400 300 200 100   0 −100−200−300−400 Hz

FIGURE 6.3   The ^{31}P NMR spectra of the phospholane molecule obtained in CDCl$_3$ by the CPMG pulse sequence at a delay time τ of 1 ms. (From Ariza-Castolo, A., *Concepts Magn. Reson. A*, 32A, 168, 2008. With permission.)

interactions. However, the magnitudes of the QCC and CSA can still be obtained by relaxation time measurements under specific conditions, considered below. The important advantage of relaxation studies in solutions is the fact that relaxation curves are mono-exponential; therefore, calculations of relaxation times are simplified.

6.2.1   QCC Values from Variable-Temperature Relaxation Experiments and T_{1MIN} Times

In general, the nuclear quadrupolar coupling constants interpreted in terms of chemical bonding are measured and studied by nuclear quadrupole resonance (NQR) spectroscopy applied for solids; however, some "inconvenient" nuclei have relatively small quadrupole moments, such as deuterium, and in some cases, such as ^{35}Cl nuclei in Cl$^-$ ions with small QCC values, NQR studies are problematic or impossible. In this situation, the NMR relaxation technique in solutions becomes indispensable for determining quadrupolar parameters via measurement of T_1 times.

Due to the small ^{2}H quadrupole moment, ^{2}H signals are easy to detect in solutions, particularly in ^{2}H-labeled compounds, so deuterium nuclei are most popular in relaxation studies. In addition, because the quadrupolar mechanism is generally dominant in ^{2}H relaxation, its preliminary evaluation is not necessary. This statement is well illustrated by, for example, ^{2}H relaxation experiments on the isotopomeric hydride/ dihydrogen complexes $PP_3Ru(DH)D$ and $PP_3Ru(D_2)D$. Here, the D–H distance in the dihydrogen ligand in the $PP_3Ru(DH)D$ complex is very short (<0.92 Å), so it can potentially provide the strongest dipole–dipole proton–deuterium interactions. According to the ^{2}H relaxation measurements performed for D_2 and RuD ligands in the completely deuterated complex $PP_3Ru(D_2)D$, the ^{2}H T_1 relaxation times are 0.0586 s and 0.0165 s, respectively, in CH_2Cl_2 at 180 K and the working frequency of 61.402 MHz. Very close values are obtained for the isotopomer $PP_3Ru(DH)D$: 0.0546 s for Ru(HD) and 0.0145 s for RuD ligands. Simple estimation shows that ^{2}H dipolar relaxation in the H–D unit contributes only 6.8%, which is close to the errors encountered in T_1 determinations (see Chapter 3).

With complete domination of the quadrupolar mechanism and fast isotropic reorientations of the electric field gradient vector eq_{ZZ} (for isotropic molecular motions), the deuterium quadrupolar relaxation rate ($1/T_1$) in a high-temperature zone ($1 \gg \omega_Q^2\tau_C^2$; see Chapter 3) can be expressed via Equation 6.1:

$$1/T_1 = 1.5\pi^2(DQCC)^2(1 + \eta^2/3)\tau_C \tag{6.1}$$

Thus, for example, at room temperature, the T_1 time directly gives the DQCC value, if parameters η and τ_C are found independently. For example, correlation times τ_C can be determined via T_1 relaxation experiments performed on the same molecule for other target nuclei such as ^{1}H or ^{13}C. It is easy to show that errors in such DQCC determinations will be ≤2.2% because deuterium relaxation in solutions is generally mono-exponential and errors in ^{2}H T_{1min} time measurements are ≤5%.

It is obvious that the asymmetry parameter η cannot be found with T_1 relaxation data; therefore, *a priori* it is unknown. However, by definition, this parameter changes from 0 to 1. In other words, the uncertainty in DQCC values calculated from ^{2}H T_1 times will be ≤10% even in the absence of accurate knowledge. In practice, however, ambiguity in the η value is significantly smaller, particularly for simple chemical bonds D–X. For example, the η parameter is <0.2 in C–D bonds (CD_3X, for $X = $ F, Cl, Br, I, CD_3CN, or D_2CO), thus reducing possible errors. Much larger errors can appear in the presence of unrecognized anisotropic reorientations (see below).

Some examples of DQCC determinations are illustrated in Figure 6.4.[8] The ^{2}H T_1 data obtained in a toluene–H_8 solution of the Os cluster at 223 K (Figure 6.4A) show the remarkable difference in ^{2}H T_1 times measured for the terminal and bridging hydride ligands. Anisotropic molecular tumbling can lead to such a difference. However, the ^{1}H T_1 times in a ^{1}H isotopomer of this compound are identical for both of the ^{1}H ligands. Therefore, molecular motions of this cluster can be accepted as

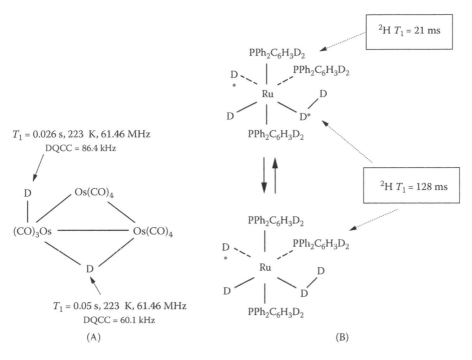

FIGURE 6.4 (A) The ^2H T_1 data obtained in a toluene–H_8 solution of the Os cluster at 223 K. (B) The ^2H T_1 time measurements in the Ru hydride performed for the signal of *ortho*-deuterons in the aromatic rings and for the $D_2/(D_2)$ signal averaged due to fast chemical exchange ($CHCl_3$).

being isotropic and the isotropic approximation can be applied. The correlation time τ_C required for DQCC determination can be obtained independently by variable-temperature ^1H T_1 measurements. Finally, Equation 6.1 gives DQCC values of 86.4 ± 1.5 and 60.1 ± 2.0 kHz for the terminal and bridging D-ligands, respectively, assuming $\eta = 0$. The latter is reasonable because the electric field gradient tensors at terminal D ligands are usually axially symmetric.

In contrast, the electric field gradient tensors at bridging D ligands show deviation from axial symmetry. For example, an asymmetry parameter η of 0.31 was found in binuclear complexes $[R_4N][DMr_2(CO)_{10}]$ (where M = Cr or W). However, even at $\eta = 0.31$, the ^2H T_1 time of the Os–D–Os ligand measured as 0.05 s in Figure 6.4A leads to a DQCC value of 63.3 kHz instead of the 60.1 kHz calculated at $\eta = 0$. The difference does not seem to be large.

In order to exclude correlation time τ_C from Equation 6.1, another approach to DQCC determinations can be applied. Double ^2H-labeled compounds are investigated, and the DQCC value for one of the deuterium nuclei can be obtained when the value for the second one is known. Figure 6.4B demonstrates this approach with deuterium T_1 time measurements in solutions of the Ru–$D_2(D_2)$ complex containing the *ortho*-deuterons in the aromatic rings. Because the quadrupole coupling constant for aromatic deuterons is well known (determined as 182 KHz by solid-state ^2H

NMR[8]) and changes only slightly from one compound to another, the DQCC(D–Ru) value can be easy calculated by excluding the τ_C term in Equation 6.1. It is clear that the reverse problem—determining molecular motion correlation times from ^{2}H T_1 times—can be easily solved when DQCC constants are known.

Because theoretically the plots of $\ln(T_1(Q))$ vs. $1/T$ go through minima at $\tau_C = 0.62/\omega_Q$, the equation for determining the deuterium QQC can be written as follows at the temperature of the T_{1min} observation:[8]

$$1/T_{1min} = 0.672(1 + \eta^2/3)(DQCC)^2/\nu_D \tag{6.2}$$

where ν_D, DQCC, and T_{1min} are measured in MHz, KHz, and seconds, respectively. Thus, at $\eta = 0$, the ^{2}H T_{1min} times directly give DQCC values only in an approximation of isotropic molecular motions. Technically, due to a relatively low deuterium working frequency, the minimal ^{2}H T_1 times are difficult to reach even at the lowest temperatures in dilute solutions of regular non-bulky molecules. However, the situation can change in concentrated viscous toluene solutions, such as toluene solutions where the ^{2}H T_{1min} times are reached due to increasing the effective τ_C and E_a values characterizing molecular motions. Figure 6.5A illustrates a T_{1min} temperature that is dependent on the molecular size of iron complexes.[9] It should be emphasized that this temperature, itself, does not play a role in DQCC calculations, which require only T_{1min} values. Finally, in the chemical context, because one interpretation of the DQCC value as a measure of the electric field gradient along X–D bonds is the ionicity of these bonds, it can be easily determined from relaxation time measurements in solutions of different compounds to characterize their reaction ability.[8,9]

By analogy with deuterium, quadrupolar coupling constants can be determined for other nuclei by measuring T_1 times in solutions. For example, such measurements have been reported for ^{35}Cl nuclei of ions Cl$^-$ in liquid chloride salts.[10] ^{35}Cl nuclei have a spin of 3/2, and their T_1 and T_2 times are given by Equations 6.3 and 6.4, where the QCC is measured in MHz:

$$1/T_1(Q) = (2/25)\pi^2(QCC/h)^2(1 + \eta^2/3)(\tau_C/(1 + \omega_Q^2\tau_C^2) + 4\tau_C/(1 + 4\omega_Q^2\tau_C^2)) \tag{6.3}$$

$$1/T_2(Q) = (1/25)\pi^2(QCC/h)^2(1+ \eta^2/3)[3\tau_C + 5\tau_C/(1 + \omega_Q^2\tau_C^2) + 2\tau_C/(1 + 4\omega_Q^2\tau_C)] \tag{6.4}$$

As follows from these equations, at $T_1 \neq T_2$ or, for example, at $T_1(500$ MHz$) \neq T_1(600$ MHz$)$, the ^{35}Cl QCC values can be directly calculated on the basis of the T_1/T_2 relaxation data. In fact, both of the T_1 and T_2 times (or the times determined at different frequencies) exclude from the equations the correlation times (τ_C) to determine the quadrupolar coupling constants if molecular motions are again isotropic. In the case of Cl$^-$ ions, the asymmetry parameter η can be taken as 0.[10] The ^{35}Cl QCC values have been interpreted in terms of bond ionicity to create an ionicity scale, an important requisite for successful identification of salt properties.[10]

The application of ^{59}Co NMR relaxation for the characterization of the cobalt carbonyl cluster in solutions is shown in Figure 6.5B.[11] The solid-state ^{59}Co NMR spectrum of the cluster exhibits two signals, which are differently broadened and belong to apical and basal cobalt atoms in the cluster structure. The relatively sharp

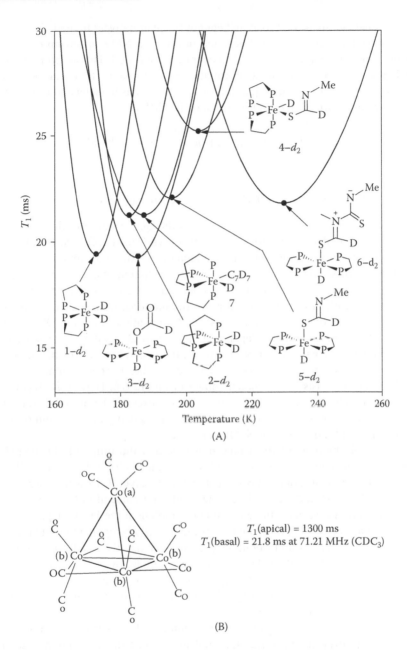

FIGURE 6.5 (A) The dependence of ^2H T_1 time (ms) on temperature (K) for iron–deuterium complexes. (From Field, L.D. et al., *J. Organomet. Chem.*, 692, 3042, 2007. With permission.) (B) Cobalt cluster and ^{59}Co T_1 times measured for basal and apical cobalt atoms.

resonance has been assigned to the apical cobalt atoms, and the ^{59}Co QCC value of 12.4 MHz and an asymmetry parameter (η_{Co}) value of 0.32 have been obtained directly from the line shape in the solid-state NMR spectrum. In contrast, the signal of basal cobalt atoms is too broad and cannot be quantitatively treated. At the same time, both of the signals are well detected in the ^{59}Co NMR spectrum of a $CDCl_3$ solution, and their ^{59}Co T_1 times can be accurately measured. The structure in Figure 6.5B is not asymmetric so the approximation of isotropic motions can be applied. Then, correlation times τ_C are equal for the apical and basal nuclei and Equation 6.5 can be used to obtain the basal QCC value of 100 MHz:

$$C_{QCC}(\text{basal}) = C_{QCC}(\text{apical}) \, [T_1(\text{apical})/T_1(\text{basal})] \tag{6.5}$$

6.2.2 T_1 Relaxation and DQCC Values in Mobile Structural Units

As mentioned above, DQCC values (as well as the QCC values of other nuclei) can be accurately calculated for errors of ≤3% from ^{2}H T_1 data collected for molecules experiencing isotropic tumbling. This statement is extremely important because much larger errors or even uncertainties in the calculations can appear for approximations of isotropic motions applied for molecular systems moving anisotropically or for systems with fast internal motions on the motional time scale of whole molecules. The effects of anisotropic motions on T_1 relaxation times and QCC calculations are not simple and greatly depend on the structural and dynamic properties of the compounds under investigation. However, the general principles behind understanding these effects for such systems and approaches to their treatments can be considered on the basis of the model shown in Figure 6.6. Here, a D_2 ligand as a structure unit in transition metal–dihydrogen complexes can be immobile (i.e., moving with the whole molecule), fast spinning, or experiencing fast 180° jumps or fast high-amplitude librations on the NMR time scale.

For isotropic motions of the whole molecule and the immobile D_2 ligand, ^{2}H NMR relaxation is obviously well treated with Equations 6.1 or 6.2 or the full equations describing quadrupolar relaxation shown in Chapter 3. However, even in this case, atoms X and D in the immobile X–D bond experience ultra-fast vibrational or librational motions. Because the electric field gradient at the D site depends on the X–D bond length (see Chapter 2) and the principal eq_{ZZ} component is situated along this bond, the vibrations and librations partially average the quadrupolar coupling constant. Thus, DQCC values determined, for example, by ^{2}H T_{1min} measurements will be effectively reduced. The theory shows that this reducing effect is not dramatic but can reach 6% of the DQCC value. Thus, corresponding corrections in DQCC calculations can be applied.[12]

It is well known that D_2 ligands in transition metal–dihydrogen complexes can remain mobile even in the solid state and experience the 180° jumps shown in Figure 6.6A at the lowest temperatures, while the whole molecules are immobile. Because the eq_{ZZ} component for deuterium in D_2 ligands deviates from the D–D direction, the ^{2}H NMR relaxation rate, $1/T_1(J)$, in the jumping D_2 unit can be expressed as follows:

$$1/T_1(J) = (9/160)(1 + \eta^2/3)(\sin 2\alpha)^2(DQCC)^2[\tau_J/(1 + \omega_D^2\tau_J^2) + 4\tau_J/(1 + 4\omega_D^2\tau_J^2)] \tag{6.6}$$

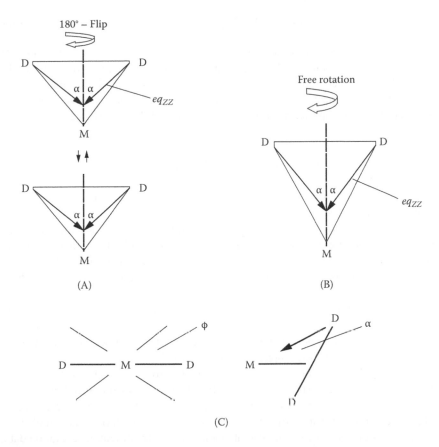

FIGURE 6.6 (A) D_2 ligands in transition metal–dihydrogen complexes experiencing 180° jumps. (B) Free rotation of D_2 ligands in transition metal–dihydrogen complexes. (C) D_2 ligand undergoing internal high-amplitude librations.

where 2α is the angle formed by principal eq_{ZZ} components at two deuterons and τ_J is the correlation time describing the jumps. As a ^2H T_{1min} time is measured, Equation 6.6 can be rewritten as follows:[13]

$$1/T_{1min}(J) = 0.0128(1 + \eta^2/3)(\sin2\alpha)^2(DQCC)^2/\nu_D \qquad (6.7)$$

This expression is obviously valid to determine DQCC in the solid state, where other motions are absent. It is interesting that, according to this model, the 180° jumps do not contribute to ^2H relaxation at $\alpha = 0$ when the eq_{ZZ} vector is exactly situated along bond D–D.

Because, topologically, the 180° jumps of the D_2 ligand correspond to displacement of the principal eq_{ZZ} component by the 2α angle, the molecules with the D_2 unite. Experiencing fast 180° jumps can be considered in solutions as two equally populated states, each of which undergoes isotropic motions together with the whole molecule. Under these conditions, the ^2H T_{1min} time can be expressed via Equation 6.8:

$$1/T_{1min}(180°) = 0.672C_1(1 + \eta^2/3)(DQCC)^2/\nu_D \qquad (6.8)$$

where $C_1 = 0.25(1 + 3\cos^2 2\alpha)$. This expression is equivalent to Equation 6.2 at $\alpha = 0$.

The case of fast-spinning dihydrogen ligands is shown in Figure 6.6B. Here, the D–D bonds experience a free rotation at correlation times much shorter than correlation times τ_C, which describe motions of whole molecules. Then, the Woessner spectral density function, applied for this case, gives the ^{2}H T_{1min}(ROT) value expressed by Equation 6.9:

$$1/T_{1min}(ROT) = 0.168(3\cos^2\alpha - 1)^2(1 + \eta^2/3) (DQCC)^2/\nu_D \qquad (6.9)$$

where α is the angle formed by the principal electric field gradient component, eq_{ZZ}, and the rotational axis.

Finally, the last case considered within the framework of this motional model is the D_2 ligand that experiences internal high-amplitude librations in solutions, as illustrated in Figure 6.6C. These fast D_2 librations can be characterized by amplitudes ϕ in a twofold potential to affect the ^{2}H T_{1min} times according to Equation 6.10:

$$1/T_{1min}(LIB) = 0.672F(\alpha,\phi)(1 + \eta^2/3)(DQCC)^2/\nu_D \qquad (6.10)$$

Here, factors $F(\alpha,\phi)$ can be calculated for variations in the values of α and ϕ using the spectral density function suggested in Reference 14.

The principal conclusion that can be drawn from the motional model is that the fast internal motions can strongly elongate the ^{2}H T_{1min} times measured experimentally, and this effect depends on the type of fast motions, their geometries, and particularly the α angle. Because the angle α shows the eq_{ZZ} orientation in the molecular coordinate system, it is a very important element in the treatment of relaxation data. In the case of D_2 ligands in transition metal complexes, the angle strongly depends on the bonding mode in the D_2 unit; the eq_{ZZ} vector is situated along bond D–D for strong D–D bonding or along bond M–D for weak D–D bonding. Generally speaking, this angle can be found by quantum chemical calculations or experimentally by solid-state NMR. The absence of this information obviously leads to uncertainty, and the DQCC values cannot be accurately determined by ^{2}H T_1 relaxation time measurements. This latter point is also valid for other molecular systems (and other nuclei) characterized by fast internal motions. However, some estimation or identification of a tendency can still be done. For example, an α angle close to the magic angle (54°) and rotational or librational motions with high amplitudes ϕ will have a maximal influence on the ^{2}H T_{1min} times. When the D_2 ligand is the 180° jumping ligand, then the ^{2}H T_{1min} will be longest at $\alpha = 45°$. With the help of the above equations, when DQCC values are constant one can also show that the maximum possible elongation of T_{1min} times by a factor of >4, by 4, and by 2.8 is expected for free rotation, 180° jumps, and librations, respectively.

Because of the uncertainty in calculations of quadrupolar constants caused by internal motions, the reverse task, directed toward determination of internal motions by relaxation time measurements, seems to be more realistic. For example, quadrupolar parameters of ^{2}H nuclei in the dihydrogen complex Ru(D_2)Cl(dppe)$_2^+$ have been obtained independently by solid-state ^{2}H NMR spectra.[15] The NMR data collected at a temperature of 5.4 K show that the tensor of the electric field gradient in the solid complex is practically axially symmetric ($\eta = 0.1$). The principal component eq_{ZZ} deviates from the D–D bond with an α angle of 45°, and the static DQCC value has been determined to be 107 kHz.

The ^{2}H T_{1min} time in the solid complex has been measured as 0.161 s at a deuterium frequency of 61.45 MHz. Then, assuming 180° D$_2$ jumps (typical of the solid state), the ^{2}H T_{1min} value of 0.42 s can be calculated on the basis of the above quadrupolar parameters. The calculated time is remarkably longer than the 0.161 s measured experimentally. Thus, two-site jumps with a combination of rotational diffusion characterize the D$_2$ motions in the solid complex. Variable-temperature relaxation experiments on the complex in solutions have given a ^{2}H T_{1min} time of 0.047 s, much shorter than that for the solid state (0.161 s) due to whole-molecule tumbling in solutions. The quadrupolar parameters obtained by solid-state NMR spectra result in a ^{2}H T_{1min} time of 0.128 s calculated with the fast-spinning model vs. 0.047 s found experimentally. In contrast, the 180° jumping model applied for solutions produces a time of 0.032 s. This time is only slightly shorter than the experimental ^{2}H T_{1min} time. Thus, internal D$_2$ motions of Ru(D$_2$)Cl(dppe)$_2^+$ in solutions can again be described as a combination of rotational diffusion and 180° jumps. Similar logic and approaches can be used for other mobile molecular fragments and other nuclei.

6.2.3 Chemical-Shift Anisotropy Values from T_1 Data

The chemical shift anisotropies $\Delta\sigma$ depend on the local environments of nuclei and the elements of molecular symmetry. Therefore, the $\Delta\sigma$ values contributing to nuclear relaxation can potentially provide structural information, particularly in the case of heavy nuclei with very large $\Delta\sigma$ magnitudes. Generally, the chemical-shift anisotropy (CSA) relaxation mechanism for heavy nuclei, such as ^{205}Tl, ^{195}Pt, ^{207}Pb, ^{57}Fe, and ^{103}Rh, is dominant. Nevertheless, even in this case the dipole–dipole and/or spin–rotation interactions can be significant; therefore, the CSA contributions should be accurately evaluated from total relaxation rates. The evaluation is based on the dependence of the CSA relaxation on magnetic field strength in the region of fast molecular motions (see Chapter 3). This dependence (i.e., T_1 measurements performed at two magnetic fields $B^{(1)}$ and $B^{(2)}$) can result in accurate evaluation of the CSA contribution, noted as $(T_1)_{CSA}^{-1}$, in Equation 6.11:

$$(T_1)_{CSA}^{-1}(B^{(1)}) = \left\{ \left[(T_1)_{OBS}^{-1}(B^{(1)}) \right] - \left[(T_1)_{OBS}^{-1}(B^{(2)}) \right] \right\} \left[1 - (B^{(2)})/(B^{(1)})^2 \right]^{-1} \quad (6.11)$$

The ^{103}Rh relaxation in solutions of the complex Rh(COD)(tetrakis(1-pyrazole) borate) demonstrates this point.[16] According to the ^{103}Rh T_1 measurements carried out for this complex, the CSA contributions to the total relaxation rate are calculated as 92 and 97% at the magnetic fields of 7.05 and 11.75 T, respectively. Then, the $\Delta\sigma(^{103}$Rh) value can be calculated via Equation 3.19 in Chapter 3 at known correlation time τ_C. In turn, this correlation time can be found from the ^{13}C T_1 time measured for olefin carbons in the Rh complex. Because the C–H distance is equal to 1.09 Å and relaxation of the olefin carbons is dominated completely by dipole–dipole C–H interactions, the motional correlation time τ_C has been determined as 5.2×10^{-11} s. Finally, this value gives an anisotropy $\Delta\sigma(^{103}\text{Rh})$ value of 6500 ppm.

The situation is slightly complicated moving from heavy nuclei to protons due to the presence of strong dipole–dipole proton–proton interactions. In this case, the CSA contributions can be evaluated by determination of the relaxation rates in partially deuterated molecules or by nuclear Overhauser effect (NOE) measurements in addition to the variable-field experiments. Examples can be found in the recommended literature.

6.3   NMR RELAXATION AND INTERMOLECULAR INTERACTIONS

In general, the simplest intermolecular interactions can be written via Equation 6.12:

$$(^1\text{H}-A + B) \leftrightarrow (^1\text{H}-A\cdots B) \tag{6.12}$$

where, for example, ^1H is the detected nucleus and $A\cdots B$ is a molecular aggregate formed in solutions. When energies of these interactions are small, Equation 6.12 is very fast on the NMR time scale. This is the case for weak non-covalent interactions such as electrostatic attraction, π–π interactions, specific solvation, complexation, hydrogen bonding, or dihydrogen bonding. Under these conditions, the registered ^1H NMR signal shows the completely averaged spectral parameters, including the ^1H T_1 time in Equation 6.13:

$$1/T_1(^1\text{H}_{obs}) = P(A)[1/T_1(^1\text{H}-A)] + P(A\cdots B)[1/T_1(^1\text{H}-A\cdots B)P(A\cdots B)] = 1 - P(A) \tag{6.13}$$

where $P(A)$ and $P(A\cdots B)$ are the corresponding mole fractions. By definition, molecular sizes and motional inertia moments of the aggregates are larger than the initial molecules. Therefore, reorientations of the aggregates in solutions will be slower, corresponding to increased motion correlation times τ_C. This effect can be useful for revealing interactions.

6.3.1   WEAK BONDING

As discussed in Chapter 3, relaxation rates $1/T_1$ are proportional to τ_C times in the region of fast molecular tumbling when the dipolar, quadrupolar, or CSA mechanisms dominate in the nuclear relaxation. Thus, even weak intermolecular interactions leading to the formation of aggregates can be detected experimentally as the effect of reducing T_1 times. It is obvious that T_1 measurements performed for nuclei situated in molecules with smaller sizes (but not *vice versa*) will be more sensitive and will provide a good T_1 test for the presence of weak interactions in solutions.

The validity of this statement is demonstrated in Figure 6.7, which shows the intermolecular interactions between small molecules $(CF_3)_2CHOH)$ and a more massive compound $(Bu_4N)_2[B_{12}H_{12}]$ which result in formation of the

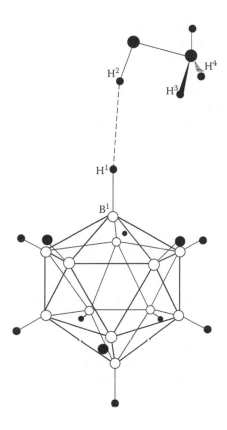

FIGURE 6.7 Weak interactions between $(CF_3)_2CHOH)$ and $(Bu_4N)_2[B_{12}H_{12}]$ in CD_2Cl_2 to form dihydrogen bonds $BH^{\delta-}\cdots^{\delta+}H$ and corresponding changes in the T_1 times accompanying this phenomenon.

$[B_{12}H_{12}]^{2-}$	$(CF_3)_2CHOH$	$[B_{12}H_{12}]^{2-} + (CF_3)_2CHOH$
^{11}B: 0.0215	^{19}F: 6.3	^{11}B: 0.0179 ^{19}F: 3.1
	^{1}HO: 5.4	^{1}HO: 1.9
	^{1}HC: 7.5	^{1}HC: 1.9

dihydrogen bonds $BH^{\delta-}\cdots^{\delta+}H$. The relaxation time measurements for a CD_2Cl_2 solution show that, at room temperature, the addition of $(CF_3)_2CHOH)$ to compound $(Bu_4N)_2[B_{12}H_{12}]$ in a 1:2 ratio results in the T_1 times being strongly reduced for ^{1}H and ^{19}F nuclei in the small alcohol molecule, while the ^{11}B T_1 time in the bulky molecule $(Bu_4N)_2[B_{12}H_{12}]$ changes insignificantly. One should remember that solutions investigated in this context should be dilute in order to avoid the effect of viscosity, which can also change when mixing reagents. The structure in Figure 6.7 shows that protons H^1 and H^2 in the dihydrogen bond are closely located. This proton–proton dipolar coupling causes an additional contribution to the total relaxation rate of HO protons. An accurate evaluation of this contribution makes determination of the H^1–H^2 bond length likely. This approach and its formalism can be found in the recommended literature.

The same idea, based on the relaxation–correlation time relationship, can be used to discover other weak non-covalent interactions in solutions; for example, the behavior of different aromatic molecules (e.g., benzene, phenol) in water solutions is of great interest for environmental chemistry. This behavior can be probed and characterized by ^2H T_1 relaxation measurements performed for deuterated molecules. Such experiments show clearly that the ^2H T_1 times of benzene-d_6 changes remarkably in the presence of humic acids. Thus, non-covalent interactions between benzene and humic acids are well established. Similarly non-covalent van der Waals and π–π interactions between fulvic acids and a β-glucosidase enzyme are detected by ^1H relaxation time measurements. It has been found that by increasing the content of fulvic acids the enzyme proton signals show reduced T_1 times, while the T_2 times increase.[17]

The direct detection of hydrogen bond formation in solutions by various spectroscopic methods is not always simple, particularly in the case of the strong, long-lived H-bonded systems. Because molecules potentially capable of H bonding can reduce their rotational reorientation rates $1/\tau_C$ in hydrogen bonding solvents, the experimental measurements of τ_C values via T_1 relaxation times can help with the detection of H-bonded complexes. Among the different nuclei, deuterium is most convenient for two reasons: (1) quadrupolar relaxation dominates completely, and (2) DQCC values for simple molecules (e.g., aromatic molecules) are well known so the ^2H T_1 times directly give τ_C values via Equation 6.1. For example, the deuterated molecule in Figure 6.8A has three ^2H labels and can be successfully used as a probe in a large set of alcohols and non-H-bonding solvents. In fact, according to ^2H T_1 relaxation time measurements, the rotational motional rates decrease in solvents capable of hydrogen bonding. Because this approach is based on ^2H NMR and registers only resonances of the probe, it can potentially be applied for studies of H bonds in supramolecular or bioorganic systems.[18]

6.3.2   ION PAIRING

Ion pairing interactions play a very important role in organic and metallorganic chemistry with regard to developing an understanding of the mechanisms of chemical reactions. For example, ion pairing interactions strongly affect the rates of proton transfers. Among the various spectroscopic methods developed for detection of ion pairing in solutions, nuclear relaxation takes a worthy place. The discussion below demonstrates the general principles behind the relaxation approach to testing ion-pairing interactions in solutions.

Figure 6.8B shows the T_1 behavior of ^{19}F nuclei in BF_4^- ions at different temperatures observed in acetone-d_6 (and CD_2Cl_2) solutions; it is unusual due to the presence and the absence of the dihydrogen complex $trans$-$[FeH(H_2)(dppe)_2]^+[BF_4^-]$.[19] It

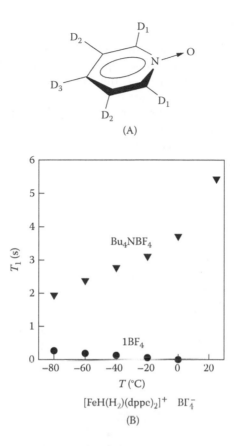

FIGURE 6.8 (A) Hydrogen-bonding probe pyridine noxide-d_5, where oxygen is a proton-accepting center (see text). (B) Temperature dependence of the ^{19}F T_1 times in an acetone-d_6 solution measured for BF_4 ions in Bu_4NBF_4 and the Fe complex noted as $1BF_4$. (From Algarra, A.G. et al., *Chem. Commun. (Camb.)*, 30, 4563, 2009. With permission.)

should be noted that the polarity of acetone assists in the dissociation necessary to form free ions. Two important features can be noted when comparing Bu_4NBF_4 and *trans*-[FeH(H$_2$)(dppe)$_2$]$^+$[BF$_4^-$]. First, the ^{19}F T_1 times of BF$_4^-$ ions in the dihydrogen complex are anomalously short, while the paramagnetic impurities that potentially shorten relaxation times are absent. Second, the ^{19}F T_1 relaxation of BF$_4^-$ ions in the compounds exhibits quite different temperature dependencies. As seen, in the absence of the dihydrogen complex, the ^{19}F T_1 time increases with increasing temperature. In contrast, the ^{19}F T_1 time decreases with increasing temperature in the dihydrogen complex *trans*-[FeH(H$_2$)(dppe)$_2$]$^+$[BF$_4^-$]. These observations strongly suggest an ion-pairing interaction for the latter case. The more accurate conclusion is that the high relaxation rate for BF$_4^-$ ions in the dihydrogen complex and the ^{19}F T_1 temperature dependence are caused by a large contribution of a spin–rotation relaxation mechanism operating in the corresponding ion pairs.

6.3.3 COMPLEXATION

The same idea, based on increasing the inertia moments and molecular motion correlation times of aggregates, can be applied for testing and studying the complexation of molecules in solutions by T_1 time measurements. Complexation of the cyclic organic molecule shown in Figure 6.9A with the massive dimeric rhodium(II) tetrakistrifluoroacetylate will obviously restrict its mobility. In turn, an increase in molecular motion correlation times can be seen experimentally as shortening T_1 times. The ^1H and ^{13}C T_1 measurements can be performed in chloroform solutions of

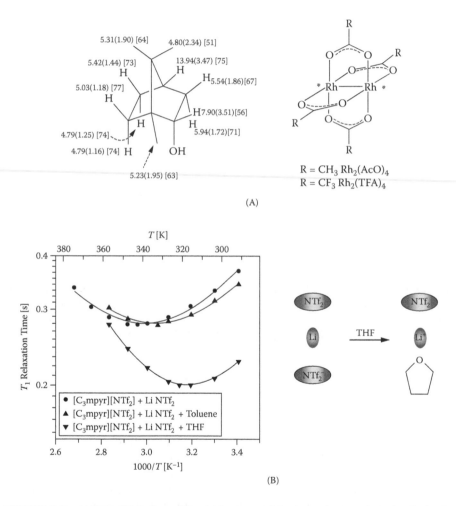

(A)

(B)

FIGURE 6.9   (A) The ^1H T_1 times (seconds) measured in the cycle organic molecule in the absence and in the presence of the Rh complex (in parentheses); the reducing T_1 effects are shown as percentages. (B) The variable-temperature ^7Li T_1 times measured for the electrolyte [C$_3$mpyr][NTf$_2$]/Li[NTf$_2$] in the absence and in the presence of toluene and THF (left); representation of interactions with participation of THF (right). (From Bayley, P.M. et al., *Phys. Chem. Chem. Phys.*, 13, 4632, 2011. With permission.)

different oxygen-containing organic molecules (alcohols, ethers, ketones, aldehydes, carboxylic acids, and esters) to test their complexation.[20] Figure 6.9A shows the ^{1}H T_1 times for the cyclic molecule measured in the absence and in the presence of two equivalents of the Rh complex. As can be seen, the ^{1}H T_1 reducing factor is very strong (between 2 and 4) for all of the protons, clearly indicating the formation of aggregates. Moreover, no ^{1}H T_1 effects were observed in the presence of rhodium tetraacetate, which is not capable of complexation.

In some cases, the simple T_1 test shown above is not sufficient to find intermolecular interactions in solutions or liquids such as complexation or intermolecular coordination. This situation occurs, for example, during studies and applications of ionic liquid (IL) electrolytes, such as (IL)/Li[NTf$_2$] in lithium batteries, which require detailed knowledge of how and why the different organic contrasting solvent additives affect the transport properties and conductivity in these systems.[21] The problem can be solved only on the basis of variable-temperature NMR experiments to observe the full T_1 temperature curves and T_{1min} times.

It has been demonstrated, for example, that toluene and tetrahydrofuran (THF) increase the conductivity in (IL)/Li[NTf$_2$] systems at low temperatures. However, it was unclear why THF acts more effectively. The effect could be explained by more pronounced THF–lithium ion interactions, a suggestion that has been tested by measurement of the ^{7}Li T_1 times to demonstrate the structuring in the electrolytes and the influence of THF and toluene on lithium coordination within the systems.[21] It has been established that room-temperature ^{7}Li T_1 measurements do not give a single-valued answer; however, a clear difference between toluene and THF can be observed in the variable-temperature relaxation data shown in Figure 6.9B, particularly in the ^{7}Li T_{1min} zone. As seen from the relaxation curves, the ^{7}Li T_1 minimum is strongly shifted toward low temperatures in the presence of THF. This result shows clearly that toluene does not interact with the lithium ions but THF does, as shown in the figure.

6.4 SOLVENT RELAXATION

Solvent relaxation is a specific field of NMR applied for the study of colloidal systems representing combinations of polymers, surfactants, and colloidal particles. The focus of such research is on relaxation time measurements performed for molecules of solvents, with the goal of distinguishing between molecules of solvents located at the surface (and interacting with the surface) and those moving in the bulk solution. Such discrimination opens the way to studying how interfacial interactions and interfacial structure affect molecular mobility, which can be characterized by motion correlation times. For example, the mobility of free water and that of water bonded to polymer chains are quite different. In water–agar systems, the difference in H$_2$O motional correlation times is 0.4×10^{-8} s for water associated with the agar, ~10^{-6} s for bonded water, and ~10^{-12} s for free water molecules.

Incorporation of paramagnetic metal ions into surfactants creates additional possibilities because the solvent relaxation rates will be enhanced due to dipolar interactions between the solvent and the paramagnetic species. In turn, this enhancement effect will determine the location of the surfactant in colloidal systems.

Both T_1 and T_2 measurements can be performed in such systems to obtain characteristic correlation times. One should remember that in accordance with the relationships between molecular motion correlation times and spin–lattice and spin–spin relaxation times, solvent relaxation in colloidal systems always corresponds to the conditions where the T_1 and T_2 times behave similarly.

The strategy in the solvent relaxation studies can be represented by ^1H T_2 measurements performed for polymers in aqueous solutions and gels with the CPMG pulse sequence. Protons of the water in these systems generally experience an exchange with OH$^-$ protons in the polymer molecules. Then, the measured relaxation can be treated as follows:

$$1/T_2 = \left[\left(1 - p_C \right) / T_{2f} \right] + \left[p_C / \left(T_{2C} + \tau(ex) \right) \right] \tag{6.14}$$

where p_C is the number of the exchanging protons related to the total number of detectable protons in the system, T_{2f} is the relaxation time of free water, T_{2C} is the relaxation time of exchangeable protons, and $\tau(ex)$ is the exchange correlation time. The $\tau(ex)$ and p_C magnitudes can be calculated using the composition of the polymer gel and the content of water. As follows from Equation 6.14, non-exchangeable protons in polymer systems are excluded from the model. This is valid because the relaxation times of these protons are generally too short to be measured under experimental conditions. Finally, the important element of this model is the absence of a remarkable difference in the chemical shifts of water protons and hydroxyl protons. In other words, the relaxation rate of the exchangeable protons is much larger than the chemical shift difference expressed in frequency units.

In practice, T_2 measurements are generally carried out for water signals using a variety of polymer concentrations.[22] According to the model, when the $\tau(ex)$ value becomes independent of the polymer concentration, the measured relaxation rate $1/T_2$ will increase linearly with increasing concentration. At the same time, a deviation from linearity can occur when the mobility of the hydrogen-bonded water changes due to volume phase transitions. This feature can serve as a good relaxation test for the presence of phase transitions in systems under investigation.

Figure 6.10 shows the results of T_2 relaxation experiments performed on a poly(N-isopropylacrylamide) (PNIPAM) microgel dispersion where the diameter of the particles (noted in the Debye equation, Equation 3.2 in Chapter 3) obtained by independent hydrodynamic measurements; the T_2 times were measured for pure water and for water in the microgel.[22] The curves show that the T_2 time in the microgel dispersion (in contrast to pure water) changes dramatically in the region of the temperature that corresponds to a volume phase transition in full accordance with hydrodynamic data. According to the data in Figure 6.10, the ^1H T_2 time in the microgel dispersion increases with temperature (between 20° and 30°C), similar to what occurs with pure water only below the PNIPAM critical solution temperature. Then, in the region of ~32°C, the T_2 values reduce strongly due to a restriction in water dynamics caused by intermolecular interactions with the polymer chains. It should be added that the solvent relaxation NMR techniques can be also applied for polymers in non-aqueous systems and polymer melts.

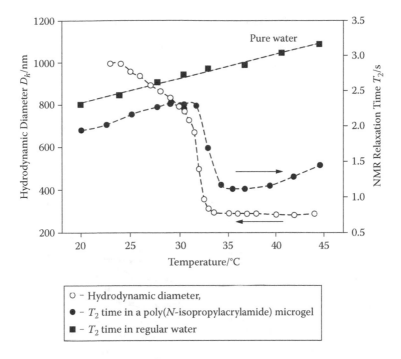

FIGURE 6.10 Temperature dependence of relaxation time T_2 of water molecules measured by the CPMG pulse sequence in a PNIPAM microgel dispersion and in pure water. (From Cooper, C.L. et al., *Soft Matter*, 9, 7211, 2013. With permission.)

As mentioned above, solvent relaxation in colloidal systems generally shows similar T_1 and T_2 times. The situation can change in non-aqueous solutions, and the experimental ^1H T_1/T_2 ratios can become too large to manifest, an unusual relaxation and dynamic behavior of solvent molecules. The large T_1/T_2 ratios are measured, for example, in the concentrated benzene solutions of isotactic poly(methyl methacrylate).[22] This is a good test for the motional anisotropy of solvent molecules. Anisotropic motions of benzene originate from a solvent association with polymer molecules. They reduce T_2 times, but T_1 times do not respond to changes in the polymer molecular weight.

6.5 RELAXATION IN MOLECULAR SYSTEMS WITH CHEMICAL EXCHANGES

The general principles of T_1 time behavior in the presence of slow chemical exchanges in solutions were considered in Chapter 4. It should be emphasized again that, in combination with saturation transfer experiments, relaxation is a unique tool for establishing a slow exchange in NMR spectra where resonance lines are sharp and show natural line widths. This section provides several examples of relaxation applications, including the measurement of selective relaxation times and spin-locking experiments to determine the type of problems solved in the presence of chemical

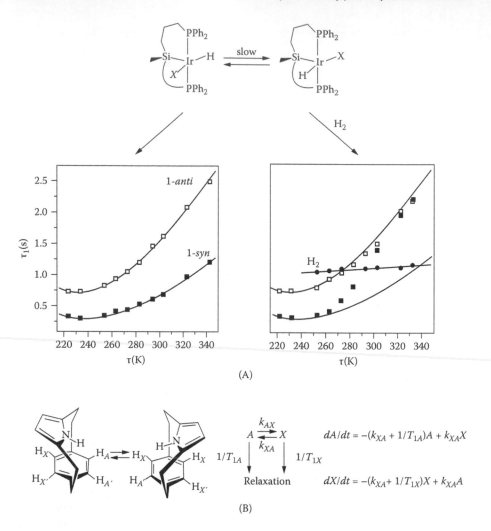

FIGURE 6.11 (A) Variable temperature ^{1}H T_1 relaxation data collected for two hydride resonances of the hydride Ir complex at X = Cl in toluene-d_8 in the absence (left) and in the presence (right) of H_2. (From Sola, E. et al., *J. Am. Chem. Soc.*, 132, 9111, 2010. With permission.) (B) Conversion of the cycle molecule in solutions and the kinetic scheme of the process. (From Dahlquist, F.W. et al., *J. Magn. Reson.*, 17, 406, 1975. With permission.)

exchange. The *anti/syn* forms of the Ir monohydride complex shown in Figure 6.11A are separately observed in room-temperature ^{1}H NMR spectra at *anti/syn* ratios of 93:7 or 83:17 in benzene-d_6 or CD_2Cl_2, respectively.[23] The resonances are sharp and show the natural line width. In accordance with the structural features of forms with different environments of ^{1}H–Ir nuclei, the ^{1}H T_1 times of the hydride resonances are remarkably different even at 340 K. In addition, the T_1 temperature dependencies are similar in accordance with the same inertia moments of both forms. Thus, no *anti/syn* transformation is observed on the ^{1}H NMR time scale. The situation changes in the presence of molecular dihydrogen. Under a H_2 atmosphere, the relaxation curves demonstrate agreement of $^1H\ T_1$ times at temperatures greater than 260 K. This tendency reveals clearly *anti/syn* transformation on the T_1 time scale that is accelerated under a H_2 atmosphere. However, it is interesting that the $^1H\ T_1$ time of the H_2 signal increases with temperature only slightly due to the small molecular size. Thus, a Ir–H/H_2 exchange is not involved into the *anti/syn* transformation.

In addition to the regular (non-selective) relaxation times, slow chemical exchanges can be quantitatively characterized by the measurement of selective T_1 times. This method is particularly important for solutions of small molecular systems where T_1 times are long. If, for nuclei A and X potentially experiencing a slow exchange, the first selective 180° pulse in the inversion–recovery sequence (see Chapter 3) excites only nuclei A, then evolution of integral intensities $I(t)$ due to the exchange during delay time t between the selective 180° pulse and non-selective 90° pulse can be written as a function of time t in Equation 6.15:

$$\left\{[I_0A - IA(t)] - [I_0X - IX(t)]\right\} \big/ \left\{[I_0A - IA(t)] + [I_0X - IX(t)]\right\} = \exp\left(-2t/\tau_A\right) \quad (6.15)$$

to calculate the lifetime τ_A.

This method has been used to probe the conversion of cycle organic molecules in solutions that follow the kinetic scheme shown in Figure 6.11B.[24] The conversion rates k_{AX} and k_{XA} in the kinetic scheme describing the process are equal. The relaxation times T_{1A} and T_{1X} are also equal (or very similar). Thus, magnetization as a function of time at sites A and X can be given by the sum of the two exponentials. This is typical when the two processes of nuclear relaxation and chemical exchange control magnetization tending to equilibrate after excitation.

By definition, chemical exchanges in solutions occurring at exchange frequencies in a diapason of tens of kHz can affect only $T_{1\rho}$ times (but not T_1 times). In fact, radiofrequency field B_1, applied to a sample as a long radiofrequency pulse, is weaker than external magnetic field B_0 by several orders. As a result, in the presence of the exchanges, the $T_{1\rho}$ and T_1 times will not be equal. Therefore, these exchanges can potentially be characterized by spin-locking experiments directed toward $T_{1\rho}$ measurements.

It should be emphasized that exchange frequencies of tens of kHz are generally sufficient to average completely the resonances in NMR spectra experiencing the exchange. Therefore, an experimental result of $T_{1\rho} < T_1$ obtained for an averaged resonance by T_1 and T_{1r} relaxation measurements is a good test for the presence of a fast chemical exchange. The exchange relaxation contribution T_{1EXCH} for two types of exchanging nuclei A and X can be expressed as

$$(T_{1EXCH})^{-1} = (T_{1\rho})^{-1} - (T_1)^{-1} \quad (6.16)$$

In turn, the rates of exchange $1/\tau$ (where τ is the lifetime of A or X) can be obtained by a variation in the spin-locking field power ($\omega_1 = \gamma B_1$):

$$T_{1EXCH} = \left[(1/2)\pi^2 \left(v^A - v^X\right)^2 \tau\right]^{-1} + \tau\omega_1^2 \left(2\pi^2 \left(v^A - v^X\right)^2\right)^{-1} \quad (6.17)$$

where $\tau = \tau_A = \tau_X$. The latter corresponds to equally populated spin states A and X.

The results of these $T_1/T_{1\rho}$ measurements can be represented as plots of the T_{1EXCH} vs. ω_1^2 to give the τ values and the chemical shift differences ($v^A - v^X$). This is a unique feature of the spin-lock experiments because the ($v^A - v^X$) differences are not detectable in regular NMR spectra due to the fast exchange.

6.6 STRUCTURAL ASPECTS OBTAINED FROM RELAXATION IN SOLUTIONS

X-ray diffraction crystallography (or neuron diffraction) is considered the most valuable tool for the structural characterization of chemical compounds ranging from small organic or inorganic molecules to very complex biochemical molecular systems. Nevertheless, its application is limited for crystalline solids (large crystals in the case of neuron diffraction), and many compounds are amorphous and heterogeneous. In such cases, the major role belongs to solid-state NMR, the structural applications of which are considered in Chapter 10. However, many problems in chemistry are related to compounds that exist only in solutions, or the compound under investigation should be structurally characterized in solutions because the structures can be different in solutions and the solid state. Because the rates of dipole–dipole nuclear relaxation, $1/T_1$, are inversely proportional to the sixth power of internuclear distances, they are an important tool for structural formulation in solutions. This section considers some relaxation approaches and examples of their application to demonstrate how to perform such experiments and how to treat the collected data to account for molecular motions.

6.6.1 ^{1}H T_1 CRITERION IN STRUCTURAL ASSIGNMENTS

Protons experience maximal dipolar coupling, and their natural abundance is 100%. Both of these circumstances explain the wide application of ^{1}H T_1 times in structural formulations or even as reliable structural criteria. Two cases, inorganic and organic molecules, can be used as practical examples. Transition metal hydrides can exist in two structural forms: classical or non-classical (or dihydrogen) complexes, as shown in Figure 6.12.A. Due to chemical bonding, H$\cdots$H distances in dihydrogen complexes can be <1 Å vs. >1.5 Å in classical forms.[25] This difference is clearly seen in ^{1}H T_1 relaxation times measured in solutions, where dihydrogen complexes show extremely short T_1 times, particularly T_{1min} values. The molecular masses of the complexes shown are similar, and the difference in their molecular motion correlation times, τ_C, is minimal. Therefore, the T_1 shortening effect is easily observed even at room temperature.

Two iridium hydride complexes, $IrH(H_2)Cl_2(PCy_3)_2$ and $IrH_5(PCy_3)_2$, with hydride environments (see Figure 6.12A) demonstrate how this T_1 effect works. The $IrH(H_2)Cl_2(PCy_3)_2$ complex exhibits two ^{1}H resonances, assigned to the H$_2$ and H ligands, while the H ligands in the $IrH_5(PCy_3)_2$ complex are equivalent, corresponding to a single ^{1}H resonance. The dihydrogen ligand in the $IrH(H_2)Cl_2(PCy_3)_2$ complex is *trans*-located to the closest hydride atom. This distance is too large to provide a significant dipolar contribution to the total ^{1}H $T_1(H_2)$ relaxation time. Each hydride ligand in the $IrH_5(PCy_3)_2$ complex neighbors two closely located hydride atoms. Nevertheless, the

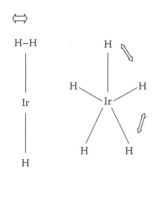

The ^{1}H T_1 times measured for hydrogen ligand signals at 250 MHz in solutions.

Hydride	T_1(ms)	Solvent/Temperature
$IrH_5(PCy_3)_2$	820	CD_2Cl_2, 193 K
$H_2Fe(CO)_4$	3000	Toluene-d_8, 203 K
$W(H_2)(CO)_3(PPr^i_3)_2$	5*	Toluene-d_8
$trans$-$IrH(H_2)Cl_2(PCy_3)_2$	4.2*	Toluene-d_8, 235 K
$[IrH(H_2)(bq)(PPh_3)_2]^+$	8*	CD_2Cl_2, 200 K

* T_{1min}

(A)

The ^{1}H T_{1min} times as a function of the r(H–H) distance immobile (IM) and fast spinning (FS) dihydrogen ligands

r(H–H) Å	T_{1min} (ms)	T_{1min} (ms)
	IM	FS
0.75	1.1	4.4
1.0	6.5	26
1.1	11.5	45.8
1.2	19.3	77.2
1.3	31.2	125
1.35	39	157
1.65	130	—

(B)

FIGURE 6.12 (A) Classical and non-classical hydride ligands in transition metal complexes and their ^{1}H T_1 times. (B) The ^{1}H T_{1min} time calculated at a working frequency of 250 MHz for immobile and fast-spinning H_2 units.

$IrH_5(PCy_3)_2$ signal shows a long ^{1}H T_1 time of 820 ms vs. the extremely short ^{1}H T_1 time of 4.2 ms measured for the H_2 signal of $IrH(H_2)Cl_2(PCy_3)_2$. Because the dipolar contacts between hydride atoms and protons located in phosphorus ligands are negligible, the relaxation data allow formulation of the empirical criterion that ^{1}H T_1(250 MHz) times of >150 ms correspond to classical transition metal hydrides, while ^{1}H T_1(250 MHz) times of <80 ms show their non-classical nature.

The reliability of this empirical rule and the reverse task, aimed at determination of r(H–H) values in the H_2 units by ^1H T_1 times, deserve special considerations. The molecular motion correlation times, τ_C, are known to affect the T_1 times; however, they can be excluded by measurement of the T_{1min} values. Then, Equation 6.18[26] can be written for proton relaxation, where r(H–H) is the distance measured in Å and ν is the working frequency (expressed in MHz):

$$T_{1min}(\text{H–H}) = [(r(\text{H–H})/5.815]^6 \nu \qquad (6.18)$$

This equation is convenient for r(H–H) calculations. However, dihydrogen ligands can experience a fast internal rotation (see Figure 4.10A in Chapter 4) on the time scale of whole molecular tumbling. Thus, H_2 motions are anisotropic. According to relaxation theory, any anisotropic molecular reorientations strongly change T_1 times measured experimentally. As mentioned in Chapter 4, fast H_2 rotation can lead to a fourfold ^1H T_{1min} elongation. This fact should be taken into account when calculating ^1H T_{1min} times at frequency ν of 250 MHz, for example, for immobile and fast-spinning proton pairs, where the r(H–H) distance changes between 0.75 and 1.65 Å (Figure 6.12B). The distance of 0.75 Å is close to that in molecular hydrogen, whereas r(H–H) = 1.65 Å is generally accepted for the classical dihydride structure. The data show that the fast-spinning elongated dihydrogen ligands (r(H–H) = 1.30 to 1.35 Å) and the classical hydride structure are not distinguishable. However, even the fast-spinning dihydrogen ligands at short H–H distances of ≤1 Å can be reliably identified by ^1H T_{1min} measurements in solutions. The same idea can be applied for other nuclei, such as ^{19}F, and other molecular fragments.

The considerations and modeling estimations discussed above show that the reverse task, calculating r(H–H) from ^1H T_{1min} times for any proton pairs or other nuclei, will suffer from uncertainties when the motional model is not established independently. The effects are clearly seen in Figure 6.12B, where, for example, the value of 1.2 Å corresponds to the T_{1min} time of 19.3 ms measured in an immobile proton pair and 77.2 ms in a fast-spinning pair. In addition, details regarding the smaller effects of low-amplitude libration motions can be found in Reference 14 and in the recommended literature.

Figure 6.13 illustrates T_1 times used as structural criteria in studies of organic molecules capable of fast transformations in solutions. 2-Guanidinobenzimidazole can experience the very fast tautomeric conversions and fast intra- and intermolecular proton exchanges. It is obvious that the structural formulation of this compound will be not precise in the absence of recognition and description of these processes.

Generally speaking, this problem is not simple because the above transformations represent low-energy equilibriums. In addition, in combination with the fast exchanges, they produce an effectively symmetrical molecule. Nevertheless, measurements of ^{13}C and ^1H T_1 times in different media can help with structural assignments.[7] In fact, the intramolecular hydrogen bond NHH···N(3) leads to a planar molecular structure characterized by the ^{13}C and ^1H T_1 relaxation data in DMSO. In the presence of sulfuric acid, this hydrogen bond can be broken due to the formation of a conjugated acid (Figure 6.13D,E), which, in turn, forms intermolecular bonds with the acid. As a result, the structure becomes non-planar, the rotation of the

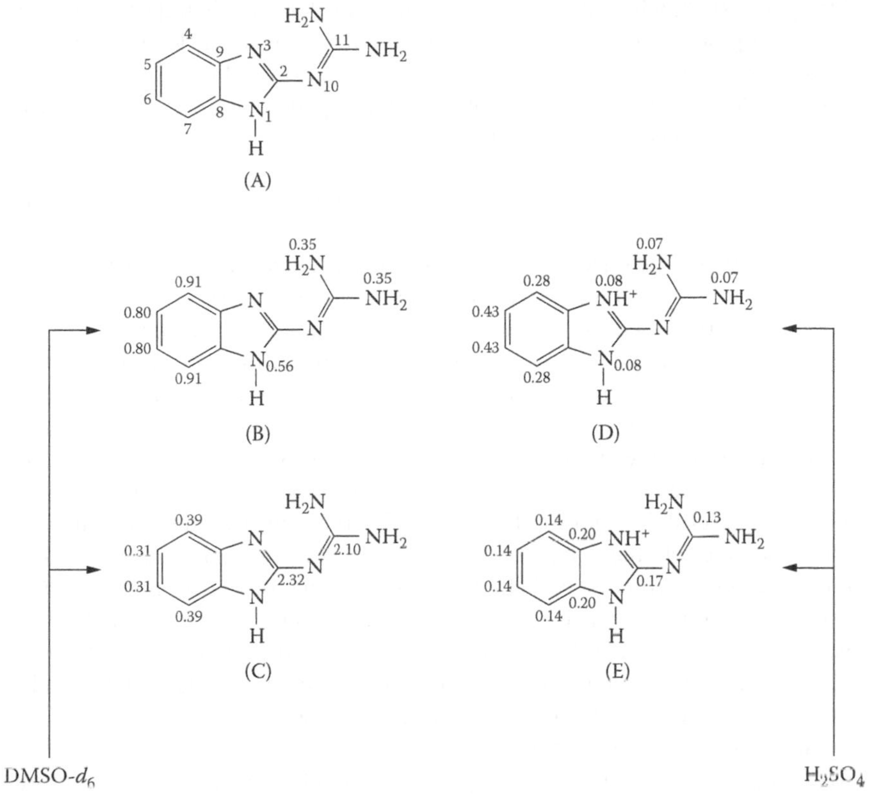

FIGURE 6.13 (A) 2-Guanidinobenzimidazole and (B–E) its structures and their T_1 time characterization: ^{13}C T_1 times (C, E) and ^1H T_1 times (B, D). (From Ariza-Castolo, A., *Concepts Magn. Reson. A*, 32A, 168, 2008. With permission.)

C_2–N_{10} bond is hindered due to steric strain, and the relaxation rates of both nuclei (^1H and ^{13}C) increase remarkably. This follows from the data obtained in DMSO and H_2SO_4.

6.6.2 INTERATOMIC DISTANCES FROM T_1 DATA IN SOLUTIONS

According to the theory of nuclear relaxation applied for any pair of nuclei, their T_1 (or T_2) times, controlled by hetero- or homonuclear dipolar coupling, allow direct determination of interatomic distances (or bond lengths) in solutions. To do so accurately, four conditions are necessary: (1) nuclear relaxation should be mono-exponential; (2) the dipolar contributions to total relaxation rates, measured experimentally, should be precisely evaluated; (3) the molecular motion correlation times, τ_C, should be found independently, or the T_{1min} times, independent of the τ_C values, should be measured; and (4) molecular reorientations should be isotropic or the types of anisotropic motions should be well defined. The methodology and strategy in such studies are represented in this section by the determinations of interatomic distances for nuclear pairs $^1H–X$, where protons are the target nuclei. This is important because, due to the largest γ value and 100% natural abundance of ^1H nuclei, the duration of the relaxation experiments is not long. This point is also valid for ^{19}F nuclei.

It is easy to show that dipolar coupling in $^1H–X$ pairs at constant H–X distances decreases proportionally to coefficient k in Equation 6.19:

$$k = 0.44(\nu_X/\nu_H)^2 I_X(I_X + 1) \tag{6.19}$$

where ν_H and ν_X are the Larmor frequencies of H and X nuclei, respectively, and I_X is the spin of X nuclei. According to the equation, accurate determination of H–X distances by ^1H T_1 times will be problematic and unreliable in, for example, $^1H–^{15}N$ or $^1H–^{103}Rh$ pairs due to the very weak dipolar interactions. In contrast, proton–boron dipolar coupling will be stronger, and ^1H or even ^{11}B T_1 measurements can be more successful in this context.

Again, due to the low natural abundance of nuclei ^{13}C and ^{15}N, the choice of ^1H nuclei as targets for the determination of distances in the pairs $^1H–^{13}C$ and $^1H–^{15}N$ will not prove useful. In contrast, the ^{13}C or ^{15}N T_1 times measured for the same pairs can indicate precisely the corresponding distances. For the same reasons, determination of carbon–carbon or carbon–nitrogen distances obviously requires ^{13}C- or ^{15}N-labeled compounds, particularly in the case of complex systems such as peptides. At the same time, non-labeled biomolecules modified by paramagnetic probes can be useful for accurate calculations of distances between target nuclei and the paramagnetic centers due to strong paramagnetic relaxation enhancements. These experiments will be considered in Chapter 7.

Referring back to the discussion on protons, it should be emphasized that in some cases even weak dipolar coupling (e.g., in ^1H and ^{195}Pt pairs) can accurately indicate bond lengths (e.g., H–Pt). The ^1H NMR spectra of such compounds will exhibit the central resonance and two satellite lines corresponding to the non-magnetic and magnetic platinum isotopes, respectively. Under this condition, the $^1H–^{195}Pt$ dipolar contributions can be well evaluated by ^1H T_1 measurements performed for the central line and the satellites. In fact, the difference $1/T_1(sat) – 1/T_1(central)$ directly gives this contribution. Finally, the large spin numbers of many nuclei (e.g., ^{93}Nb, ^{51}V, ^{55}Mn, ^{187}Re, ^{59}Co, ^{181}Ta) can compensate for their low γ values; thus, the H–X distances can be determined by ^1H T_1 relaxation measurements in solutions after accurate evaluations of the corresponding dipolar contributions.

Different methods and procedures can be applied for distance determinations in solutions that are based on the ^1H T_1, ^1H T_{1sel}, ^1H T_{1bis}, and ^1H T_{1min} measurements (or their combinations) and NOE experiments. The approaches and technical details can be found in the recommended literature; however, one of the methodologies is considered in this section to demonstrate in practice that the H–X distances can be determined in solutions at an accuracy even better than that of x-ray analysis in the solid state.

In general, the contributions of metal hydride dipolar interactions to total ($^1H–$ metal) T_1 relaxation rates in molecules, shown in Figure 6.12A, can be found by experiments on compounds with deuterated phosphorus ligands. However, without deuteration, they can be accurately evaluated by non-selective (^{1}H T_1) and selective (^{1}H T_{1sel}) relaxation measurements performed for hydride signals. Then, due to the additivity of the relaxation rates, Equations 6.20 and 6.21 can be written as

$$1/T_1 = 1/T_1(\text{H–}M) + 1/T_1(\text{H}\cdots\text{H}) \tag{6.20}$$

$$1/T_{1sel} = 1/T_1(\text{H–}M) + 1/T_{1sel}(\text{H}\cdots\text{H}) \tag{6.21}$$

where the $1/T_1(\text{H–}M)$ and $1/T_1(\text{H}\cdots\text{H})$ relaxation rates correspond to metal hydride dipolar coupling and dipolar interactions between the hydride ligand and all of the protons in a molecule, respectively. Following relaxation theory, the hydride–proton and metal hydride contributions can be expressed via Equations 6.22 to 6.24:

$$1/T_1\,(\text{H}\cdots\text{H}) = 0.3\gamma_{\text{H}}^4\hbar^2 r(\text{H}\cdots\text{H})^{-6}\left[\left(\tau_C/\left(1+\omega_{\text{H}}^2\tau_C^2\right)\right)+\left(4\tau_C/\left(1+4\omega_{\text{H}}^2\tau_C^2\right)\right)\right] \tag{6.22}$$

$$1/T_{1sel}\,(\text{H}\cdots\text{H}) = 0.3\gamma_{\text{H}}^4\hbar^2 r(\text{H}\cdots\text{H})^{-6}\left[\begin{array}{l}\left(\tau_C/\left(1+\omega_{\text{H}}^2\tau_C^2\right)\right)\\ +\left(2\tau_C/\left(1+4\omega_{\text{H}}^2\tau_C^2\right)\right)+\tau_C/3\end{array}\right] \tag{6.23}$$

$$1/T_1\,(\text{M–H}) = (2/15)\gamma_{\text{H}}^2\gamma_{\text{M}}^2\hbar^2 I(I+1)r(\text{M–H})^{-6} \tag{6.24}$$
$$\times\left[\begin{array}{l}\left(3\tau_C/\left(1+\omega_{\text{H}}^2\tau_C^2\right)\right)+\left(6\tau_C/\left(1+\left(\omega_{\text{H}}+\omega_{\text{M}}\right)^2\tau_C^2\right)\right)\\ +\left(\tau_C/\left(1+\left(\omega_{\text{H}}-\omega_{\text{M}}\right)^2\tau_C^2\right)\right)\end{array}\right]$$

When the T_1 and T_{1sel} time measurements are performed in a high-temperature zone (i.e., at $\omega_{\text{H}}^2\tau_C^2 \ll 1$), then a new constant, k, in Equation 6.25, can be introduced to express the relationship between the proton–hydride and metal hydride relaxation contributions, where $f = T_{1sel}/T_1$:[27]

$$k = \left[0.3\gamma_{\text{H}}^4\hbar^2 r(\text{H}\cdots\text{H})^{-6}\right]\Big/\left[(2/15)\gamma_{\text{H}}^2\gamma_{\text{M}}^2\hbar^2 I(I+1)r(\text{M–H})^{-6}\right] \tag{6.25}$$
$$= (f-1)/(0.5-(f/3))$$

Finally, combining these equations leads to Equation 6.26, which is appropriate for distance calculations:

$$1/T_1 = (4/30)r(\text{M–H})^{-6}\gamma_{\text{H}}^2\gamma_{\text{M}}^2\hbar^2 I(I+1) \tag{6.26}$$
$$\times\left[\begin{array}{l}\left((3+k)\tau_C/\left(1+\omega_{\text{H}}^2\tau_C^2\right)\right)+\left(4k\tau_C/\left(1+4\omega_{\text{H}}^2\tau_C^2\right)\right)\\ +\left(6\tau_C/\left(1+\left(\omega_{\text{H}}+\omega_{\text{M}}\right)^2\right)\tau_C^2\right)+\left(\tau_C/\left(1+\left(\omega_{\text{H}}-\omega_{\text{M}}\right)^2\tau_C^2\right)\right)\end{array}\right]$$

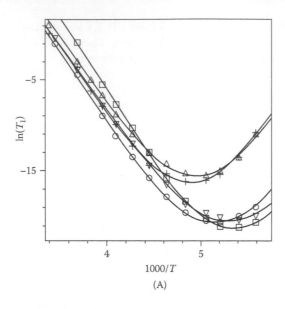

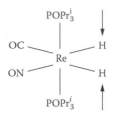

200 MHz: E_a = 3.0 kcal/mol,
τ_0 = 2.0 10^{-13} s, r(Re–H) = 1.70 ± 0.08 Å
300 MHz: E_a = 3.2 kcal/mol,
τ_0 = 1.6 10^{-13} s, r(Re–H) = 1.71 ± 0.03 Å

200 MHz: E_a = 3.2 kcal/mol,
τ_0 = 1.5 10^{-13} s, r(Re–H) = 1.72 ± 0.05 Å
300 MHz: E_a = 3.3 kcal/mol,
τ_0 = 1.4 10^{-13} s, r(Re–H) = 1.69 ± 0.03 Å

(A)   (B)

FIGURE 6.14 (A) Variable-temperature ^{1}H T_1 data for hydride resonances in solutions of ReH$_2$(CO)(NO)(POPr$^i_3$)$_2$. 200 MHz: O, ReH with δ = –1.51 ppm; ∇, ReH with δ = –4.78 ppm. 300 MHz: +, ReH with δ = –1.51 ppm, △, ReH with δ = –4.78 ppm. (From Gusev, D.G. et al., *Inorg. Chem.*, 32, 3270, 1993. With permission.) (B) Data obtained by fitting procedures.

In the methodology context, this theory implies the ^{1}H T_{1sel} times to be measured only in the high-temperature region, whereas the full variable-temperature ^{1}H T_1 curves obtained experimentally are fitted to Equation 6.26. It is obvious that a good fit results in metal hydride distance r(M–H), activation energy E_a of the molecular reorientations, and the correlation constant τ_0.

Figure 6.14A shows the fitted experimental data obtained in a toluene-d_8 solution of the transition metal hydride complex ReH$_2$(CO)(NO)(POPr$^i_3$)$_2$.[27] The agreement between the theory and the experimental data is very good, and the results of the treatments give the Re–H bond length and the E_a and τ_0 values, which are practically identical for experiments performed at frequencies of 200 and 300 MHz (Figure 6.14B).

When a ^{1}H T_{1min} time is well determined by variable-temperature relaxation experiments and the ^{1}H T_{1sel} values have been accurately measured again in a high-temperature zone, then computer fitting procedures are not obligatory and Equation 6.27 can be applied for determination of the bond length:

$$r(\text{M–H}) = C\left[(1.4k + 4.47)\left(T_{1min}/\nu\right)^{1/6}\right]$$ (6.27)

$$C = 10^7\left(\gamma_H^2\gamma_M^2\hbar^2 I(I+1)/15\pi\right)^{1/6}$$

In this equation, r(M–H), T_{1min}, and ν are measured in Å, seconds, and MHz, respectively, and the dipolar constant C for nuclei Re, Mn, Nb, and Ta, for example, is calculated as 4.20, 4.31, 5.11, and 3.74, respectively.

In practice, two factors restrict application of the ^1H T_{1sel}/T_{1min} method. The inverting 180° pulse should actually be selective. This is particularly important in the case of ^1H NMR spectra, where, for example, two resonances are closely located but only one of them would be excited. Crude estimation leads to identification of 25 to 30 Hz as the minimal chemical shift difference, expressed in Hz, at which the acting pulse remains still selective. The second restriction is connected with the nature of molecules involved in the experiments: the T_{1sel}/T_{1min} approach cannot be applied when a resonance, selectively excited, participates in a slow chemical exchange.

A similar methodology can be used to measure proton–proton distances in solutions. The approach to evaluation of dipole–dipole proton–proton contributions is based on measurement of the ^1H T_1, ^1H T_{1sel}, and ^1H T_{1bis} times to obtain the differences between the $1/T_{1sel}$ and $1/T_{1bis}$ rates.[27,28] These differences give the so-called cross-relaxation rate, σ_{ij}, in Equation 6.28:

$$\sigma_{ij} = \left(1/T_{1sel} - T_{1bis}\right)_i = \left(1/T_{1sel} - 1/T_{1bis}\right)_j \tag{6.28}$$

$$\sigma_{ij} = 0.1\left(\mu_0/4\pi\right)^2 \gamma_H^4 \hbar^2 r\left(H_i \cdots H_j\right)^{-6} \left(6\tau_C/\left(1+4\omega_H^2\tau_C^2\right) - \tau_C\right)$$

where i protons are coupled by j protons. It is obvious that distances $H_i \cdots H_j$ can be calculated at the known molecular motion correlation times, τ_C.

Sometimes CSA relaxation masks the dipolar contribution necessary for calculating interatomic distances. Figure 6.15 shows such an example, where the hydride ligands in the osmium cluster relax via the dipolar and CSA interactions.[29] Because

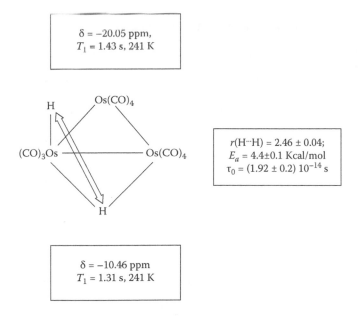

FIGURE 6.15 Hydride–hydride distance and characteristics of molecular reorientations obtained for the Os carbonyl cluster in solutions.[29]

protons other than hydrides in the molecule are absent, both of the relaxation mechanisms provide the corresponding contributions to the total ^{1}H T_1 relaxation rates measured for the H–Os signals in Equation 6.29:

$$1\big/T_1^{TOT} = 0.3\left(\mu_0/4\pi\right)^2 \gamma_H^4\,\hbar^2 r(\mathrm{H}\cdots\mathrm{H})^{-6} \tag{6.29}$$

$$\times \left[\begin{array}{l}\left(\tau_C\big/\left(1+\omega_H^2\tau_C^2\right)\right)+\left(4\tau_C\big/\left(1+4\omega_H^2\tau_C^2\right)\right) \\ +(2/15)\gamma_H^2 B_0^2(\Delta\sigma)^2\left(\tau_C\big/\left(1+\omega_H^2\tau_C^2\right)\right)\end{array}\right]$$

Because the CSA relaxation rate is proportional to the magnetic field strength (B_0^2), one can expect that the CSA mechanism will actually be effective for the highest magnetic fields because proton $\Delta\sigma$ values are generally small, which leads to the following methodology for the measurement of proton–proton distances in solutions.

At a low magnetic field corresponding, for example, to the ^{1}H Larmor frequency of 90 MHz, the $r(\mathrm{H}\cdots\mathrm{H})$, E_a, and τ_0 values can be calculated by fitting the variable-temperature ^{1}H T_1 measurements to the first contribution in Equation 6.29. Then, these values (found and represented in Figure 6.15) are applied for computer fitting of the variable-temperature ^{1}H T_1 data collected at a high magnetic field (e.g., 400 MHz) to Equation 6.29, giving the $\Delta\sigma$ values. For example, $\Delta\sigma$ values of 22.6 ± 2.0 and 20.0 ± 1.9 ppm are calculated for bridging and terminal hydride ligands, respectively. As seen, the errors in such determinations are quite satisfactory.

A general analysis of the errors in measurements and quantitative interpretations of relaxation times in solutions can be found in the recommended literature.

REFERENCES AND RECOMMENDED LITERATURE

1. Mann, M.D. and Chmelka, B.F. (2000). *Anal. Chem.*, 72: 5131.
2. Bakhmutov, V.I., Vorontsov, E.V., and Antonov, D.Y. (1998). *Inorg. Chim. Acta*, 278: 122.
3. Doddrell, D. and Allerhand, A. (1971). *Proc. Nat. Acad. Sci. USA*, 68: 1083.
4. Gusev, D.G., Vymenits, A.B., and Bakhmutov, V.I. (1991). *Inorg. Chim. Acta*, 179: 195.
5. Rong, Q., Espanol, M., de Freitas, D.M., and Geraldes, C.F.G.C. (1993). *Biochemistry*, 32: 13490.
6. Allerhand, A., Clouse, A.O., Rietz, R.R., Roseberry, T., and Schaeffer, R. (1972). *J. Am. Chem. Soc.*, 94: 7.
7. Ariza-Castolo, A. (2008). *Concepts Magn. Reson. A*, 32A: 168.
8. Bakhmutov, V.I. (2002) Deuterium spin lattice relaxation and deuterium quadrupole coupling constants: a novel strategy for characterization of transition metal hydrides and dihydrogen complexes in solution. In: *Unusual Structures and Physical Properties in Organometallic Chemistry* (Gielen, M., Willem, R., and Wrackmeyer, B., Eds.), pp. 145–165. West Sussex: John Wiley & Sons.
9. Field, L.D., Shaw, W.J., and Clentsmith, G.K.B. (2007). *J. Organomet. Chem.*, 692: 3042.
10. Ingmana, P. and Driver, G.W. (2012). *Phys. Chem. Chem. Phys.*, 14: 13053.
11. Sizun, C., Kempgens, P., Raya, J., Elbayed, K., Granger, P., and Rose, J. (2000). *J. Organomet. Chem.*, 604: 27.
12. Henry, E.R. and Szabo, A. (1985). *J. Chem. Phys.*, 82: 4653.
13. Bakhmutov, V.I. (2004). *Magn. Reson. Chem.*, 42: 66.
14. Morris, R.H. and Witterbort, R. (1997). *J. Magn. Reson. Chem.*, 35: 243.

15. Wehrmann, F., Fong, T.P., Morris, R.H., Limbach, H.H., and Buntkowsky, G. (1999). *Phys. Chem.Chem. Phys.*, 1: 4033.
16. Socol, S.M. and Meek, D.V. (1985). *Inorg. Chim. Acta*, 101L: 45.
17. Mazzei, P. and Piccolo, A. (2013). *Eur. J. Soil Sci.*, 64: 508.
18. Szymczak, N.K., Oelkers, A.B., and Tyler, D.R. (2006). *Phys. Chem. Chem. Phys.*, 8: 4002.
19. Algarra, A.G., Fernandez-Trujillo, M.J., Lledos, A., and Basallote, M.G. (2009). *Chem. Commun. (Camb.)*, 30: 4563.
20. Gaszczka, R. and Jazwinski, J. (2013). *J. Mol. Struct.*, 1036: 78.
21. Bayley, P.M., Best, A.S., MacFarlanec, D.R., and Forsyth, M. (2011). *Phys. Chem. Chem. Phys.*, 13: 4632.
22. Cooper, C.L., Cosgrove, T., van Duijneveldt, J.S., Murray, M., and Prescott, S.W. (2013). *Soft Matter*, 9: 7211.
23. Sola, E., Garca-Camprub, A., Andres, J.L., Martn, M., and Plou, P. (2010). *J. Am. Chem. Soc.*, 132: 9111.
24. Dahlquist, F.W., Longmuir, K.J., and Du Vernet, R.B. (1975). *J. Magn. Reson.*, 17: 406.
25. Kubas, G.J. (2001). *Metal Dihydrogen and σ-Bond Complexes*. New York: Kluwer Academic/Plenum.
26. Desrosiers, P.J., Cai, L., Lin, Z., Richards, R., and Halpern, J. (1991). *J. Am. Chem. Soc.*, 113: 4173.
27. Gusev, D.G., Nietlispach, D., Vymenits, A.B., Bakhmutov, V.I., and Berke, H. (1993). *Inorg. Chem.*, 32: 3270.
28. Bakhmutov, V.I., Howard, J A K., Keen, D.A. et al. (2000). *J. Chem. Soc. Dalton Trans.*, 1631.
29. Aime, S., Dastru, W., Gobetto, R., and Viale, A. (2001). NMR relaxation studies of polynuclear hydride derivatives. In: *Recent Advances in Hydride Chemistry* (Poli, R. and Peruzzini, M., Eds.), p. 351. Amsterdam: Elsevier.

Recommended Literature

Bakhmutov, V.I. (2005). *Practical NMR Relaxation for Chemists*. Chichester: Wiley.
Buntkowsky, G., Limbach, H.H., Wehrmann, F., Sack, I., Vieth, H.M., and Morris, R.H. (1997). ^{2}H NMR theory of transition metal dihydrides: coherent and incoherent quantum dynamics. *J. Phys. Chem. A*, 101: 4679.
D'Averqne, E. (2008). *Protein Dynamics—A Study of the Model-Free Analysis of NMR Relaxation Data*. Saarbrücken, Germany: VDM Publishing.
Harris, R.K. (1986). *Nuclear Magnetic Resonance Spectroscopy: A Physicochemical View*. Avon: Bath Press.
Kaplan, J.H. and Frenkel, G. (1980). *NMR of Chemically Exchanging Systems*. New York: Academic Press.
Schulz, F., Sumerin, V., Heikkinen, S., Pedersen, B., Wang, C., Atsumi, M., Leskel, M., Repo, T., Pyykko, P., Petry, W., and Rieger, B. (2011). Molecular hydrogen tweezers: structure and mechanisms by neutron diffraction, NMR, and deuterium labeling studies in solid and solution. *J. Am. Chem. Soc.*, 133: 20245.
Smerald, A. (2013). *Theory of the Nuclear Magnetic $1/T_1$ Relaxation Rate in Conventional and Unconventional Magnets*. Heidelberg: Springer.

7 Special Issues in Solution NMR

Due to wide applications of NMR in chemistry, geochemistry, biochemistry, biology, etc., NMR experiments and the problems and solutions sought are quite varied. In spite of the standardization and development of NMR techniques and software, in some sense any complex NMR experiment and its interpretation can seem unique, particularly to inexperienced researchers who have dealt primarily with routine one-dimensional (1D) and two-dimensional (2D) NMR spectra recorded for the analysis of reaction mixtures. Nevertheless, some problems require either the application of particular techniques and approaches or special knowledge and strategies. Some of these issues are considered in this chapter, the choice of which is obviously subjective.

7.1 OPTICAL ISOMERS IN NMR SPECTRA

Molecular symmetry, or its absence, and the type of symmetry play a central role in the origin of magnetic non-equivalency observed in NMR spectra. Symmetry is directly connected with the number of resolved signals and their assignments in NMR spectra and hence with determination of molecular structure by solution NMR. Moreover, the concept of molecular symmetry is one of the main factors behind the automatic assignment of signals in ^{1}H and ^{13}C NMR spectra of organic compounds carried out by various computer programs.[1]

Among the different elements of molecular symmetry, the chirality of molecules, or optical asymmetry, in many biochemical and biological processes is key to molecular recognition. Therefore, the manifestation of chirality in NMR spectra is worthy of special attention. It is generally accepted that protons or other nuclei situated in molecular groups can be heterotopic, diastereotopic, homotopic, or enantiotopic. By definition, for example, heterotopic and diastereotopic protons should differ in high-resolution ^{1}H NMR spectra. In fact, because they have different environments their chemical shifts and spin–spin coupling constants should also be different. The effect obviously depends on nuclei, compounds, temperature, and solvents. In contrast, homotopic protons (or other homotopic nuclei) are identical in NMR spectra and show the same chemical shifts and spin–spin coupling constants. For example, the central carbon atom in the two structures in Figure 7.1A has only three different substituents, and these structures are not distinguishable in ^{1}H NMR spectra. Due to their symmetry, the CH proton gives one signal, the CH_3 groups give one signal, and, for example, the *ortho* protons of the aromatic ring also give one signal due to an extremely fast rotation around the C–C bond.

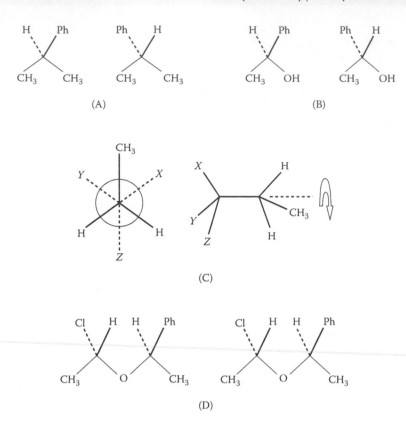

FIGURE 7.1 (A) Structures with homotopic protons. (B) Optical (mirror) isomers with enantiotopic protons. (C) Projections of a $(Y)(X)(Z)C-C(CH_3)H_2$ molecule with diastereotopic protons. (D) Diastereomers in the presence of two asymmetric carbons.

When four substituents at a carbon are different substituents, then the carbon becomes an asymmetric center, giving rise to the phenomenon of *optical isomerism*. Figure 7.1B illustrates chirality, where two optical (mirror) isomers cannot be converted into each other. These isomers can become distinguishable in the presence of a polarized (anisotropic) light. The protons attached to the asymmetric carbon in the structures are enantiotopic. In contrast to polarized light, the magnetic field is isotropic so neither of the isomers will be distinguishable in ^{1}H NMR spectra to give the same chemical shifts and spin–spin coupling constants.

Nevertheless, the presence of an asymmetric center in a molecule can be established in ^{1}H NMR spectra by observing two different resonances for two chemically identical groups neighboring the asymmetric carbon. Figure 7.1C shows the projections of a molecule $(Y)(X)(Z)C-C(CH_3)H_2$, where the protons neighboring with the asymmetric carbon have different environments even at very fast (free) rotation around the C–C bond. As a result, a ^{1}H NMR spectrum can show two ^{1}H resonances for the CH$_3$ groups with a chemical shift difference dependent on compounds, temperature, and solvents. Similarly, the CH_2 protons in the amino acid C_6H_5- $CH_2-C(NH_2)H(COOH)$ can appear in ^1H NMR spectra as the AB part of an ABX spin system at commensurable values of the $J(A-B)$ constant and the $A-B$ chemical shift difference. However, the form of a NMR spectrum depends on many factors, including the magnetic field strength. In other words, chirality corresponds to an increased number of NMR signals, and this phenomenon in the early days of NMR often resulted in confusing interpretations of NMR spectra.

In contrast to the structures in Figure 7.1A, the CHCl (as well as CHPh) protons attached to the asymmetric carbon will be distinguishable in a ^1H NMR spectrum of the molecule shown in Figure 7.1D due to the presence of a second asymmetric carbon giving rise to the phenomenon of diastereomerism. In fact, the molecule in Figure 7.1D gives four diastereomeric forms (++, −−, +−, and −+) in terms of their absolute configuration, an observation that can be used for the analysis of molecular chirality by solution NMR.

Because the magnetic field is isotropic, the diastereotopic conditions important for registration of NMR spectra can be created by the application of chiral solvents, such as optically active amines or optically active alcohols.[2] However, generally, the chemical shift differences of enantiotopic protons in such solvents are quite small and change from 1 to 10 Hz as a function of the magnetic field strength, optical purity of the solvent, temperature, and concentrations. Nevertheless, the optical purity of compounds or even their absolute configuration can still be determined. Currently, optically active solvating reagents, such as (R)-(-)2,2,2,-trifluoro-1-(9-anthryl)etha-nol,[3] are widely applied for the analysis of chiral drugs.

The addition of paramagnetic optically active chemical shift reagents can also be used to study molecular chirality in solutions. These reagents, containing lantha-nide cations Eu^{3+}, Yb^{3+}, or Pr^{3+}, are capable of intermolecular interactions with the coordination centers in molecules under investigation (e.g., oxygen, nitrogen). Upon the addition of relatively small amounts of the reagents, the equilibrium between free and coordinated molecules is fast on the NMR time scale and the resonances observed are averaged. The nuclei located closer to centers Eu^{3+}, Yb^{3+}, or Pr^{3+} will experience larger paramagnetic shifts. In turn, the induced chemical shifts depend on the equilibrium constants and the molar reagent-to-compound ratios. Generally, the equilibrium constants are very large and a 1:1 ratio is sufficient to observe clear effects. Moreover, under these conditions, the ^1H NMR spectra of even complex spin systems can be remarkably simplified. It should be emphasized that Pr^{3+} ions provide minimal Fermi-contact shifts; therefore, they are preferable in experiments where resonance lines experience minimal broadenings.

It is clear that, in the presence of a chiral shift reagent such as $Pr(HFC)_3$, its com-plexation with optically asymmetric molecules will result in ^1H NMR spectra where the optic isomers of the molecules will be resolved. For example, if the shift reagent has a (R) configuration, its complexation with (R) isomers of molecules under inves-tigation will cause larger chemical shifts, which can reach values of 0.5 to 0.7 ppm. For this reason, (R) and (S) isomers and their ratios can be well determined. The various methods and approaches to determining chirality in solutions are summa-rized in References 4 and 5.

An intriguing phenomenon observed in enantiomeric mixtures is the statistically controlled anisochronism of diastereomeric association.[6] This phenomenon consists of the appearance of NMR signals in achiral media belonging to optical isomers due to their mutual associative interactions, an observation that can be used for enantiomeric analysis. Rates of these association processes are very fast on the NMR time scale.

Again, diastereomerism is the reason for NMR signals being distinguished in solutions of optic isomers. In fact, in terms of the asymmetric center configurations, the associations ++ and +− lead to two short-lived diastreomeric forms. In addition, equilibrium constants characterizing this association are also different. The important point with regard to the mathematical treatment of the phenomenon is the fact that the diastereomeric non-equivalency in NMR spectra can be observed only in enantiomerically enriched solutions.

This effect is shown in Figure 7.2, where the diastereomeric non-equivalency in the $^{31}P\{^1H\}$ NMR spectra can reach 0.1 to 0.5 ppm. Such a difference obviously provides an accurate integration of the ^{31}P signals (Figure 7.2A) and determination of enantiomeric forms. As earlier, the effect depends on solvents and temperature; however, isomeric ratios now play a major role and the chemical shift difference goes to zero at equal concentrations of optic isomers (Figure 7.2B,C). Similarly, determination of the enantiomeric excess by NMR has been reported for phosphoric acids due to formation of their diastereomeric salts.[7]

Molecular chirality can be determined by application of a prochiral host molecule, which can interact with an optically active guest molecule, as shown in Figure 7.3.[8] The prochiral host molecule should have prochiral groups to be used as convenient NMR reporters. One such host molecule is N,N'-disubstituted oxoporphyrinogen, for which enantiotopic CH protons (marked with •) are the reporters. Due to the initial symmetry of this rigid host molecule, the CH protons show a singlet in ^1H NMR spectra. However, with the addition of optically pure guest molecules to a solution of the host its symmetry changes and the CH protons become non-equivalent. They then evolve to show an AB system and then an AX system at increasing concentrations of one of the optical isomers. It should be emphasized that a driving force behind the spectroscopic effects is the host–guest complexation via H bonding. Because the cycle containing nitrogen is rigid, the distance between the CH reporters and the binding site plays a major role.

7.2 SOLUTION NMR OF BIOMOLECULES: GENERAL PRINCIPLES

High-resolution NMR plays a key role in studies of biomolecules in solutions, such as proteins and protein complexes, to determine their structure or dynamics on a very large time scale from nanoseconds to milliseconds.[9–11] Because these systems are very complex, their investigation by NMR requires a detailed understanding of the theoretical basis and practical aspects of NMR experiments, from pulse calibrations and correctly setting parameters to artifact phenomena in final spectra. Also required is extensive experience, aided by numerous computer supplements to simplify structural analyses. In addition, some experiments may be based on specific NMR instrumentations that often are not available in the laboratory. This is why bio-NMR is considered in reviews and even books as a specific field of NMR

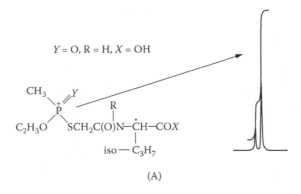

$Y = O, R = H, X = OH$

(A)

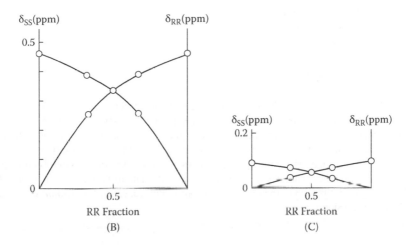

(B)

(C)

FIGURE 7.2   (A) The $^{31}P\{^1H\}$ NMR spectrum of phosphorus-containing compounds show-ing the non-equivalency of enantiomers when their concentrations are not equal. (B, C) Chemical shift difference for ^{31}P signals from SS and RR enantiomers of phosphorus-con-taining compound A vs. a fraction of RR isomer at $-60°$ and $-20°C$, respectively, where solid lines are obtained theoretically and points correspond to the experimental data.[6]

applications. This field is too large to be covered in this brief text; therefore, this section gives an account of only the main principles and strategies used in structural studies of protein molecules to show how their structure can be established in solu-tions at an accuracy comparable to x-ray diffraction.

Currently, around 15% of protein structures are determined by solution NMR, where the main limitations are associated with the size of the protein molecules. In fact, increasing size leads to decreasing molecular mobility and increasing line widths in NMR spectra. The latter result reduces spectral resolution; therefore, proteins with molecular mass up to 30 to 35 kDa are generally routinely analyzed, particularly when these molecules are labeled with ^{13}C, ^{15}N, and ^2H nuclei. The labeled compounds reduce the total experimental time and increase the volume

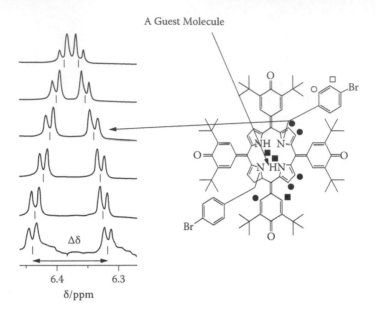

FIGURE 7.3   Structure of the host molecule, *N,N′*-disubstituted oxoporphyrinogen, where enantiotopic CH protons, marked with the symbol •, are the ¹H NMR reporters. These signals experience an evolution in the presence of a chiral guest from a singlet to the *AB* system and the *AX* system (from top to bottom) due to an increasing Δδ value. The evolution is observed when the concentration of one of the optical isomers increases. The other symbols correspond to protons that also show changes in their chemical shifts due to H bonding; however, their signals do not mask the area of the reporters. (From Labuta, J. et al., *Nat. Commun.*, 4, 2188, 2013. With permission.)

of NMR parameters, named *restraints*, in structural bio-NMR. Nevertheless, in spite of such labeling, only few structures of proteins with molecular mass up to 70 kDa have been reported.[9,12] Typically, in order to completely solve the structure of proteins, NMR data should provide 10 to 20 independent interatomic distances to account for one amino acid unit, generally obtained by 2D or three-dimensional (3D) nuclear Overhauser effect spectroscopy (NOESY) NMR experiments. Then, dihedral angles and average atom–atom vector directions are added. These restraints can be identified by direct measurements of ^3J coupling constants in NMR spectra and by studies of residual dipolar coupling extracted from experiments performed on proteins partially oriented in the external magnetic field.

Proteins containing paramagnetic centers represent special cases for NMR studies. On the one hand, paramagnetic centers in proteins cause additional line broadenings, which prevent the collection of highly resolved NMR spectra. However, on the other hand, paramagnetic cations can provide an additional number of NMR restraints needed to solve the structure. In fact, due to their presence, protein molecules can show pseudocontact chemical shifts and relaxation enhancements, which can be combined to determine the structure more accurately.

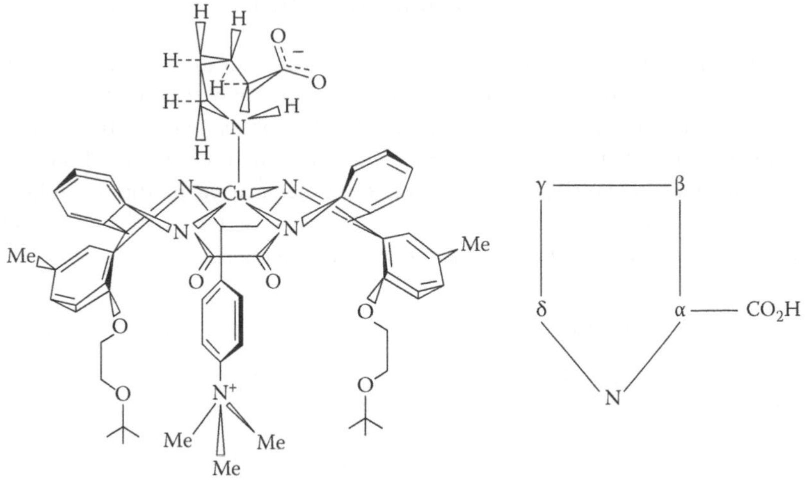

(A)

Cu⋯Proton Distances (Å)

α	β	γ	δ	
3.29 ± 0.25	4.62 ± 0.35	4.66 ± 0.35	2.92 ± 0.3	3.49 ± 0.25

Cu⋯Carbon Distances (Å)

α	β	γ	δ	CO_2^-
3.1	4.0	3.7	3.2	4.0

(B)

FIGURE 7.4 (A) The association formed by a proline molecule with a paramagnetic Cu(II) complex in water. (B) The distances Cu⋯H and Cu⋯C obtained by ^{1}H and ^{13}C T_1 time measurements.[13]

Relaxation enhancements, measured and interpreted for ^{1}H and ^{13}C nuclei in a water solution of proline containing the paramagnetic Cu(II) complex shown in Figure 7.4A, demonstrate how the relaxation restraints can be used for accurate structural descriptions in solutions even in the presence of fast reversible complexation. The water solutions have been prepared at various concentrations of the proline/Cu(II) complex (N_P/N_S in Equation 7.1), and the relaxation times T_1 and T_2 have been accurately measured for protons and carbons of proline molecules.

As expected, the ^{1}H and ^{13}C T_1 times of proline molecules shorten remarkably after addition of the paramagnetic Cu(II) complex. For example, the ^{1}H T_1 times of a pure proline solution have been measured as 2.02 to 3.44 s vs. 0.2 to 1.07 s, when N_S/N_P is 5×10^{-4}. Due to fast exchange between the free and associated proline molecules, the relaxation rates $1/T_{1M}$ represent the weighted averages of two environments corresponding to the bulk solution and the bound state. In the presence of a large proline excess, the Cu coordination sphere is obviously completely saturated and the relaxation rates follow Equation 7.1:

$$1/T_{1M} = (N_P/N_Sq)(1/T_1 - 1/T_{1F}) \qquad (7.1)$$

where $1/T_{1M}$ and $1/T_{1F}$ are the relaxation rates measured in the presence and in the absence of paramagnetic centers, respectively; N_P is the molar concentration of the proline; N_S is the molar concentration of the Cu(II) complex; and q is the proline/Cu ratio, which can be taken as 1 in accordance with the structure shown in Figure 7.4A.

Under the above conditions, the T_{1M} and T_{2M} times caused by dipolar relaxation due to electron–nucleus interactions can be written in terms of the Solomon–Bloembergen equations, Equations 7.2 and 7.3 (see Chapter 3):

$$1/T_{1M} = (2/15)\gamma_I^2 g^2 S(S+1)\beta^2 / r^{-6} \left[3\tau_C / (1+\omega_I^2\tau_C^2) + 7\tau_C / (1+\omega_S^2\tau_C^2) \right] \qquad (7.2)$$
$$+ (2/3)S(S+1)(A/\hbar)^2 \left[\tau_e / (1+\omega_S^2\tau_e^2) \right]$$

$$1/T_{2M} = (1/15)\gamma_I^2 g^2 S(S+1)\beta^2 / r^{-6} \left[\begin{array}{c} 4\tau_C + 3\tau_C / (1+\omega_I^2\tau_C^2) + \\ +13\tau_C / (1+\omega_S^2\tau_C^2) \end{array} \right] \qquad (7.3)$$
$$+ (1/3)S(S+1)(A/\hbar)^2 \left[\tau_e + \tau_e / (1+\omega_S^2\tau_e^2) \right]$$

Then, taking correlation time τ_C as 10^{-10} s (typical of solutions), the copper–proton and copper–carbon distances in Figure 7.4B can be calculated.[13] Finally, on the basis of the distances obtained, the structure and conformation of the proline fragment can be determined.

A similar approach can be applied for studies of more complex molecular systems, such as paramagnetic peptides.[14] It has been found that distances between protein NH protons and the paramagnetic nitroxide-based probe can be obtained by proton relaxation enhancements in good agreement with the x-ray data only at a distance between 10 and 20 Å. Shorter distances cannot be determined accurately due to large broadenings in ^1H NMR spectra observed for signals of protons located near the nitroxide probe.

The methodology applied for the structural study of proline in Figure 7.4 illustrates only a single cycle, corresponding to the path from NMR restraints measured experimentally to the structural parameters calculated in their base. For complex biomolecular systems, this cycle will be repeated many times to obtain the final structure when a large ensemble of conformers in biomolecules must be taken into account. Generally, the calculations can be considered complete when agreement is reached between restraints measured experimentally and those calculated for a conformer family to reach a good root mean square deviation (RMSD).

An important problem in the determination of protein structures by solution NMR is the correct and full assignment of resonances in very complex NMR spectra. This step obviously depends on the skills of the researcher, as incorrectly assigned resonances will give incorrect inter-atomic distances, resulting in the wrong structural information. Sometimes a structure cannot be determined due to such incorrect assignments. Various computer programs and tools have been developed to assign NMR resonances automatically. Moreover, on the basis of statistical procedures, these programs allow determination of the reliability of the structure identified. It should be emphasized that, as in the case of x-ray analysis, details regarding programs for structural calculations of proteins are available in the literature for NMR users.[9]

The precision of structural determinations obviously depends on the number of NMR restraints obtained experimentally. Here, NOE distances play a major role. The data collected by NOE experiments are generally considered in terms of allowed distance ranges,[9] where the lower limit of 2 Å is generally accepted as the sum of two hydrogen atomic radii, while the upper limit is between 2 and 5 Å. Because, for example, in folded proteins, interatomic distances of 5 Å are typical for many protons, their NOESY spectra remain unresolved. To work around this problem, four-dimensional (4D) NOESY NMR experiments have been developed that are usually performed on ^{13}C- and ^{15}N-labeled protein molecules. Another strategy is based on studies of deuterated compounds, which can increase the NOE range for amide protons up to 8 Å.

In addition to the NOE and ^3J coupling constants, the number of NMR restraints can be enlarged by residual dipolar couplings collected in orienting media by the addition of hydrogen bonds, mutual orientations of internuclear vectors obtained by cross-correlations, backbone angles obtained by an analysis of the database for chemical shifts, and paramagnetic restraints. In turn, the paramagnetic restraints can be obtained by studying chemically modified protein molecules synthesized with the replacement of diamagnetic cations by paramagnetic cations.

The calcium cations in the protein calbindin D_{9k}, for example, can be substituted for various lanthanide paramagnetic ions.[12] Experiments performed on Ce protein have provided the NMR restraints of 1823 NOEs, 191 dihedral angles, 15 hydrogen bonds, 769 pseudocontact shifts, 64 residual dipolar couplings, 26 T_1 relaxation rates, and 969 values of pseudocontact shifts. These NMR restraints are quite sufficient to solve the protein structure more precisely (Figure 7.5A). The RMSD factor is 0.25 Å compared to the value of 0.69 Å obtained for the structure in Figure 7.5B solved for the initial calcium-containing protein molecule.

Figure 7.5C shows the calculations applied for the determination of paramagnetic protein structures on the basis of regular NMR restraints (e.g., NOE), residual dipolar couplings (RDCs), and pseudocontact chemical shifts (PCSs).[12] The procedures involving paramagnetic parameters are added to the well-known programs Xplor–NIH and Cyana, where the χ components of the magnetic susceptibility tensor characterizing the paramagnetic metal ion are used to calculate the coordinates of the atoms in the proteins. It should be noted that the cycle is initiated by estimation of the magnetic susceptibility anisotropy parameters or by a model protein structure. A similar strategy is applied for NMR studies of protein–protein interactions in solutions even for the formation of relatively weak protein complexes. Descriptions of the various approaches can be found in the recommended literature.

Finally, signal assignments and the identification of spin systems in the NMR spectra of ^{13}C-labeled proteins can be performed with experiments based on direct detection of ^{13}C nuclei. This detection is often used for structural studies of ^{13}C-labeled proteins, where the negative effect caused by the large proton–proton dipolar interactions is reduced by deuteration. In the case of paramagnetic protein molecules, experiments involving the direct detection of ^{13}C nuclei are particularly useful. Due to the lower γ value, paramagnetic effects on ^{13}C relaxation are

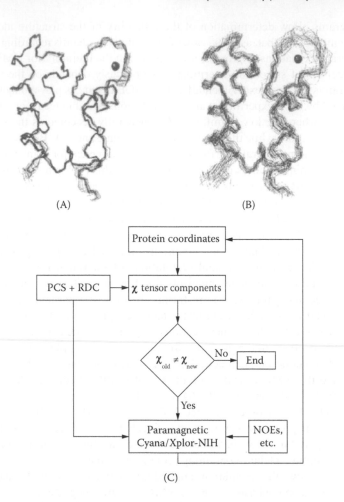

FIGURE 7.5 (A) Structure of the Ce–calbindin peptide solved with paramagnetic-based restraints. (B) Structure of the calcium-containing calbindin protein found by NMR. (C) Cycle showing the calculations of protein structures in solutions on the basis of regular NMR restraints, residual dipolar couplings (RDCs), and pseudocontact chemical shifts (PCSs). (From Bertini, I. et al., *Dalton Trans.*, 2008, 3782, 2008. With permission.)

remarkably smaller than those on ^{1}H relaxation, leading to lines that are too broad. To avoid the limit of sensitivity for nuclei with low γ values, specially designed NMR probes have been developed.

7.3   DYNAMICS OF LIQUIDS BY NMR

The microscopic behavior of liquids at different temperatures and pressures strongly affects chemical reactions in homogeneous and heterogeneous media on an industrial scale due to changes in the transport properties of their molecules and ions. Therefore, the dynamics of liquids is a fundamental problem in modern chemistry. Among the different systems, liquids capable of hydrogen bonding (e.g., alcohols) are of the greatest interest. In fact, they have unique properties, particularly under supercritical conditions. Different physical methods have been used to describe the structure and properties of systems containing hydrogen bonds: x-ray and neutron diffraction, infrared absorption, Raman scattering, and molecular simulations. However, these methods describe rather static structural properties and cannot explain the complex behavior of liquids on various time scales, particularly the supercritical states appearing at high temperatures and high pressures. In contrast, NMR can show the dynamics of liquids, including H-bonded liquids, where knowledge of the lifetimes of hydrogen bond structures and rates of molecular motions is particularly valuable. This section considers the principles and applications of diffusion NMR spectroscopy, relaxation experiments, and high-pressure NMR.

7.3.1 Diffusion NMR Spectroscopy in Liquids

Transport properties of molecules and ions in liquids can be directly characterized by diffusion NMR spectroscopy. As mentioned earlier in Chapter 4, this method introduces a new dimension to detect effects on a molecular level, where nuclei serve as convenient probes for the measurement of translation diffusion acting equally on all of the nuclei in molecules. This is in contrast to nuclear relaxation, where the relaxation times T_1 (or T_2), dependent on molecular motion correlation times, can be different for different nuclei in the same molecule due to specific magnetic interactions and the presence of local intramolecular motions, such as rotational reorientations or segmental motions. According to the Debye–Einstein theory, diffusion coefficient D is a measure of transport properties and depends on a friction factor:

$$D = k_B T / f_T \tag{7.4}$$

where k_B is the Boltzmann constant, T is the absolute temperature, and f_T is the friction factor. In the case of spherical molecules, the f_T factor is expressed via Equation 7.5:

$$f_T = 6\pi\eta r \tag{7.5}$$

where η is the solvent viscosity and r is the hydrodynamic radius. Generally speaking, D coefficients cannot be measured by regular NMR techniques because the external magnetic field is isotropic and homogeneous. In contrast, their measurements are possible by incrementing the areas of gradient pulses (q) in pulsed field gradient (PFG) experiments. These experiments, including the Hahn echo pulse sequence, result in diffusion NMR spectra obtained by transforming the signal amplitudes with respect to q^2.

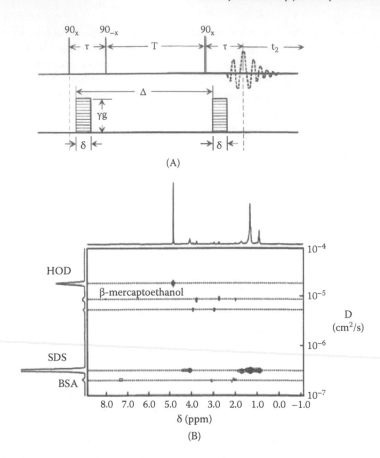

FIGURE 7.6 (A) The stimulated Hahn echo pulse sequence applied for DOSY experiments including pulsed field gradients δ. (B) The 2D DOSY NMR spectrum obtained for a water solution of sodium dodecylsulfate (SDS), bovine serum albumin (BSA), and β-mercaptoethanol in phosphate buffer at pH = 7.2. (From Johnson, C.S., *Prog. Nucl. Magn. Reson. Spectrosc.*, 34, 203, 1999. With permission.)

Besides simple Hahn echo pulses, stimulated Hahn echo pulses are often used in diffusion NMR. As seen in Figure 7.6A, a second 90°_{-x} pulse is added to the standard pulse sequence. Then, the registering pulse acts to collect the echo signal after time τ, leading to diffusion-ordered NMR spectroscopy (DOSY) NMR spectra. Because the magnetization created in a sample follows the Bloch equations, incorporation of the diffusion effects into these equations gives quantitative descriptions of diffusion experiments. It should be added that PFGs affect transverse components of the total magnetization.[15,16]

Generally, DOSY NMR experiments are carried out with automated collection of the data based on a programmed set of gradient areas to generate 1D, 2D, or 3D DOSY NMR patterns, where the diffusion dimension is represented by D values calculated with standard software. Technically, such experiments require a high-quality gradient NMR probe and a computer to control the so-called gradient drivers. Finally, these experiments are easy to interpret when the collected data, Fourier transformation, phasing, and baseline corrections all lead to undistorted NMR spectra. Two-dimensional DOSY experiments performed to rationalize equilibriums in solutions containing sodium dodecylsulfate (SDS) and bovine serum albumin (BSA) are examples of such measurements.[15] Because these equilibriums are important for understanding SDS–protein interactions, the aim of a DOSY experiment is to determine the fraction of surfactant bound to a protein vs. the total surfactant concentration.

Figure 7.6B shows a room-temperature DOSY NMR spectrum recorded in a D_2O solution of BSA and SDS containing 0.01-M β-mercaptoethanol in phosphate buffer at pH = 7.2. The diffusion dimension exhibits resonances of BSA belonging to the BSA–SDS complex, a single resonance for SDS, non-intense lines for the denaturing compound, and resonances belonging to the reaction product $HOCH_2CH_2SSCH_2CH_2OH$. According to the measurements, the BSA–SDS complex experiences slow diffusion. Further interpretation of the measurements addresses the single resonance of SDS detected in the spectra. This observation reveals a fast SDS exchange between monomers (*mon*), micelles (*mic*), and the BSA–SDS complex (*prot*) on the NMR time scale. Then, diffusion coefficient D, measured experimentally for SDS, can be written as follows:

$$D = P_{prot}D_{prot} + P_{mon}D_{mon} + P_{mic}D_{mic} \tag{7.6}$$

where

$$P_{prot} + P_{mon} + P_{mic} = 1$$

and P_{prot}, P_{mon}, and P_{mic} are the corresponding mole fractions, and D_{prot}, D_{mon}, and D_{mic} are self-diffusion constants, respectively. It is obvious that calculation of fraction P_{prot} is possible when all of the diffusion coefficients are known. The P_{mon} or P_{mic} values can be taken from the NMR spectrum directly, whereas the D_{mon} and D_{mic} values should be measured independently.[15]

Three-dimensional DOSY versions of diffusion NMR spectroscopy have also been described in the literature, where the diffusion dimension is added to the regular 2D NMR experiment in order to avoid overlapped resonances, thus increasing the accuracy of the measurements.[15]

Similarly, NMR chromatography is applied for the separation of small molecules in mixtures by diffusion coefficients. Separation in suspensions suggests the application of DOSY NMR techniques to provide a two-dimensional contour map, where the chemical shift is plotted on the horizontal axis and the diffusion rate is plotted on the vertical axis. Again, each component in the mixture produces a separate 1D NMR spectrum corresponding to its diffusion constant.[17]

The disadvantage of DOSY NMR in solutions is the very limited capability of separation in the diffusion axis. In fact, the free induction decay in a regular DOSY experiment represents a sum of sinusoidal decays in the acquisition dimension, while the diffusion dimension contains a sum of Gaussian-type decays. Under these conditions, Fourier transformation can provide high resolution in the frequency domain. At the same time, the diffusion dimension is treated with Laplace transformation, leading to a low resolution for two or more overlapped resonances.

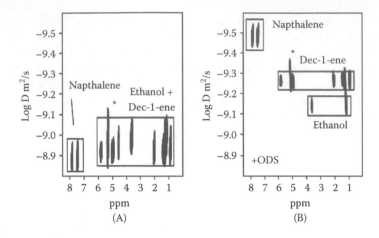

FIGURE 7.7 (A) Regular DOSY NMR spectrum recorded for a mixture of molecules dissolved in deuterated ethanol which are poorly separated. (B) The addition of functionalized silica gives excellent spectral separation. (From Caldarelli, S., *Magn. Reson. Chem.*, 45, S48, 2007. With permission.)

The results of DOSY experiments can be enhanced by performing them in structured media; for example, silica gel can be added to solutions for enhancement of separation in the diffusion dimension. Whereas DOSY experiments separate compounds by their molecular weight (diffusion constants), experiments in the presence of silica can add a factor of hydrogen-bonding affinity to silica. A negative effect of structured media is that the NMR signals from liquids broaden greatly up to hundreds or thousands of hertz. Therefore, under these conditions, the use of solid-state magic-angle spinning (MAS) NMR is preferable. Here, solid components (potentially also observed in solid-state MAS NMR spectra) can relax completely during the diffusion pulse sequence prior to acquisition. Figure 7.7 shows two DOSY NMR spectra recorded in a mixture of compounds in a solution and with the addition of silica. The data obtained in the presence of the silica demonstrate excellent separation in the diffusion coordinate.[18] In addition to DOSY experiments, different diffusion-based NMR techniques can be applied to study ligand binding in protein solutions. When the mixtures of ligands are quite complex, the spectral selectivity in NMR diffusion coefficient measurements can be improved. The best approach to improving the selectivity is based on gradient modified spin echo (GOSE) experiments to analyze any complex mixtures.[19] The GOSE pulse sequence (simple or modified) allows rejection of resonances belonging to coupled spins so only singlet lines in the ^{1}H NMR spectra are observed. It should be noted that such a selection of singlet magnetization can be incorporated into different standard pulse sequences, such as NOESY.

7.3.2 HIGH-PRESSURE NMR: STRUCTURE AND DYNAMICS OF LIQUIDS

As mentioned earlier, the effects of pressure and temperature on hydrogen-bonded liquids (e.g., water, alcohols) can lead to a better understanding of intermolecular interactions and aid in the creation of a model for liquids under normal and supercritical conditions. These effects play an important role even in biochemical systems, where, for example, studies of protein solutions at subzero temperatures or of the phase behavior and dynamics of phospholipid membranes at high pressure are particularly interesting. Finally, pressure effects on chemical equilibriums and reaction rates are very valuable for studies of the reaction mechanisms.

Typically, the pressure applied for simple liquids or complex biochemical molecular systems varies from 0.1 MPa to 1 GPa. It should be emphasized that even the highest pressure of 1 GPa does not affect the lengths of covalent bonds. The angles between the bonds also remain unchanged. At the same time, however, intermolecular distances and molecular conformation states can change with pressure, and such changes play a major role in the interpretation of observed results.

High-pressure NMR requires specially designed NMR probes and high-pressure NMR cells. Figure 7.8A shows the equipment typically needed for the generation of pressure, including pumps, high-pressure tubing, and valves.[20] High spectral resolution, an important element of high-pressure NMR experiments, particularly for complex molecular systems, can also be achieved through the use of modern high-pressure NMR probes. As in the case of regular variable-temperature NMR experiments, the temperature in high-pressure NMR probes should be well calibrated, usually by utilizing reference thermocouples, where the reproducibility of temperature measurements should be well verified for various temperature diapasons.

The methodology of studying the structure and dynamics of liquids under high pressure can be well illustrated by high-pressure NMR experiments on methanol,[21,22] which is the simplest alcohol and has unique properties and microscopic behavior that play an important role in reactions on an industrial scale. The properties of methanol at different temperatures and pressures can be characterized by three NMR parameters: the proton chemical shift difference ($\delta(OH) - \delta(CH_3)$), the ^1H T_1 times of OH groups, and the ^1H T_1 times of CH_3 groups. It should be emphasized that a capillary high-pressure NMR cell and a NMR probe applied for these studies should provide minimal line widths in the ^1H NMR spectra needed for accurate measurements. Generally, line widths of 4 to 8 Hz are sufficient to achieve good accuracy.[21]

Figure 7.8B,C summarizes the collected NMR data. As seen, the ($\delta(OH) - \delta(CH_3)$) difference measured at pressures between 50 and 3500 bar and temperatures between 298 and 773 K consistently changes, including the sign, to show a dramatic jump near the critical point (512.6 K). Thus, the hydrogen bond network in methanol experiences significant changes. Nevertheless, even at the critical point hydrogen bonding is still present. The tendencies shown in Figure 7.8B suggest that the hydrogen-bonding network of methanol increases with increasing pressure at constant temperature and decreases with increasing temperature at constant pressure.

The effects of pressure and temperature on ^1H T_1 times for CH_3 and OH groups are also significant and can clearly be seen in Figure 7.8C. An analysis of the data in terms of dipole–dipole and spin–rotation relaxation, which differ in their dependence on molecular motion correlation times (see Chapter 3), has shown that both of the mechanisms contribute to the total T_1 times.[21] However the contributions are different at high and low temperatures. Moreover, the dominant mechanism changes when high pressure is applied. Evaluation of the contributions shows that dipole–dipole

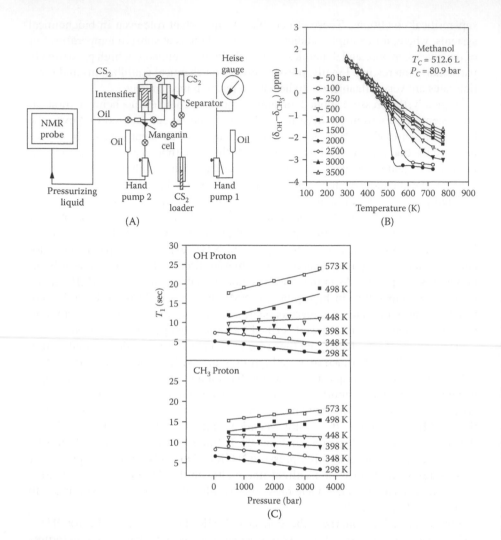

FIGURE 7.8 (A) Diagram showing high-pressure-generating equipment for NMR experiments. (From Jonas, J. et al., *J. Magn. Reson. B*, 102, 299, 1993. With permission.) (B) Chemical shift difference for OH and CH₃-protons in a methanol sample as a function of pressure and temperature. (C) ¹H spin–lattice relaxation times measured for OH and CH₃ protons in methanol as a function of pressure. (From Bai, S. and Yonker, C.R., *J. Phys. Chem. A*, 102, 8641, 1998. With permission.)

interactions are dominant in the temperature range of 300 to 400 K, while spin–rotation interactions are dominant above 500 K. Within the temperature range of 400 to 500 K, similar contributions from both relaxation mechanisms are observed.

As generally accepted, a hydrogen-bonded network of liquids, such as alcohols or water, can be characterized quantitatively by the number of hydrogen bonds per one molecule. When intramolecular dipole–dipole interactions completely govern ¹H relaxation, its rate is given by

$$1/T_1 \text{ (intra)} = (3/2)\hbar^2\gamma^4 \sum \left(1/r_{ij}^6\right)\tau_C \tag{7.7}$$

where $i \neq j$. Internuclear distance r can be calculated at a molecular motion correlation time (τ_C) found independently. In turn, a hydrogen-bond model can be written as

$$\sum \left(1/r_{ij}^6\right) = \langle n_{HB}\rangle / \langle r_{ij}^6\rangle \tag{7.8}$$

where $i \neq j$ and n_{HB} is accepted as the average number of neighbors. According to the experiments on methanol,[21] this number changes from 1.2 to 0.5 with an increase in temperature from 300 K to 400 K.

7.4 PARA-HYDROGEN AND NMR SPECTROSCOPY IN SOLUTIONS

The initial magnetization polarization created in a sample placed in an external magnetic field is low because of the low energy difference for Zeeman levels and their low population difference. From a practical standpoint, this circumstance is not desirable, as it leads to low NMR sensitivity in contrast to infrared and ultraviolet methods. There are several approaches to enhancing the sensitivity based on cryogenic NMR probes, optical pumping, and dynamic nuclear polarization (considered in Chapter 1). However, in the chemical context, para-hydrogen-induced polarization is of greatest interest.[23]

Molecular hydrogen exists as four nuclear spin isomers. According to the principle of symmetrization accepted in quantum mechanics, the overall wave function of the fermion dihydrogen should be anti-symmetric with respect to the exchange of nuclei. Anti-symmetric rotational states in a dihydrogen molecule will correspond to symmetric nuclear spin states (ortho), while symmetric rotational states will correspond to anti-symmetric nuclear spin states (para). Thus, the term ortho-hydrogen is associated with hydrogen molecules existing in one of the threefold degenerate spin isomers: $\alpha\alpha$, $\alpha\beta+\beta\alpha$, and $\beta\beta$. In contrast, only one isomer ($\alpha\beta-\beta\alpha$) will correspond to para-hydrogen molecules. Because the ortho and para isomers can exist in different rotational states, their ratio will depend on the temperature; however, special chemical procedures allow the isolation of pure para-hydrogen. This para-hydrogen can be involved in chemical reaction products that can be observed by NMR during hyperpolarization.[24]

This effect can be illustrated in practice by the addition of para-hydrogen to the transition metal complexes Ir(CO)(dppe)Br or Ir(CO)(dppe)(CN). The addition leads to reaction products that show anti-phase ^{1}H–Ir resonances in the ^{1}H NMR spectra shown in Figure 7.9A,B. The intensities of these signals are increased significantly. The energy level distribution for the nuclei shown in Figure 7.9A,B explains this effect. When a compound formed due to a reaction with regular H_2 contains two non-equivalent hydrogen atoms, its ^{1}H NMR spectrum exhibits an AX spin system. This system corresponds to four resonances characterized by the spin–spin coupling constant $J(1-2) = J(3-4)$ in accordance with Boltzmann populations of the energy levels shown in Figure 7.9A. When the para-hydrogen is introduced into

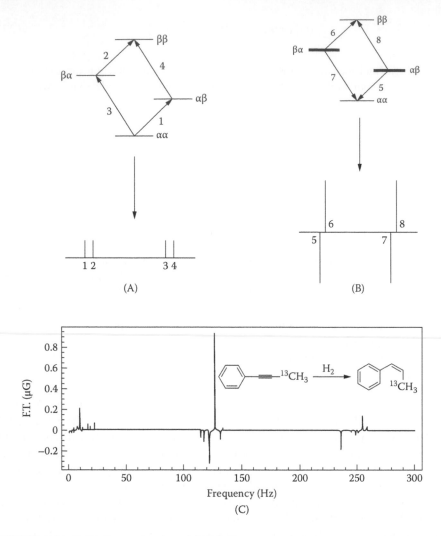

FIGURE 7.9 (A, B) Energy levels and ¹H NMR spectra of an *AX* system expected at normal (Boltzmann) populations and *para*-hydrogen-derived populations, respectively. (From Duckett, S.B. and Mewis, R.E., *Acc. Chem. Res.*, 45, 1247, 2012. With permission.) (C) Zero-field NMR spectrum of 1-phenyl-1-propyne after the addition of *para*-hydrogen. The resulting *Z* magnetization is recorded by the atomic magnetometer. (From Theis, T. et al., *Nat. Phys.*, 7, 571, 2011. With permission.)

the reaction and the symmetry of its spin state does not change during the reaction, then spin states αβ and βα corresponding to this product will be more populated, as shown by the line thickness in Figure 7.9B. Again, the signals correspond to the *AX* spin system; however, now the signals separated by proton–proton spin–spin coupling constants are anti-phase and their intensities are strongly increased due to hyperpolarization.

Two factors are important for observation of the phenomenon: (1) the addition of *para*-hydrogen to a solution should be fast on the relaxation time scale, and (2) the initial magnetic symmetry of ^{1}H nuclei in the H_2 molecule should be broken in a reaction product. Under these conditions, the *para*-hydrogen effect can be applied for mechanistic investigations of different reactions (e.g., hydrogenation, catalysis, hydroformylation) to detect low-populated intermediates that are spectrally invisible when regular hydrogen is added.[24] Here, in addition to single-pulse experiments, the correlation spectroscopy (COSY), heteronuclear single quantum coherence (HSQC), heteronuclear multiple quantum coherence (HMQC), and nuclear Overhauser effect spectroscopy (NOESY) techniques can also be used to characterize *para*-hydrogen polarized species.

For the *para*-hydrogen-induced polarization method, increasing signal-to-noise ratios in NMR spectra can be applied even for experiments in zero magnetic fields, which differ significantly from the application of regular zero-field NMR using thermal nuclear polarization. Details regarding a zero-field device and the principles of detection for *para*-hydrogen-induced polarization can be found in Reference 25, which describes bubbling *para*-hydrogen through a solution to achieve polarization. Figure 7.9C shows the zero-field NMR spectrum of 1-phenyl-1-propyne (^{13}C enriched in the CH_3 group) after bubbling of *para*-hydrogen.[25] The spectrum demonstrates a signal-to-noise ratio comparable to the spectra obtained with high-field NMR spectrometers. Thus, zero-field *para*-hydrogen polarization is a powerful tool for the identification of chemical compounds as products of hydrogenation reactions.

7.5 SOLUTION NMR SPECTROSCOPY AND HETEROGENEOUS MOLECULAR SYSTEMS

Radiofrequency pulses used in solution NMR spectrometers are generally low in power, and signals from solids cannot be recorded, on principle. For samples containing solid/liquid phases, only the mobile fractions (liquids) will be represented in the NMR spectra. This condition is key to studies of heterogeneous molecular systems by solution NMR techniques, such as nanoparticles dispersed within bulk liquid samples or suspensions.

Applications of regular solution NMR for such objects can be particularly interesting when solvents (e.g., water) are used to modify organic or inorganic systems, synthetic polymers, biopolymers, and foodstuffs, exploiting the influence of water on their properties. Studies of the solvent dynamics and behavior of dissolved substances in frozen systems have resulted in the development of specific scientific fields such as cryochemistry and cryobiology. Here, the focus of researchers is on the formation of heteroporous thermoreversible polymer hydrogels in frozen water solutions and the behavior of water molecules at temperatures above and below 0°C.

The strategy and methodology of such investigations can be illustrated by samples of an aqueous (D_2O) solution of poly(vinyl alcohol) (PVA), which has a melting point of 3.5°C.[26] PVA is a high-polarity, water-soluble polymer used for modeling the properties of biopolymers when the temperature zone directly preceding the melting point is of greatest interest.

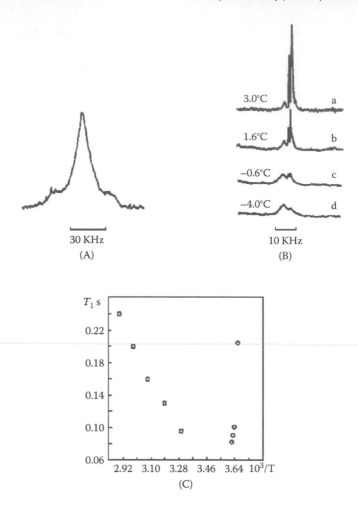

FIGURE 7.10 (A) ^{2}H NMR spectrum of a 10% solution of PVA in D_2O after two-step thermostating (1 hr at –20°C and 0.5 hr at –66°C. (B) Temperature dependence of the ^{13}C NMR spectra in the same solution. (C) Temperature dependence of ^{13}C T_1 times measured for groups CH_2 of PVA: □, an unfrozen solution; O, a solution frozen and thermostated for 1 hr at –20°C and 1 hr at the shown temperature. (From Gusev, D.G. et al., *Magn. Reson. Chem.*, 28, 651, 1990. With permission.)

Two target nuclei can serve as probes, ^{2}H and ^{13}C, to characterize D_2O and PVA, respectively. Because the line widths of D_2O in the solid and liquid states differ dramatically (more than 100-fold), the ^{2}H NMR spectra show only deuterons belonging to the mobile water molecules (Figure 7.10A) but not in the ice structure. By analogy, the ^{13}C NMR spectra indicate the behavior of PVA chains at freezing (Figure 7.10B), while the ^{13}C relaxation time measurements (Figure 7.10C) can trace their segment mobility.

As seen in Figure 7.10B, the ^{13}C NMR spectra in a frozen solution show the ^{13}C resonances, which are quite apparent and experience fundamental changes between –1° and –4°C. These changes indicate that the formation of the hydrated polymer

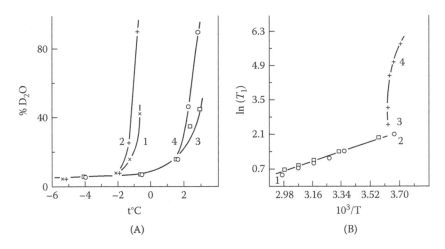

FIGURE 7.11 (A) Integral ^{2}H intensity in D$_2$O solutions of PVA (10%): 3 and 4, 98.8% of ^{2}H; 1 and 2, 8.5% of ^{2}H. Solutions were frozen at –20°C (1 h) and thermostated at the shown temperature for 0.5 hr (1 and 3) and 3 hr (3 and 4). (B) Patterns for ln(^{2}H T_1) vs. $1/T$ in a 99.8% D$_2$O solution of PVA (10%); curves 1 and 2 correspond to an unfrozen sample (O) and a sample thawed after freezing at –20°C followed by thermostating for 3 hr at 2°C (□). Curves 3 and 4 correspond to samples pre-frozen and thermostated for 1 hr at –20°C. (From Gusev, D.G. et al., *Magn. Reson. Chem.*, 28, 651, 1990. With permission.)

phase was complete. Thus, a point lying between –1 and –4°C can be regarded as the inhibition point for fast motions in the liquid microphase of the frozen PVA solution. At the same time, the ^{2}H NMR spectra of the frozen solution exhibit the isotropic ^{2}H signal in the absence of quadrupolar splitting even at very low temperatures (Figure 7.10A). This result reveals the fast isotropic rotational and translational diffusion of the water molecules under these conditions.

The standardized integral ^{2}H intensities measured in the frozen D$_2$O–PVA and D$_2$O/H$_2$O–PVA solutions in Figure 7.11A allow determination of the amount of the water not included in the ice structure. Even the fine effects of isotropic enrichment in the mobile microphase with ^{1}H$_2$O molecules can be seen when the solution, frozen at –20°C, is placed in a temperature range of 1 to 5°C before the melting point.

The molecular motion correlation times τ_C for water molecules and the PVA chain segments can be calculated on the basis of ^{13}C (Figure 7.10C) and ^{2}H (Figure 7.11B) T_1 relaxation times, which are governed by dipolar and quadrupolar relaxation mechanisms, respectively. This step helps to rationalize the mechanism of the cryo-structurization, an important factor in the formation of gels in frozen solutions. Similar strategies can be used to study other heterogeneous systems using solution NMR techniques without specific modifications.

7.5.1 Low-Field NMR in Heterogeneous Samples

It is generally accepted that low-field NMR (Chapter 4) requires the use of compact NMR spectrometers equipped with permanent magnets to create the fields of ≤2 T that correspond to the proton frequency of ≤85 MHz.[27] These devices have a low sensitivity, so most low-field NMR spectrometers are used primarily for the detection of ^{1}H and ^{19}F nuclei with large γ constants and 100% abundance. Nevertheless, even ^{13}C low-field NMR experiments can be performed for hyperpolarization of nuclei, such as *para*-hydrogen-induced polarization. For heterogeneous samples, high homogeneity of the magnetic field is not necessary, and NMR experiments can be performed without lock stabilization (i.e., in the absence of deuterated solvents). Therefore, low-field NMR devices offer the advantages of low maintenance costs and simplicity.

A traditional approach based on low-field NMR techniques and used widely in industry is monitoring chemical reactions while experiments are performed in NMR tubes that act as reactors and are placed into conventional NMR probes. In addition, bypass technology and the special design of NMR probes and other units allow monitoring in either continuous-flow or stopped-flow modes.

Heterogeneous industrial samples are usually characterized by measurement of signal amplitudes, which are proportional to the number of spins in the volume of the probe. For example, low-field NMR determination of proton content in fuels is now an internationally accepted standard procedure for their characterization. This determination obviously requires an initial calibration of the signal amplitude against the number of ^{1}H nuclei. Details regarding the quantification of low-field NMR experiments can be found in References 27 to 29.

Industrial samples generally contain various species that can be discriminated in terms of ^{1}H signals. This is possible via Equation 7.9 when the species show different relaxation properties, particularly ^{1}H T_2 times, connected with the line widths, $\Delta\omega$:

$$1/T_2 = (3/20)(\mu_0/4\pi)^2\left(\gamma^4\hbar^2/4\pi^2 r^6\right)\left[\begin{array}{c}3\tau_C/\left(1+\Delta\omega^2\tau_C^2\right)+5\tau_C/\left(1+\omega_0^2\tau_C^2\right)\\+2\tau_C/\left(1+4\omega_0^2\tau_C^2\right)\end{array}\right] \qquad (7.9)$$

which is valid for rigid molecules. Finally, the important parameters of heterogeneous systems are molecular motion correlation times, which can be found via low-field relaxation (T_1 and T_2) time measurements.[27–29] Here, in addition to the accurate determination of times τ_C, various empirical relationships can be successfully exploited; for example, the well-established correlations between ^{1}H T_2 times and macroscopic parameters (viscosity, melting point, or crystallinity) are often utilized. A similar approach is used for determination of the solid fat content in fats and their blends.

Low-field NMR experiments on samples of sweet corn at a ^{1}H frequency of 22.6 MHz can be used to demonstrate relaxation time characterization of heterogeneous samples in food chemistry.[30] The ^{1}H T_2 times are measured with a standard Carr–Purcell–Meiboom–Gill (CPMG) pulse sequence. Because the relaxation is not exponential (as is the case for most heterogeneous foodstuffs), relaxation curves are treated by a multi-exponential function to obtain T_2 times for the corresponding component. Figure 7.12 shows the data obtained for samples of fresh sweet corn and dried sweet corn. In Figure 7.12A, the T_2 times of 550 to 650 ms, 200 to 250 ms, 50 to 60 ms, and 5 to 10 ms correspond to vacuolar water, cytoplasmic water, extracellular

Fresh Corn – 78% Water

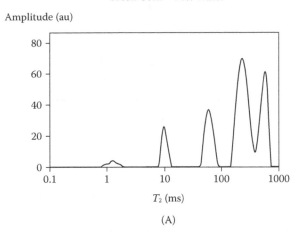

(A)

Dried Corn - 8% Water

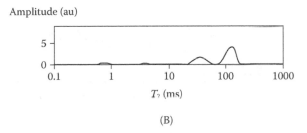

(B)

FIGURE 7.12 (A) The CPMG pattern obtained in a sample of fresh sweet corn. (B) The CPMG pattern obtained in a sample of dried sweet corn.

water, and cell wall protons, respectively.[30] In addition to allowing determination of their relative amounts, the relaxation data demonstrate how the water fractions changed in the sample of dried sweet corn (Figure 7.12B).

In a more academic context, an important application of low-field NMR is probing the molecular mobility of polymers in solutions, because T_1 relaxation times can give detailed information on the molecular motion correlation times in a MHz diapason. Chemical shift resolution in low static magnetic fields has currently improved remarkably and thus has broadened the domain of low-field NMR applications. In this context, the comparison of molecular dynamics data obtained in both high and low magnetic fields is of great interest. The relaxation data collected for polystyrene solutions in $CDCl_3$ in the ^{1}H frequency region between 20 and 600 MHz allows such a comparison (see Figure 7.13).[31] The spectral resolution reached for the low magnetic field (20 MHz) is sufficient to observe separately the signals of aromatic and aliphatic protons (Figure 7.13A). Their [1]H T_1 times can be accurately measured and analyzed as a function of parameters c_V (the solution concentration) and M_n (the molecular mass of the polymer).

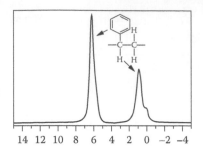

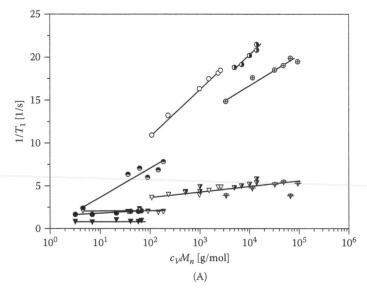

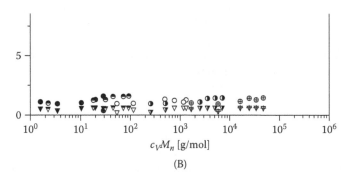

FIGURE 7.13 (A, top) The ^{1}H NMR spectrum of polystyrene in CDCl$_3$ at 20 MHz. (A, bottom) The ^{1}H T_1 times measured at a frequency of 20 MHz for the aliphatic (circles) and aromatic (triangles) protons in polystyrene solutions as a function of their concentration (c_V) and the molecular mass of the polymer (M_n). (B) The ^{1}H T_1 times measured for the same samples at a frequency of 600 MHz. (From Cudaj, M. et al., *Macromol. Chem. Phys.*, 213, 1833, 2012. With permission.)

As follows from relaxation theory, under some conditions relaxation rate T_1^{-1} decreases with increasing NMR working frequency (see Chapter 3). Therefore, the difference between samples, as a function of parameter $c_V M_n$, can clearly be observed in a low magnetic field (Figure 7.13A) but is dramatically reduced in a high magnetic field (600 MHz) (Figure 7.13B). This result illustrates the advantage of using low-field NMR. It is also interesting that the relaxation rate of aromatic protons at a low working frequency of 20 MHz depends on the $c_V M_n$ parameters only slightly, in contrast to the T_1^{-1} value for aliphatic protons (Figure 7.13A), where no discrimination is observed in a high magnetic field (600 MHz) (Figure 7.13B).

The relaxation data were treated with the Lipari–Sabo motional model and Cole–Davidson distributions to characterize quantitatively the motions of the phenyl rings and CH_2 groups by correlation times τ_C.[31] It has been found that experiments in a low magnetic field (20 MHz) better describe molecular mobility than those performed at high magnetic fields (200 MHz and higher).

The ^1H relaxation time measurements can be carried out in ultra-low magnetic fields corresponding to the ^1H frequencies between 10 Hz and 100 kHz.[32] These experiments cover the large diapason of motions connected with diffusion, intramolecular motions, chemical reactions, and biological processes, such as protein folding, catalysis, and ligand binding.

7.6 FREE RADICALS IN NMR SPECTRA OF SOLUTIONS

The large magnetic moment of an unpaired electron causes paramagnetic chemical shifts of nuclei neighboring with this electron. Low- or high-field displacements of resonances in NMR spectra are dictated by the sign of the isotropic electron/nucleus hyperfine coupling constants. In addition, strong electron–nucleus dipolar interactions can result in significant broadening of resonances, particularly for nuclei located closer to unpaired electrons. All of these factors complicate the observation of paramagnetic species; however, in spite of the difficulties, even simple one-pulse NMR experiments can be successfully performed to characterize free radicals in solutions.

When NMR signals of free radical species (e.g., stable verdazyl radicals[33]) are well observed in ^1H NMR spectra, the ^1H NMR spectra are recorded in a wide range of δ (generally from +100 to –120 ppm) and concentrations of the radicals. The chemical shifts measured experimentally include both paramagnetic and diamagnetic components, and the paramagnetic chemical shifts can be calculated on the basis of the concentration dependencies. Then, if the paramagnetic shift values are well determined, the corresponding isotropic electron–nucleus hyperfine coupling constants, A, can be calculated. Finally, spin density ρ on a neighboring carbon atom can be found via Equation 7.10:

$$A = \rho Q \tag{7.10}$$

using the known parameter Q; for example, $Q = -23.7$ G for aromatic CH fragments.

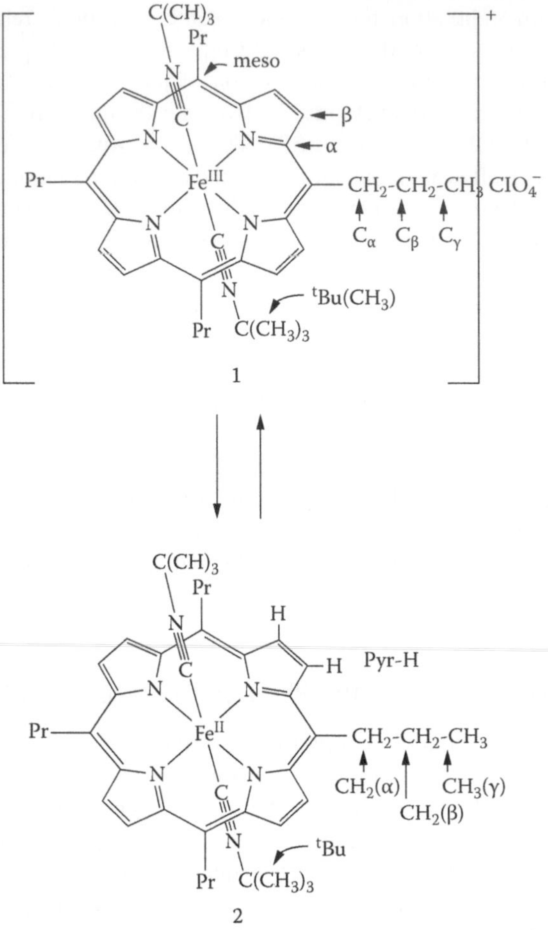

FIGURE 7.14 Self-exchange electron transfer between diamagnetic Fe(II) complex 2 and paramagnetic Fe(III) complex 1 in solutions.[34]

Sometimes, the resonances of radical species are not visible in NMR spectra due to exclusively fast relaxation. This situation occurs for the low-spin bis(*tert*-butylisocyanide) complex of chloro(*meso*-tetrapropylporphyrinato)iron(III) (complex 1 in Figure 7.14), which shows no resonances in the ^{13}C and ^1H NMR spectra.[34] These conditions change the methodology of studies such that the problem of determining the chemical shifts for the invisible carbon and hydrogen atoms is solved on the basis of the self-exchange electron transfer between the diamagnetic Fe(II) complex (complex 2 in Figure 7.14) and the paramagnetic Fe(III) complex (complex 1 in Figure 7.14). In fact, when the paramagnetic compound is added to a solution of the diamagnetic complex, its ^1H and ^{13}C chemical shifts follow Equation 7.11, where $X = $ (complex 1)/(complex 2):

$$\delta(obs) = \delta_{Fe(II)}(1 + X)^{-1} + \delta_{Fe(III)}X(1 + X)^{-1} \qquad (7.11)$$

Thus, the ^1H and ^{13}C values can be found for complex 1.

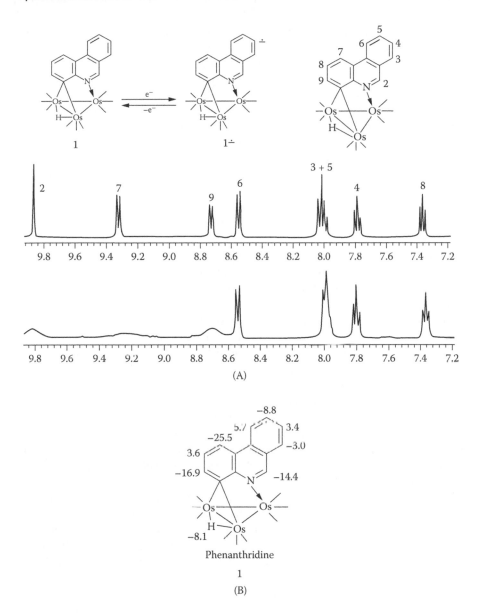

FIGURE 7.15 (A) Room-temperature ¹H NMR spectra of cluster 1 before (top) and after (bottom) the addition of cobaltocene in CD₂Cl₂. (B) The scaled distribution of unpaired spin density in cluster 1. (From Nervi, C. et al., *Chem. Eur. J.*, 9, 5749, 2003. With permission.)

A similar approach has been used to characterize radical anions generated by the addition of cobaltocene to solutions of diamagnetic Os cluster 1 in Figure 7.15A.[35] Again, due to the fast electron transfer exchange, the ¹H NMR signals of three aromatic hydrogens (H², H⁷, and H⁹) and the resonance of the Os–H hydrogen broaden greatly. However, it can clearly be seen that the broadening effect is different for each resonance observed. At a lower temperature (–80°C), all of the proton resonances transform into the baseline due to the very large line widths. The ^{13}C NMR spectrum of cluster 1 in the presence of cobaltocene shows the resonances of carbons C^2, C^7, and C^9. In full accordance with the ^{1}H NMR data, they are also selectively broadened to give the distribution of unpaired electron density shown in Figure 7.15B.

Free radicals formed in chemical and biological processes can be successfully studied by solution NMR with so-called spin trapping techniques.[36] These studies are based on the ability of some compounds (e.g., 5-diisopropoxy-phosphoryl-5-methyl-1-pyrroline-N-oxide) to react with short-lived free radical species. The reactions result in stable radical adducts that can be conveniently characterized by regular NMR techniques, such as ^{31}P NMR spectra for spin trapping.

Another methodology is applied to study radical species using complementary electron paramagnetic resonance (EPR)/NMR experiments, where the electron–nucleus double resonance (ENDOR) technique plays a major role. An ENDOR experiment is based on the simultaneous application of two electromagnetic irradiation fields, working in the microwave region and in the usual radiofrequency range, in order to drive EPR and NMR transitions of unpaired electrons and nuclei that are hyperfine coupled.

Because the relaxation times of electrons and nuclei in solutions (generally between 10^{-5} and 10^{-7} s) are much shorter than those in the solid state, solution ENDOR experiments in continuous wave (CW) mode require very powerful saturating microwave and radiofrequency fields. For this reason, these experiments are more demanding technically; however, they can directly give the isotropic hyperfine coupling constants, $A(iso)$, which are needed to identify the radicals. In fact, the $A(iso)$ values (e.g., via Equation 7.10) lead to a spin-density distribution of unpaired electrons to describe the electronic structure of the radical. A disadvantage of the ENDOR technique is its sensitivity, which is lower by 1 to 2 orders relative to EPR spectroscopy. An electron–nucleus–nucleus triple resonance technique, which is more sensitive, and a pulse ENDOR version are also available.[37]

The hyperfine coupling constants in radical intermediates in fast radical reactions can be measured by the chemically induced dynamic nuclear polarization (CIDNP) effect observed in solution NMR spectra with the spin-selective recombination of radical pairs due to spin conservation during a chemical reaction.[38,39] The vector model in Figure 7.16A explains the CIDNP effect for a radical pair, marked as 1 and 2, containing one proton in radical 1. The model includes the electron Zeeman interaction in combination with a hyperfine interaction. As in the case of nuclei, the Zeeman frequency of electrons is a function of factors g, the Bohr magneton, and external magnetic field B_0. Because g factors can be different for two radicals, in this case g_1 is assumed to be lower than g_2. In the absence of the hyperfine electron–proton interaction, the two vectors are parallel at starting time t_0. However, the vectors will not be in phase and will be moving differently, as shown in this vector diagram. For example, the vector of electron 1 moves counterclockwise, while electron 2 can move clockwise to reach a singlet state. As result, the vectors become anti-parallel.[40]

In the presence of the hyperfine electron–proton interaction, the a constant gives the contribution $+a/2\hbar$ to the Larmor frequency of radical 1 at the proton spin state, shown as α>. For the β> proton spin state, this contribution will obviously be $-a/2\hbar$.

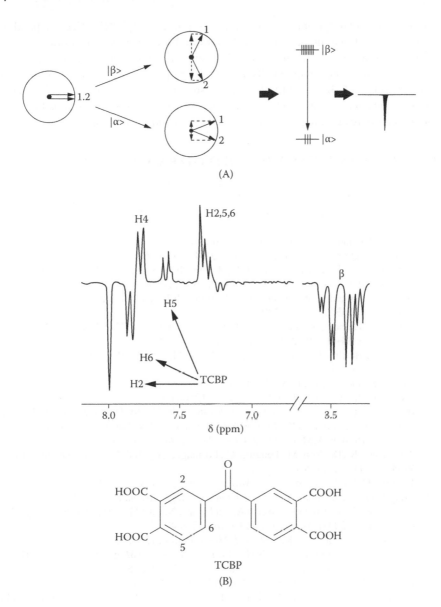

FIGURE 7.16 (A) The CIDNP effect in a radical pair with one proton in radical 1. (From Goez, M., *Concepts Magn. Reson.*, 7, 69, 1995. With permission.) (B) The ¹H CIDNP NMR spectrum obtained after the photoreaction of compound TCBP with L-tryptophan in D₂O. (From Morozova, O.B. et al., *Phys. Chem. Chem. Phys.*, 13, 6619, 2011. With permission.)

If the a constant is positive, the Zeeman and hyperfine effects will be opposite one another for the pairs with proton spin states $\alpha>$. The nuclear spin states $\beta>$ will be overpopulated in the singlet pairs, leading to the observation of an emission signal in Figure 7.16A.

When molecular systems contain many protons, the CIDNP effect, depending on the signs and magnitudes of the g factors and a constants, can result in different phases of proton signals in the ^{1}H NMR spectra. This effect, observed in the ^{1}H NMR spectrum of the compound TCBP and tryptophan in a D_2O solution, initiated by the presence of short-lived radicals,[40] is shown in Figure 7.16B. The hyperfine coupling constants, a, are easily determined on the basis of the well-established proportionality relation between the CIDNP effect and the a values.

REFERENCES AND RECOMMENDED LITERATURE

1. Quirt, A.R. and Martin, J.S. (1971). *J. Magn. Reson.*, 5: 318.
2. Pirkld, W.H., Burlingsme, T.G., and Beare, S.D. (1968). *Tetrahedron Lett.*, 56: 5849.
3. Berdnarek, E., Dobrowolski, J.C., and Mazuriek, A.P. (2001). *Acta Polon. Pharm. Drug Res.*, 58: 9.
4. Wenzel, T.J. and Wilcox, J.D. (2003). *Chirality*, 15: 256.
5. Seco, J.M., Quiñoá, E., and Riguera, R. (2004). *Chem. Rev.*, 104: 17.
6. Kabachnik, M.I., Mastryukova, T.A., Fedin, E.I., Vaisberg, M.S., Morozov, L.L., Petrovskii, P.V., and Shipov, A.E. (1978). *Russ. Chem. Rev.*, 47: 9.
7. Hulst, R., Kruizinga, W., Wynberg, H., Feringa, B.L., and Kellogg, R.M. (1995). *Recl. Trav. Chim. Pays-Bas.*, 114: 220.
8. Labuta, J., Ishihara, S., Sikorsky, T., Futera, Z., Shundo, A., Hanykova, L., Burda, J.V., Ariga, K., and Hill, J.P. (2013). *Nat. Commun.*, 4: 2188.
9. Banci, L., Bertini, I., Luchinat, C., and Mori, M. (2010). *Prog. Nucl. Magn. Reson. Spectrosc.*, 56: 247.
10. Ishima, R. and Torchia, D.A. (2000). *Nat. Struct. Biol.*, 7: 740.
11. Akke, M. (2002). *Curr. Opin. Struct. Biol.*, 12: 642.
12. Bertini, I., Luchinat, C., Parigi, G., and Pierattelli, R. (2008). *Dalton Trans.*, 2008: 3782.
13. Belokon, Y.N., Pritula, L.K., Tararov, V.I., Bakhmutov, V.I., Gusev, D.G., Saporovskaya, M.B., and Belikov, V.M. (1990). *J. Chem. Soc. Dalton Trans.*, 1873.
14. Alphonsa, J.R., De Zotti, M., Peggion, C., Formaggio, F., Toniolo, C., and De Borggraeve, W.M. (2011). *J. Pept. Sci.*, 17: 377.
15. Johnson, C.S. (1999). *Prog. Nucl. Magn. Reson. Spectrosc.*, 34: 203.
16. Koay, C.G. and Özarslan, E. (2013). *Concepts Magn. Reson. Part A*, 42A: 116.
17. Pemberton, C., Hoffman R., Aserin A., and Garti, N. (2011). *J. Magn. Reson.*, 208: 262.
18. Caldarelli, S. (2007). *Magn. Reson. Chem.*, 45: S48.
19. Otto, W.H. and Larive, C.K. (2001). *J. Magn. Reson.*, 153: 273.
20. Jonas, J. (2007). High-pressure NMR. In: *Encyclopedia of Magnetic Resonance* (Harris, R.K. and Wasylishen, R., Eds.). Chichester: John Wiley & Sons.
21. Bai, S. and Yonker, C.R. (1998). *J. Phys. Chem. A*, 102: 8641.
22. Tsukahara, T., Harada, M., Tomiyasu, H., and Ikeda, Y. (2008). *J. Phys. Chem. A*, 112: 9657.
23. Bargon, J. and Natterer, J. (1977). *Prog. Nucl. Magn., Reson. Spectrosc.*, 31: 293.
24. Duckett, S.B. and Mewis, R.E. (2012). *Acc. Chem. Res.*, 45: 1247.
25. Theis, T., Ganssle, P., Kervern, G., Knappe, S., Kitching, J., Ledbetter, M.P., Budker, D., and Pines, A. (2011). *Nat. Phys.*, 7: 571.
26. Gusev, D.G., Lozinsky, V.I., Vainerman, E.S., and Bakhmutov, V.I. (1990). *Magn. Reson. Chem.*, 28: 651.
27. Dalitz, F., Cudaj, M., Maiwald, M., and Guthausen, G. (2012). *Prog. Nucl. Magn. Reson. Spectrosc.*, 60: 52.
28. Kantzas, A., Bryan, J.L., Mai, A., and Hum, F.M. (2005). *Can. J. Chem. Eng.*, 83: 145.

29. Ramos, P.F.O., Toledo, I.B., Nogueira, C.M., Novotny, E.H., Vieira, A.J.M., and Azeredo, R.B.V. (2009). *Chemom. Intell. Lab. Sys.*, 99: 121.
30. Shao, X. and Li, Y. (2013). *Food Bioprocess Technol.*, 6: 1593.
31. Cudaj, M., Cudaj, J., Hofe, T., Luy, B., Wilhelm, M., and Guthausen, G. (2012). *Macromol. Chem. Phys.*, 213: 1833.
32. Zotev, V.S., Matlashov, A.N., Volegov, P.L., Savukov, I.M., Espy, M.A., Mosher, J.C., Gomez, J.J., and Kraus, R.H. (2005). *J. Magn. Reson.* 175: 103.
33. Lang, A., Naarmann, H., Walker, N., and Dormann, E. (1993). *Synthetic Met.*, 53: 379.
34. Niibori, Y., Ikezaki, A., and Nakamura, M. (2011). *Inorg. Chem. Commun.*, 14: 1469.
35. Nervi, C., Gobetto, R., Milone, L., Viale, A., Rosenberg, E., Rokhsana, D., and Fiedler, J. (2003). *Chem. Eur. J.*, 9: 5749.
36. Zoiaa, L. and Argyropoulos, D.S. (2010). *J. Phys. Org. Chem.*, 23: 505.
37. Möbius, K., Lubitz, W., and Savitsky, A. (2013). *Prog. Nucl. Magn. Reson. Spectrosc.*, 75: 1.
38. Goez, M. (1995). *Concepts Magn. Reson.*, 7: 69.
39. Pine, S.H. (1972). *J. Chem. Educ.*, 49: 664.
40. Morozova, O.B., Ivanov, K.L., Kiryutin, A.S., Sagdeev, R.Z., Kochling, T., Vieth, H.M., and Yurkovskaya, A.V. (2011). *Phys. Chem. Chem. Phys.*, 13: 6619.

Recommended Literature

Belton, P.S., Gil, A.M., Webb, G.A., and Rutledge, D., Eds. (2003). *Magnetic Resonance in Food Science: Latest Developments*. London: Royal Society of Chemistry.
Bertini, I., Luchinat, C., Parigi, G., and Pierattelli, R. (2005). NMR spectroscopy of paramagnetic metalloproteins. *ChemBioChem.*, 6: 1536.
Bertini, I., Molinari, H., and Niccolai, N., Eds. (1991). *NMR and Biomolecular Structure*. Deerfield Beach, FL: VCH Publishers.
Clore, G.M, and Iwahara, J. (2009). Theory, practice and applications of paramagnetic relaxation enhancement for the characterization of transient low-population states of biological macromolecules and their complexes. *Chem. Rev.*, 109: 4108.
Damoense, L., Datt, M., Green, M., and Steenkamp, C. (2004). Recent advances in high-pressure IR and NMR techniques for the determination of catalytically active species in rhodium- and cobalt-catalyzed hydroformylation reactions. *Coord. Chem. Rev.*, 248: 2393.
Hills, B.P. (2006). Applications of low-field NMR to food science. *Annu. Rep. NMR Spectrosc.*, 58: 177.
Rule, G.S. and Hitchens, T.K. (2005). *Fundamentals of Protein NMR Spectroscopy*. New York: Springer Verlag.
Wenzel, T.J. (2007). *Discrimination of Chiral Compounds Using NMR Spectroscopy*. New York: Wiley.
Williamson, M.P. (2009). Applications of the NOE in molecular biology. *Annu. Rep. NMR Spectrosc.*, 65: 77.
Wüthrich, K. (1995). *NMR in Structural Biology*. Singapore: World Science.
Yonker, C.R. and Linehan, J.C. (2005). Use of supercritical fluids as solvents for NMR spectroscopy. *Prog. Nucl. Magn. Reson. Spectrosc.*, 47: 95.

8 Solid-State NMR Spectroscopy
General Principles and Strategies

Solid-state NMR spectroscopy is widely applied in modern chemistry and biochemistry to characterize insoluble molecular systems ranging from simple inorganic and organic molecules to complex bio-organic molecules and proteins and their complexes. Objects of solid-state NMR studies in materials science are inorganic/organic aggregates in crystalline and amorphous states, composite materials, heterogeneous systems including liquid and gas components, suspensions, and molecular aggregates with dimensions on the nanoscale, where different nuclei can be used as NMR probes. Solid-state NMR is capable of solving various problems at the atomic level, including simple descriptions of molecular fragments and structural units in various systems and the accurate determination of their structure and mobility on a very large time scale. In contrast to NMR in solutions, solid-state NMR spectroscopy allows measurement of three-dimensional (3D) chemical shifts that potentially better describe the electron structure of compounds.

Solid-state NMR techniques have experienced ongoing development aimed at improving the resolution, sensitivity, and registration of inconvenient nuclei that have high quadrupolar interactions and are traditionally very popular in materials chemistry. A number of reviews describing modern one-dimensional (1D) and two-dimensional (2D) NMR techniques and approaches can be found in the recommended literature.

The physical basis of NMR in liquids and solids is the same; however, there are certain features that can cause difficulties in the application of solid-state NMR spectroscopy even for researchers familiar with NMR in solutions. These include the excitation of nuclear spins (using very powerful pulses) and registration of NMR signals in solids (direct and indirect). Also, the high spinning rates of samples applied in magic-angle spinning (MAS) experiments differently affect the line shapes of nuclei experiencing dipolar and quadrupolar interactions. In addition to the main spin transitions, quadrupole nuclei often show satellite transitions, which affect both line shape and line broadening. All of these factors play an important role in the interpretation of NMR spectra and even in the design of studies adequate for the task at hand.

8.1   DETECTION OF NMR SIGNALS IN SOLIDS

Molecular motions in solutions are generally isotropic, high amplitude, and very fast on the NMR time scale. Because of these motions, resonances in NMR spectra are very sharp and line widths depend primarily on the homogeneity of the magnetic field applied. In solids, because of limited molecular mobility and non-averaged dipolar, chemical shift anisotropy (CSA), and quadrupolar interactions, the resonances are very broad. For example, line width in the ^1H NMR spectrum of liquid water is <1 Hz, while the resonance in ice is broadened up to ~10^5 Hz. This is a major feature of solid-state NMR, which requires special methods and techniques for the excitation and registration of signals.

Observation of NMR signals in solids is based on the excitation of nuclear spins by various radiofrequency (RF) pulse sequences that can be divided into two groups implying direct and indirect excitation. Direct excitation is generally applied for simplest solid-state NMR experiments similar to solution NMR. Here, a single RF pulse excites the necessary frequency range. Commercial solid-state NMR spectrometers routinely produce the powerful RF pulses that uniformly excite the frequency area of 50 to 100 kHz. Generally speaking, such an area is sufficient to detect undistorted resonances for nuclei with spins of 1/2 in diamagnetic molecular systems even in the presence of strong dipolar coupling or chemical shift anisotropy interactions. The situation changes in the case of quadrupolar nuclei, where extremely strong quadrupolar interactions can produce inhomogeneous line broadening of up to several MHz; thus, uniform excitation will be impossible.

Under these conditions, a RF pulse applied at the Larmor frequency of quadrupolar nuclei will effectively excite only the central spin transitions of these nuclei, while satellite transitions will remain off resonance. As a result, the collected free induction decays (FIDs) and NMR spectra after Fourier transformation will be remarkably distorted. The distorted FIDs can appear even for nuclei with smaller quadrupolar properties; for example, such distortions can be seen in NMR spectra of directly detected ^{14}N nuclei for quadrupole coupling constants of >200 kHz. To reduce FID distortions, experiments can be carried out by collecting a number of sub-spectra, the individual widths of which are determined by incrementing the RF frequency applied. On the other hand, in the absence of sub-spectra, the FID distortions can be partially avoided by decreasing pulse lengths or by selective irradiation. The composite and shaped pulses, or the frequency-stepped adiabatic half-passage (FSAHP) method, can also be applied to record such spectra.[1] Minimization of baseline distortions in NMR spectra after Fourier transformation can be achieved by using Hahn echo pulse sequences ($90^\circ_X \rightarrow \tau \rightarrow 180^\circ_Y \rightarrow \tau \rightarrow$ FID) or solid echo pulse sequences ($90^\circ_X \rightarrow \tau \rightarrow 90^\circ_Y \rightarrow \tau \rightarrow$ FID), where τ is the echo delay time. However, it should be emphasized that Hahn echo and solid echo experiments performed on spinning samples should be synchronized with rotor periods to reach maximal peak intensities. The main disadvantage of a technique based on direct excitation is that the low sensitivity requires a large number of scans to achieve good signal-to-noise ratios in the NMR spectra, particularly in static samples where line widths are extremely large. This disadvantage is intensified for nuclei with a low natural abundance and long relaxation times.

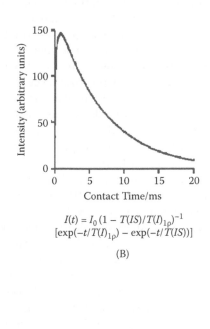

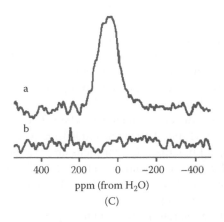

^1H $90°_x$ - Contact Pulse$_{-Y}$ ^1H Decoupling

^{13}C - Contact Pulse$_{-Y}$ - ^{13}C FID

Hartmann–Hahn Conditions

$$y_H B_1(^1H) = y_c B_1(^{13}C)$$

(A)

$$I(t) = I_0 (1 - T(IS)/T(I)_{1\rho})^{-1}$$
$$[\exp(-t/T(I)_{1\rho}) - \exp(-t/T(IS))]$$

(B)

a

b

400   200   0   −200   −400

ppm (from H_2O)

(C)

FIGURE 8.1 (A) Cross-polarization (CP) pulse sequence and Hartmann–Hahn conditions for $^{13}C\{^1H\}$ CP NMR experiments in solids. (B) Kinetics of ^1H–^{31}P CP in solid $CaHPO_4 \cdot 2H_2O$. (C) The ^{27}Al–^{17}O CP MAS NMR spectrum in a static sample of α-alumina (a) and the ^{17}O NMR spectrum obtained by direct excitation (b). (From Bakhmutov, V.I., *Chem. Rev.*, 111, 530–562, 2011. With permission.)

Detection via cross-polarization (CP) is typical for solids. It is realized by polarization transfer from abundant spins (^1H, ^{19}F, etc.) to dilute spins (^{13}C, ^{15}N, ^{29}Si, etc.) to complete two important tasks. First, CP improves the sensitivity of experiments by significant enhancement of the signal-to-noise ratios in the NMR spectra of nuclei with a low natural abundance and low γ values. Second, relaxation delays in CP experiments are optimized to account for spin–lattice relaxation times of abundant spins (^1H, ^{19}F, etc.). Because these relaxation times are remarkably shorter than those of dilute spins, the duration of the experiments is reduced. The cross-polarization pulse sequence used for $^{13}C\{^1H\}$ CP NMR experiments in solids, shown in Figure 8.1A, is directed toward a transfer of magnetization from the ^1H spins to the ^{13}C spins via dipolar coupling between them. The amplitudes of two contact pulses applied at the proton and carbon frequencies should be adjusted in accordance with the Hartmann–Hahn conditions.[2] Under these conditions, the energy gaps between the ^1H and ^{13}C spin states become equal in the rotating frame, and transitions of the ^1H spins will be exactly compensated energetically by transitions of the ^{13}C spins.

The intensity of signals in CP NMR spectra will depend on the accuracy of adjustments of the Hartmann–Hahn conditions, requiring proper calibration of 90° pulses and correct adjustment of the carrier ^1H frequency, distances between coupled nuclei (i.e., their close space proximity), molecular mobility in a solid sample, and ^1H T_1 times. Thus, fast molecular motions break the contact between ^1H and ^{13}C spin systems, and intensities in the ^{13}C CP NMR spectra are reduced. By analogy, the efficiency of cross-polarization decreases with unusually short ^1H T_1 times. The very short relaxation times of abundant spins violate the polarization transfer conditions, as has been observed, for example, in paramagnetic samples, where direct detection is preferable. On the basis of the above statements, one can conclude that the signal intensities in CP NMR spectra are not proportional to the absolute amounts of corresponding nuclei in samples.

Figure 8.1B shows the ^{31}P signal intensity as a function of contact time in the cross-polarization ^1H–^{31}P NMR experiments in solid $CaHPO_4 \cdot 2H_2O$.[3] The CP ^{31}P intensity initially increases with contact time to reach a maximum and then decreases to zero. This experiment clearly demonstrates that contact times should be optimized to provide the best sensitivity in CP experiments. The kinetics of the cross-polarization in Figure 8.1B can be treated quantitatively via the equation shown, where $T(IS)$ is the cross-polarization constant and $T(I)_{1\rho}$ is the ^1H spin–lattice relaxation time corresponding to the rotating frame. Generally speaking, the $T(IS)$ constants are specific and characterize functional groups involved in the cross-polarization process. These constants are governed by internuclear distances r (via a factor of $1/r^3$), the number of neighboring protons, and the mobility of these groups.

In the case of rigid solids, the cross-polarization rate $1/T(IS)$ can be expressed as

$$1/T(IS) = (3/2)(2\pi M_2(IS)/5(M_2(II))^{1/2})^{1/2} \qquad (8.1)$$

where $M_2(IS)$ and $M_2(II)$ are the second moments (personifying line widths) that depend on the strength of the dipolar interactions between spins I and S or I and I, respectively. The CP rate decreases with decreasing second moments, due to the more intense molecular (or group) motions.

Factor $\gamma(^1H)/\gamma(S)$, corresponding to the Hartmann–Hahn conditions, shows the potential signal enhancements achieved by ^1H–S cross-polarization experiments for S nuclei with spins of 1/2. In turn, the contact time necessary to provide the polarization transfer increases with decreasing γ constants.[4] Thus, for example, ^{13}C proton–carbon and ^{29}Si proton–silicon CP NMR experiments require different contact times of 1.5 and 6 ms, respectively. ^{109}Ag proton–silver CP MAS NMR experiments can be carried out by increasing the contact time to 50 ms, whereas ^{43}Ca proton–calcium CP MAS NMR requires contact times of up to 70 ms. It should be noted that the effect of putting a high-power RF field on the NMR probe can be catastrophic for such long contact times.

Cross-polarization is particularly necessary in solid-state NMR (MAS or static) studies of molecular systems where the target nuclei show extremely large chemical shift anisotropy. For example, signal enhancements can help remarkably with recording solid-state NMR spectra of ^{199}Hg or $^{119,117}Sn$ nuclei, for which anisotropies reach values of several thousands ppm. Such magnitudes dramatically broaden line widths in static samples or lead to very large sideband patterns in MAS experiments, effectively reducing intensities in NMR spectra.

The cross-polarization sections can be added to various pulse sequences, or they can be used in experiments where the standard cross-polarization is modified by polarization inversion sections. The latter modification is very useful for signal assignments in ^{13}C proton–carbon CP MAS NMR spectra to distinguish ^{13}CH and ^{13}CH$_2$ groups. NMR experiments for the observation of quadrupolar nuclei via proton–nucleus cross-polarization require special Hartmann–Hahn matching conditions. In fact, the polarization transfer can be effective only when the energy level splitting for both of the nuclei is the same in the rotating coordinate system. In reality, however, only a small fraction of the quadrupolar nuclei can be well matched with protons; therefore, signal enhancements in the CP NMR spectra are generally insignificant or even poor. The cross-polarization of ^{1}H–^{27}Al and ^{1}H–^{2}H, for example, is not sufficient. Nevertheless, CP ^{1}H–^{2}H experiments can be utilized to record ^{2}H NMR spectra where baselines are less distorted in comparison with the method of direct excitation.

When a pair of quadrupolar nuclei can be well matched in terms of the Hartmann–Hahn conditions, the cross-polarization NMR technique becomes a very powerful tool for studies of these nuclei in solids.[5] Figure 8.1C shows the ^{27}Al–^{17}O CP NMR spectrum and an attempt to record the ^{17}O NMR spectrum upon direct excitation.[6] The advantage of the cross-polarization is obvious. Similarly, cross-polarization transfer from quadrupolar nuclei to nuclei with a spin of 1/2 can significantly increase signal-to-noise ratios in their NMR spectra. Here, the cross-polarization comes from the central transition of a quadrupolar nucleus, such as ^{27}Al or ^{23}Na to observe ^{29}Si nuclei. In the case of ^{27}Al-to-^{29}Si cross-polarization, the signal-to-noise ratio in the ^{29}Si NMR spectra can be enhanced by a factor of 5.[7] Although this factor is not large, the faster repetition rate for the ^{27}Al-to-^{29}Si CP cycles plays an important role in 2D NMR experiments. In fact, a fivefold increase in signal enhancement will correspond to a 25-fold decrease in experimental time.

8.1.1 MAS NMR vs. Wide-Line NMR

According to the theory of NMR, dipolar and quadrupolar interactions, as well as chemical shift anisotropy or their combinations, lead to very broad resonances in static powder samples, particularly amorphous ones, where various environments of nuclei result in large chemical shift dispersions. When dipole–dipole interactions in rigid solids are dominant, interpretation of such broad lines in the structural context can be based on the Van Fleck approach,[8] where dipolar second moment M_2 (represented as a square width) is measured and calculated in terms of dipolar coupling via Equation 8.2, which represents the sum of all nuclei pairs in crystals or accounts for Σr_{jk}^6 in a polycrystalline sample:

$$\sum \left(3\cos^2\theta_{jk} - 1\right)^2 / r_{jk}^6$$

$$(8.2)$$

Because the r_{jk}^6 factor is strongly dominant only for the closest dipole–dipole contacts, internuclear distances in molecules, groups, or ions can be obtained by computing and comparing the experimental and calculated wide-line NMR spectra. A similar approach can be used for molecular systems for which the nuclei experience quadrupolar and chemical shift anisotropy interactions completely dominating the NMR line shape.

In contrast to crystalline and polycrystalline samples, nucleus–nucleus directional information is losing in amorphous systems. Here, line widths increase due to the presence of chemically and structurally non-equivalent nuclei, and Van Fleck's analysis is much more complex or even impossible. Therefore, for amorphous systems, wide-line solid-state NMR experiments are used to extract information on molecular (or group) motions or phase transitions that can be characterized by relaxation time measurements and/or line-width analysis. In the context of dynamics, static experiments are preferable. They can be performed in a much larger temperature range compared to MAS NMR experiments, which require accurate temperature control and temperature calibration and specific NMR rotors (see the recommended literature).

Because dipolar, quadrupolar, and/or chemical shift anisotropy interactions lead to strong line broadenings, the sensitivity in wide-line NMR experiments is very low. All of the second-rank anisotropic interactions can be reduced or completely removed by MAS NMR (see Chapters 1 and 2) in a fast spinning regime. Two important advantages of MAS NMR should be emphasized. MAS experiments increase sensitivity and result in highly resolved resonances in MAS NMR spectra similar to solution NMR (Figure 8.2A). However, with intermediate spinning rates, which can be well compared to the spread in Larmor frequencies, MAS NMR spectra will show sets of sidebands, which again reduce sensitivity. It should be noted that the sideband patterns reproduce the line shape in a static sample, thus maintaining information about internuclear distances or chemical shift anisotropies, which can still be calculated.

The distances between the sidebands, measured in Hz, are equal to the spinning rates. Their dispositions change as a function of the adjusted rate, allowing for identification of a central resonance that corresponds to the isotropic shift. It should be emphasized that, in spite of the presence of sidebands, MAS NMR provides the unique possibility to study molecular systems (crystalline, polycrystalline, and amorphous) containing multiple sites due to increased spectral resolution.

The sidebands and isotropic resonances in the MAS NMR spectra can be easily discriminated by experiments performed at various spinning rates, although there is a risk in such a strategy, as the duration of the experiments can become too long. This is particularly important for nuclei that relax slowly and thus require long relaxation delays. To avoid this problem and to eliminate completely or partially the sideband patterns, specific pulse sequences can be applied. In general, such pulse sequences include the actions of strong 180° pulses with a timing that has been carefully chosen.

The sidebands can be completely eliminated from MAS NMR spectra by the total suppression of spinning sideband (TOSS) pulse sequence. TOSS experiments result in MAS NMR spectra where only signals corresponding to the isotropic chemical shifts are observed (see Figure 8.2A). In some sense, the TOSS solid-state NMR spectra and high-resolution NMR spectra in solutions are equivalent. Other sequences for sideband manipulations are also available.[9] It is remarkable that in

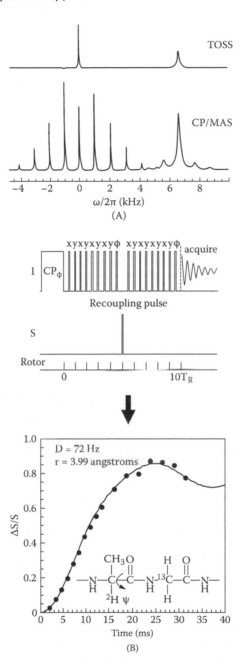

FIGURE 8.2 (A) The regular $^{13}C\{^{1}H\}$ CP MAS (bottom) and TOSS (top) NMR spectra for a sample of glycine spinning at a rate of 1030 Hz.[9] (B, top) REDOR MAS NMR experiments including a single recoupling pulse in the deuterium channel. (B, bottom) Evolution of the ^{13}C–^{2}H REDOR $\Delta S/S$ fraction, observed over time, for a ^{13}C-labeled molecule (Ala–Gly)$_{15}$.[13] (Adapted from Bakhmutov, V.I., *Solid-State NMR in Materials Science: Principles and Applications*, CRC Press, Boca Raton, FL, 2011.)

contrast to wide-line NMR, MAS NMR spectroscopy can be more successfully applied for studies of paramagnetic solids. For example, nitronyl–nitroxide radicals can be exhaustively characterized by ^{13}C CP MAS NMR spectra in spite of the line broadening effects, the large variation in isotropic chemical shifts from −670 to 1250 ppm, and unobservable carbons containing unpaired electrons.[10]

8.2 GENERAL APPROACHES AND STRATEGIES OF NMR STUDIES IN SOLIDS

The methodology and experimental approaches to solid-state NMR studies are different and depend on the following factors: the nature of the objects (crystalline or amorphous, homogeneous or heterogeneous, suspensions or powders), the nature of the target nuclei (quadrupolar or non-quadrupolar), and the type of problem, such as identification of compounds, the determination of their structure, or describing the qualitative/quantitative characteristics of their dynamics. These factors play a major role in the choice of experiments (e.g., static or MAS, spectroscopic or relaxation) and even the method of signal detection. Because of the extremely large variety of objects and problems involved in solid-state NMR (e.g., more than 600 scientific articles published yearly) and because the approaches are often modified for particular objects, only the most common principles and strategies are considered below.

8.2.1 NUCLEI WITH SPINS OF 1/2

At the present time, the focus of solid-state NMR studies of individual compounds and complex systems or materials is directed toward nuclei with spins of 1/2, whereas earlier solid-state NMR experiments were mainly focused on quadrupolar nuclei due to the absence of high-resolution techniques at the time. In solids, nuclei with spins of 1/2 generally have long relaxation times, and only one transition (+1/2 → −1/2) can be easily excited by conventional methods. Currently, high-resolution MAS NMR experiments on ^{1}H, ^{19}F, ^{13}C, ^{15}N, ^{31}P, ^{29}Si, ^{117,119}Sn, ^{199}Hg, or ^{129}Xe nuclei are most popular for structural studies because they provide detailed information on the nature and number of chemically and structurally different sites in samples containing organic and inorganic components or pharmaceutical products. However, one should remember that the spectral resolution achieved in MAS experiments is limited by the nature of the samples; their crystallinity, homogeneity, and molecular mobility play a major role. For example, a crystalline glycine shows sharp ^{13}C resonances in the ^{13}C{^{1}H} MAS NMR spectrum (Figure 8.2A), where line widths are measured as small as 5 Hz in the presence of high-power ^{1}H decoupling. In contrast, the line widths of ^{13}C resonances increase to 50 Hz and much greater in amorphous samples in spite of the high-power ^{1}H (or ^{19}F) decoupling and extremely high spinning rates provided by modern NMR devices.

For diamagnetic samples, nuclei with spins of 1/2 experience mainly homonuclear and/or heteronuclear dipole–dipole interactions expressed quantitatively via dipolar coupling constants (D; see Equation 2.23 in Chapter 2). In turn, because the strength of these interactions depends on orientation of the internuclear vector relative to static magnetic field B_0 via factor $(3\cos^2(\theta) − 1)/r^3$, decreasing the angle term $(3\cos^2(\theta) − 1)$,

MAS conditions can reduce dipolar interactions, resulting in highly resolved NMR spectra. In practice, MAS experiments result in maximally narrowed resonances when a spinning rate is larger than the line width of resonances observed in static samples. Technically, modern MAS devices make this possible by creating spinning rates up to 30 to 60 kHz. Even for ^{1}H nuclei characterized by the highest dipolar coupling constants, highly resolved MAS NMR spectra can be obtained in fast spinning samples (see below). However, it should be emphasized that, even at extremely large spinning rates, resonances of homonuclear-coupled nuclei will still be broadened due to the remaining time-dependent terms. In contrast, MAS NMR spectra of molecular systems with dominant heteronuclear dipolar coupling (e.g., ^{13}C–^{1}H, ^{15}N–^{1}H, ^{13}C–^{19}F) and chemical shift anisotropy interactions show sharp lines even at intermediate spinning rates. The addition of CP detection to MAS conditions remarkably decreases the duration of experiments on nuclei such as ^{13}C or ^{15}N. Finally, one should remember that high molecular mobility decreases CP signal enhancement. This is particularly important for samples containing liquid components. These components cannot be observed by CP detection but they can be revealed by direct detection.

8.2.2   NMR Experiments on Dipolar Recoupling

Because an important element in structural analysis is determination of chemically and structurally different sites, narrowing the lines in fast spinning samples is the greatest advantage of solid-state MAS NMR. At the same time, under MAS conditions, information on the dipolar coupling constants (D) needed for determination of interatomic separations (r) is obviously losing. However, to maintain the advantage of MAS NMR, this information can be recovered by dipolar recoupling experiments performed on spinning samples.[11–13]

These experiments use various echo pulse sequences aimed at homonuclear or heteronuclear dipolar recoupling. The former can be represented by ^{1}H–^{1}H back-to-back (BaBa) MAS NMR spectroscopy based on double quantum NMR pulse sequences with time sections:[12]

$$\text{Excitation} \rightarrow t(1) \rightarrow \text{Reconversion} \rightarrow \text{Acquisition } (t(2)) \qquad (8.3)$$

A double quantum coherence (see Chapter 1), excited in the first time section, evolves during incremented time $t(1)$ and then converts into the detectable single quantum coherence collected during acquisition time $t(2)$. When the increment of time $t(1)$ becomes equal to a rotor period, then the rotor-synchronized 2D NMR spectrum can be recorded. A decrease in time $t(1)$ corresponding to increasing the DQ spectral width results in the observation of a DQ MAS spinning sideband pattern. In turn, the pattern can be analyzed quantitatively in terms of dipolar coupling and internuclear distances.

Recovery of heteronuclear dipolar coupling in spinning samples is generally realized by rotational-echo, double-resonance (REDOR) MAS NMR experiments. For nuclear pairs with spins of 1/2 (e.g., ^{13}C and ^{15}N), these experiments can potentially result in accurate measurement of the distances between them. Experiments on quadrupolar nuclei are slightly more complex, requiring a better understanding of the theoretical and experimental details. In the case of ^{2}H and ^{13}C nuclei in samples with enriched ^{13}C–^{2}H pairs, the deuterium quadrupolar frequency area is very large; therefore, the REDOR technique experiences a problem with uniform excitation of the deuterium spectral range. In ^{13}C-labeled $(Ala–Gly)_{15}$ molecules, for example, the quadrupolar frequency of ^{2}H nuclei at C–H chemical bonds covers a frequency region of 240 kHz. This rather technical problem can be minimized by the application of special composite radiofrequency pulses,[13] shown in Figure 8.2B, because the action of a single dipolar recoupling pulse in the deuterium channel is not satisfactory. Here, instead of the 180° pulse, a ^{2}H composite 90° pulse is used. The pulses in the observed channel (i.e., the channel for ^{13}C registration) are applied at the midpoint and at the end of each rotor cycle. After Fourier transformation, the NMR data indicate the ^{13}C intensity as a function of evolution time. This pattern is illustrated in Figure 8.2B, where the solid line represents calculation of the ^{13}C–^{2}H dipolar coupling (72 kHz) and an internuclear distance (r) of 3.99 Å, which depends on the dihedral angle ψ. The importance of the REDOR MAS NMR technique is difficult to overestimate. Sometimes, it can be a single physical method that allows determination of interatomic distances in solids.[14]

Discussion regarding the theoretical basis for determining internuclear distances in diamagnetic solids and the techniques applied can be found in Reference 15 and the recommended literature. These techniques, based on various pulse sequences such as Carr–Purcell–Meiboom–Gill (CPMG), separated local field (SLF), polarization inversion spin exchange at magic angle (PISEMA), polarization inversion time-averaged nutation spin exchange at magic angle (PITANSEMA), rotational-echo double-resonance (REDOR), and transferred-echo double-resonance (TEDOR), have been successfully applied for accurate determination of F–F and F–Si distances, among others. Some approaches have been specially developed for static samples; for example, static NMR experiments on isotropically labeled glycine achieve an accuracy in measurements of C–C and C–N internuclear distances comparable to that provided by x-ray and neutron diffraction methods. The C_α–N distance in the glycine molecule has been determined as 1.496 ± 0.002 Å by solid-state NMR and as 1.476 ± 0.001 Å by neutron diffraction.[15]

8.2.3 Quadrupolar Nuclei

As noted earlier, the spectral behavior of quadrupolar nuclei in solutions is similar to that of nuclei with spins of 1/2, excluding the difference in relaxation properties. In contrast, quadrupolar nuclei and nuclei with spins of 1/2 differ strongly in the solid state. Quadrupolar nuclei interact with external and local magnetic fields to result in Zeeman splitting of energy levels; however, these levels split additionally due to strong interactions of the nuclei with electric field gradients at these nuclei. Thus, besides the chemical shift anisotropy typical of such nuclei, quadrupole coupling constants and the asymmetry parameters η_q strongly affect the line shapes of quadrupolar nuclei in solid-state NMR (see Chapter 2).

Quadrupolar interactions of nuclei perturb their Zeeman energy by first- and second-order energy corrections. The first-order term is anisotropic and includes an angular dependence similar to dipole–dipole interactions. Thus, this term can be remarkably reduced under MAS conditions. In contrast, MA spinning does not affect the second-order correction, which is isotropic. However, this term is dependent on the Larmor frequency via Equation 8.4:

$$E_m = -(e^2qQ/(4I(2I-1)))^2m/\omega_0 \qquad (8.4)$$

where m is the magnetic quantum number and ω_0 is the Larmor frequency. Thus, the second-order term decreases with increasing external magnetic field strength. This is a positive effect. However, for high magnetic fields, the magnetic anisotropy contribution increases. Thus, a reasonable compromise is needed to optimize both of the effects. It should be remembered that powerful radiofrequency pulses, usually applied for direct detection of spins in solids, simultaneously excite the central ($+1/2 \rightarrow -1/2$) transitions and satellite transitions of quadrupolar nuclei, further complicating interpretation of their solid-state NMR spectra.

Because MAS NMR experiments even at the highest spinning rates can reduce only first-order interactions, the observed resonances will still be broad due to second-order (isotropic) interactions. Therefore, in many cases, NMR experiments on static samples will be preferable, particularly at different magnetic fields. Figure 8.3A shows such experiments performed on a static sample of LaF_3 at various magnetic fields.[16] The ^{139}La NMR spectra demonstrate the positive influence of the highest fields that reduce the second-order terms. Under these conditions, the line shape is again dictated by a combination of quadrupolar and chemical shift anisotropy interactions. Nevertheless, the shapes can be simulated to obtain the quadrupolar and CSA parameters.

Because the strength of quadrupolar interactions depends on the nature of the nuclei, they can be divided into two important categories reflecting the remarkably different spectral behavior in solid-state NMR: (1) nuclei with integer spins (e.g., ^2H, ^6Li, ^{10}B, ^{14}N, ^{50}V) and (2) nuclei with non-integer spins (e.g., ^7Li, ^{11}B, ^{27}Al, ^{23}Na, ^{35}Cl, ^{37}Cl, ^{47}Ti, ^{49}Ti, ^{51}V, ^{81}Br). The first category corresponds to very strong first-order quadrupolar broadenings in static NMR spectra, which can be significantly reduced or even eliminated entirely by NMR experiments on high-spinning samples. In contrast, nuclei with non-integer spins do not experience strong first-order quadrupolar broadenings for their central transitions and therefore can be conveniently probed even by static NMR experiments. In fact, the excited central transitions appear to be the most intense resonances in static powder NMR spectra, while the satellite transitions are spread over a frequency range of many MHz, which is an advantage. At the same time, however, the resonances of the central transitions for non-integer spins will be broad due to strong second-order quadrupolar effects. These broadening can reach tens of kHz; thus, the resonances belonging to structurally different sites can remain unresolved in NMR spectra recorded in static powder samples, a disadvantage. In addition, the regular MAS NMR technique cannot help significantly because the second-order quadrupolar broadenings are reduced only partially under MAS conditions. Nevertheless, resolution in MAS NMR spectra of nuclei with non-integer spins can be improved by special approaches based on 2D NMR, complex sample reorientations during the experiments, and excitation of multiple quantum transitions.

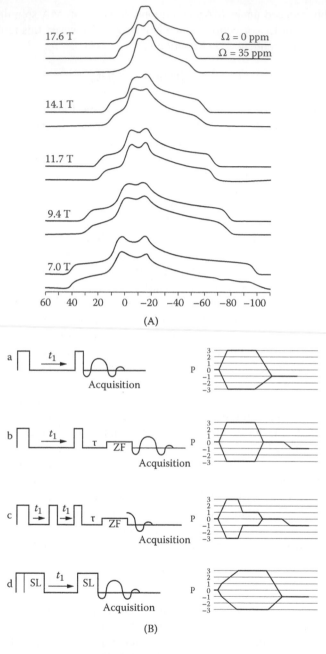

FIGURE 8.3 (A) The ^{139}La NMR spectra recorded in a static sample of LaF$_3$ at different magnetic fields. The experimental NMR spectra are shown below the simulated spectra; the top spectrum for a field of 17.6 T is computed without a magnetic shielding anisotropy contribution.[16] (B) Multi-quantum MAS NMR experiments (see symbols in the text).[17] (Adapted from Bakhmutov, V.I., *Solid-State NMR in Materials Science: Principles and Applications*, CRC Press, Boca Raton, FL, 2011.)

8.2.3.1 Multiple Quantum NMR of Quadrupolar Nuclei in Solids

By definition, multiple quantum (MQ) NMR experiments direct radiofrequency excitation of multiple quantum transitions that are normally forbidden. To be precise, the information on multiple quantum transitions is encoded in a one-quantum NMR spectrum collected during detection period $t(2)$. This information comes from the MQ coherences excited earlier during spin evolution period $t(1)$. Therefore, variations in the duration of time $t(1)$ can indicate the behavior of the forbidden coherences in time. If the number of non-equivalent spins is N, then the number of possible quantum transitions (P) will vary from N to $N - P$, where $1 \leq P \leq N$. Therefore, a smaller number of resonance lines will appear at the same spectral width, and the multiple quantum spectrum will be remarkably simplified and much better resolved.

As mentioned above, quadrupolar interactions strongly affect the regular solid-state NMR spectra of quadrupolar nuclei, masking more weak interactions, such as the effects of chemical shift anisotropy, which are also important for the characterization of systems under investigation. In contrast, multiple quantum NMR spectra do not suffer from quadrupolar broadenings and thus the CSA information becomes accessible.

Figure 8.3B illustrates the 2D MQ MAS NMR experiments often applied to characterize quadrupolar nuclei in the solid state.[17] The simplest experiment, marked (a), shows a MQ transition excited by the action of a single high-power RF pulse resulting in the MQ coherence during time $t(1)$. After time $t(1)$, a second pulse is used to convert this MQ coherence into coherence $p = -1$ collected during time $t(2)$. Then, the echo signals will be detected at time $t(2)$, which is taken to be equal to the quadrupole anisotropy multiplied by time $t(1)$. After 2D Fourier transformation, the resonances show up as ridges located along the quadrupolar–anisotropy axis, and the isotropic spectrum appears as a projection of the entire 2D spectrum.

The experiment marked (b) was accompanied by the so called Z-filter pulse, which is selective and capable of a symmetrical coherence transfer pathway; in contrast, the simplest experiment (a) produces an asymmetrical coherence transfer pathway (see the diagram). In the more complex $t(1)$ split experiment (c), the isotropic spectrum was obtained directly and without shearing the data; however, due to the wide distribution in chemical shifts, which are typical of amorphous systems, part of the signal intensity has been lost. This is an obvious disadvantage of the $t(1)$ split experiment. Finally, the experiment marked (d) used a pulse sequence where the excitation and the conversion of the coherences were performed by spin-locking (SL) sections. This experiment shows the great advantage in the collection of quantitative information. Some applications of these experiments are considered below and technical details can be found in Reference 17 and the recommended literature.

8.2.4 HIGH-PRESSURE NMR IN DISORDERED SOLIDS

As in the case of liquid-phase NMR, the high-pressure NMR technique can be applied for the characterization of solids. This statement is valid only for disordered solids, such as intercalation molecular systems that are relatively compressible even at a moderate pressure in the diapason between 1 and 5 kbar. In fact, the application of pressure to an intercalation compound can result in a decrease in the van der Waals gap sizes, which in turn can affect intercalated molecules.[18] Similar to the case of solutions, high-pressure NMR experiments require pressure-generating equipment, measuring units, and tools.[18] Special attention has been focused on the design of NMR probes, which should be equipped with a high-sensitivity solenoid coil to provide short RF pulses at high power. Specially designed sample cells are also necessary. One of them is shown in Figure 8.4A. Because this NMR technique is not fitted to spinning samples, high-pressure solid-state studies do not play a role in the structural context. At the same time, they are very valuable at probing molecular mobility (considered in Chapter 9). In this context, the static deuterium solid-state NMR spectra of guest molecules (Figure 8.4A, top) are most illustrative.

8.3   ENHANCING SPECTRAL RESOLUTION IN SOLID-STATE NMR

One of the tasks solved by solid-state NMR is the structural description of solids—either full determination of a structure or determination of different sites. The latter is particularly important for amorphous solids, for which the single-crystal x-ray method is not helpful. The simulation of x-ray powder diffraction patterns requires independent data, generally taken again from solid-state NMR spectra. It is obvious that at higher spectral resolutions this information can be extracted more precisely.

Spectral resolution is (1) dictated by the homogeneity of the external magnetic field, (2) principally dependent on the nature of samples (crystalline or amorphous, homogeneous or heterogeneous), (3) controlled by the nature of target nuclei and paramagnetic impurities, and (4) capable of allowing the discrimination of structurally different units in solid-state NMR spectra. Because chemical shift dispersions in solids (particularly amorphous) are large, the homogeneity of the external magnetic field plays a minor role, in contrast to solution NMR.

In many cases, the line widths of nuclei with spins of 1/2 are dominated by strong dipolar couplings. Because line widths reduce approximately linearly with increasing magic-angle spinning rates, spectral resolution can be improved by applying the ultra-fast spinning technique for an additional narrowing effect. It should be noted that improvements in MAS technology and applications of small NMR rotors (2.5-mm diameter or less) have allowed experiments at spinning frequencies up to 65 kHz. This is particularly important for ^1H (or ^{19}F) nuclei, where such a MAS technique can result in the highly resolved ^1H NMR spectra of simple organic solids, mesoporous silica nanoparticles, and microcrystalline proteins. However, one should remember that homonuclear ^1H–^1H (or ^{19}F–^{19}F) dipolar interactions are particularly strong; therefore, ^1H (or ^{19}F) resonances can remain broad even under MAS conditions of ultra-fast spinning.

These residual broadenings in ^1H MAS NMR spectra can be reduced by isotropic dilution with deuterium or eliminated by a homonuclear dipolar decoupling technique known as combined rotation and multiple pulse spectroscopy (CRAMPS) performed at ultra-fast spinning rates. In addition to the fast rotation of a sample in the coordinate space to reduce the factor $(1 - 3\cos^2\theta)$ to zero, the principal idea of the CRAMPS technique is based on coherent averaging of dipolar coupling in the spin space via resonant cyclic and periodic multiple pulse excitations over cycle times.[19] When the times are short relative to the inverse of the homogeneous dipolar coupling,

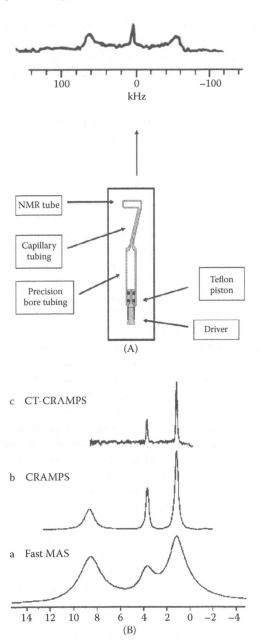

FIGURE 8.4 (A) Sample cell and its components used in high-pressure solid-state NMR studies (bottom) and the solid-state NMR spectrum of an intercalated molecule showing isotropic signals and a powder pattern (top). (From Jonas, J. et al., *J. Magn. Reson. B*, 102, 299, 1993. With permission.) (B) The ^{1}H NMR spectra of L-alanine in a powder sample: MAS at a spinning rate of 30 kHz (a), CRAMPS at a spinning rate of 12 kHz (b), and improved CRAMPS at a spinning rate of 12 kHz (c). (From Lesage, A. et al., *J. Am. Chem. Soc.*, 123, 5747, 2001. With permission.)

the resulting signal will be registered in the windows between pulses in the cyclic and periodic excitations to attenuate homogeneous dipolar interactions. Figure 8.4B shows the ^{1}H NMR spectra of L-alanine recorded in a powder sample under various conditions. As seen, the line widths decrease dramatically from regular MAS NMR to CRAMPS and improved CRAMPS experiments.[20] Other pulse sequences, such as decoupling using mind-boggling optimization (DUMBO) or phase-modulated Lee–Goldburg (PMLG),[21] or 2D ^{1}H–^{1}H double quantum MAS NMR experiments are also available to narrow the ^{1}H resonances to a few Hz.

As mentioned above, the principal problem limiting actual spectral resolution in NMR spectra of quadrupolar nuclei by the MAS techniques is connected with the second-order quadrupolar interactions that cannot be completely removed even at the highest spinning rates. Thus, the discrimination of various sites in solids by the simple MAS NMR technique is limited by the quadrupolar nature of the nuclei.

In such cases, spectral resolution can be significantly enhanced by the 2D multiple quantum MAS NMR experiment, which does not suffer from second-order quadrupolar interactions. Figure 8.5A illustrates a typical situation. The single-pulse ^{27}Al MAS NMR spectrum recorded in a sample of solid AlPO$_4$-14 is poorly resolved and valuable conclusions are not possible. In contrast, resolution is remarkably improved in the projection of the 2D triple quantum MAS NMR spectrum and shows the presence of four different aluminum sites.

Alternatively, a broadening effect caused by second-order quadrupolar interactions can be reduced by performing experiments at the highest magnetic fields or by the application of a technique based on double rotation (DOR)—that is, by sample rotation at two different angles.[22] An experiment with rotation at two different angles obviously requires a specially designed NMR probe. Finally, satellite-transition MAS NMR can be also applied to avoid second-order quadrupolar interactions.

Another strategy for enhancing spectral resolution is based on the so-called relaxation-assisted separation of signals observed in solid-state NMR spectra. Even partially relaxed NMR spectra obtained by standard inversion–recovery pulse sequences can potentially resolve overlapped lines if they have different relaxation T_1 times. This effect, applied for solutions in Chapter 6, can also be useful for solids. For example, the single-pulse ^{1}H MAS NMR spectrum recorded at a spinning rate of 10 kHz in a sample of the material $Zn_{2.5}(H)_{0.4-0.5}(C_6H_3O_9P_3)(H_2O)_{1.9-2.0}(NH_4)_{0.5-0.6}$ shows a broad ^{1}H resonance centered at 7.3 ppm. In contrast, this line is actually a superposition of three resonances at 8.0, 7.0, and 5.6 ppm in the partially relaxed ^{1}H NMR spectra which can be assigned to protons of the C_6H_3 aromatic rings, NH_4 ions, and H_2O molecules absorbed by the surface of the material.[23] The relaxation-assisted approach resolves the signals in solid-state NMR spectra coming from crystal and amorphous fractions of samples (e.g., cellulose). Generally, such signals remain unresolved due to small chemical shift differences.

The effect of enhanced resolution is based on the spin–lattice relaxation times of protons in the rotating coordinate frame which are different in amorphous and crystal phases. Figure 8.5B shows the pulse sequence applied for an analysis of cellulose samples. Here, in the proton channel, the spin-locking delay time acts between the preparation and contact time pulses. Because ^{1}H $T_{1\rho}$ times are different in amorphous and crystal fractions, this spin-locking delay time provides the cross-polarization

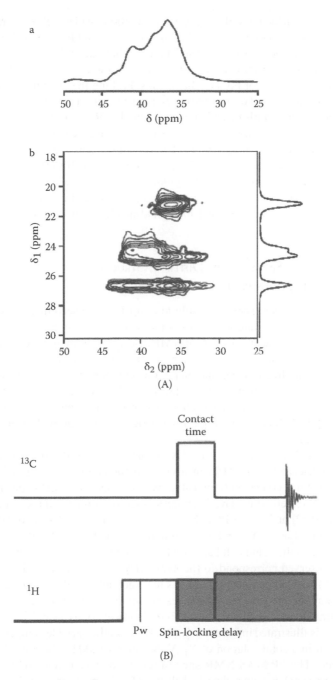

FIGURE 8.5 (A) The ^{27}Al MAS (a) and 2D 3Q MAS (b) NMR spectra and isotropic projection obtained in a sample of a material of the AlPO$_4$-14 family.[21] (B) Spin-locking pulse sequence applied for probing the crystallinity in amorphous/crystal samples.[24] (Adapted from Bakhmutov, V.I., *Solid-State NMR in Materials Science: Principles and Applications*, CRC Press, Boca Raton, FL, 2011.)

from ^1H to ^{13}C nuclei for only the crystal fraction, which relaxes more slowly. Therefore, the NMR spectrum recorded under these conditions will show only ^{13}C resonances corresponding to the crystal phase. Then, according to the protocol in such studies, this spectrum can be compared with the regular ^{13}C CP MAS spectrum exhibiting carbons for both of the phases. Probing the crystallinity in samples containing amorphous and crystal parts can also be carried out by dipolar dephasing pulse sequences, the details of which can be found in Reference 24.

Sometimes, even a simple variation in contact times carried out in CP NMR experiments can resolve signals with close chemical shifts. For example, porous coordination polymers based on bis(phosphine)$PtCl_2$ show $^{31}P\{^1H\}$ NMR spectra and $^{31}P\{^1H\}$ CP NMR spectra recorded at regular contact times, where the two phosphorus resonances expected for such systems are not resolved. In contrast, they are well observed at a short contact time of 100 s, yielding chemical shift values of 46.1 and 40.6 ppm.[25]

8.3.1 Resolving Signals from Different Structural Units in Solids via Paramagnetic Effects

This approach considers spectral resolution capable of discriminating structurally different units by solid-state NMR. In some sense, it is ideologically close to the application of shift reagents in solution NMR. Because paramagnetic shifts are typically larger than chemical shifts in diamagnetic molecular systems, host diamagnetic structures modified by paramagnetic ions can potentially give more structural details in NMR spectra than solids containing only diamagnetic ions. It is obvious that the replacement of diamagnetic ions for paramagnetic ones will be only partial because at high concentrations of paramagnetic ions the line widths of resonances will be too broad.

This idea has been probed by ^{31}P MAS NMR.[26] In fact, the 100% natural abundance of ^{31}P nuclei and high NMR sensitivity can potentially provide rich information on minor structural components invisible by less sensitive NMR experiments. The ^{31}P MAS NMR experiments have been performed on solid solutions of $La_{1-x}Ce_xPO_4$ ($x = 0.027$ to 0.32) and $Y_{1-x}M_xPO_4$ (M = V^{n+}, Ce^{3+}, Nd^{3+}; $x = 0.001$ to 0.014), where diamagnetic ions La^{3+} or Y^{3+} and paramagnetic centers Ce^{3+} and Nd^{3+} are distributed randomly through the volume. It has been found that the number and intensities of the resonances observed correspond to the symmetry and structure of the diamagnetic host phase. Moreover, paramagnetic shifts have actually resulted in better resolution to detect even low-intensity ^{31}P resonances belonging to PO_4 fragments having more than one paramagnetic neighbor, two or four chemical bonds away.[26]

The effect is illustrated in Figure 8.6, which shows the variable-temperature NMR data collected in a solid solution of $Y_{1-x}V_xPO_4$ at $x = 0.002$, where the V^{n+} ions are paramagnetic. The ^{31}P MAS NMR spectra recorded at 47°C and 162°C exhibit the most intense resonance with a chemical shift δ of -11.4 ppm, which is independent of temperature. This resonance obviously belongs to phosphorus atoms that are remote from paramagnetic ions. However, in addition, at least nine low-intensity ^{31}P resonances can be observed between -25 and $+8$ ppm, indicated in Figure 8.6 with the letters A to I. With increasing temperature, the low-intensity resonances experience

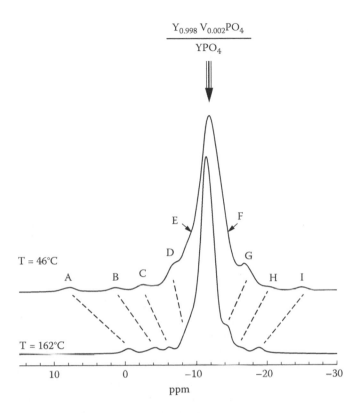

FIGURE 8.6 Variable-temperature ^{31}P MAS NMR spectra of a solid solution containing Y and V ions. (From Palke, A.C. et al., *Inorg. Chem.*, 52, 12605, 2013. With permission.)

a remarkable displacement toward the main line, demonstrating the paramagnetic nature of their shifts. Thus, due to the increased resolution, one can observe the splitting for specific ^{31}P resonances caused by local small-scale distortions.

8.4 ASSIGNMENTS OF SIGNALS

The reliable assignment of signals observed in 1D or 2D NMR spectra is the primary task for their interpretation in any context. In turn, NMR experiments should be performed at the correct parameter settings, including reasonable relaxation delays, to avoid artifacts. If artifact signals still appear, they should be accurately recognized. After Fourier transformation, the NMR spectra should be phased appropriately and baseline corrected. Phasing and baseline corrections are particularly important for the observation of wide lines in solid-state NMR.

Similarly to solutions, signal assignments should be based on isotropic chemical shifts measured experimentally through comparison with already known chemical shifts that are typical of the chemical group, oxidation state of the investigated atoms, or their coordination numbers. In other words, well-established relationships between spectrum and structure can be very useful in this context. Because the majority of spectral structural relationships have been established for solution NMR, one should remember that when moving from solutions to the solid state chemical shift values can change remarkably. For example, due to the effect of environmental closeness in solids, nuclei other than protons show a δ difference of up to 10 ppm. In addition, the nuclei in the crystallographically different fragments can show several lines instead of the single line expected in solutions.

Sometimes, isotropic spin–spin coupling constants are also valuable for signal assignments in solid-state NMR spectra. Such coupling constants, particularly in the case of nuclei with spins of 1/2, are often observed directly in NMR spectra. For example, the ^{13}C CP MAS NMR spectrum obtained in a solid mercury complex (Figure 8.7A) shows the ^{13}C resonance to be ~160 ppm, accompanied by two satellites. These satellites correspond to the $^1J(^{13}C-^{199}Hg)$ constant of 1483 Hz. This observation helps with assignment of the signal, and the $^1J(^{13}C-^{199}Hg)$ constants can be compared in solutions and in the solid state. As for solution NMR, J-resolved NMR spectroscopy can be also applied for the observation and measurement of small scalar coupling constants in the solid state. For example, the $J(^{13}C-^{13}C)$, $J(^{15}N-^{15}N)$, or $J(^{29}Si-^{29}Si)$ constants are very valuable for carrying out signal assignments.[27] The homonuclear scalar coupling constants $J(P-P)$ and $J(C-C)$ are seen even in disordered solids by NMR experiments combining spin echo and 2D INADEQUATE pulse sequences. For example, the solid-state ^{31}P NMR spectra of N,N-bis(diphenylphosphino)-N-((S)-α-methylbenzyl)-amine show the average $^2J(P-P)$ constant in the P–N–P fragment to be between 22.6 and 27.9 Hz. These $^2J(P-N-P)$ coupling constants depend on the local structural parameters;[28] therefore, they can be useful not only in the context of signal assignments but also for the study of structural distortions. Finally, when NMR spectra are recorded by direct excitation, integral intensities can also be an important element in signal assignments. The analysis of integral intensities often includes standard deconvolution procedures for overlapped resonances.

Sometimes the unusually sharp lines observed in solid-state NMR spectra obtained with direct excitation belong to liquid components (or very mobile molecular fragments) in solids. This situation occurs, for example, in heterogeneous porous molecular systems capable of gas/liquid absorption. The disappearance of these signals in CP NMR experiments will confirm the above assignment.

In the case of complex NMR spectra or controversial situations, the assignment of signals can be made on the basis of several NMR experiments that complement each other. For example, the assignment of signals in NMR spectra of X nuclei will be more reliable when experiments with both direct X excitation and ^1H-X cross-polarization are performed. In fact, the CP experiment will show the intensities of signals dependent on the number of closest protons and the H–X distances. In this sense, CP experiments at various contact times can help with signal assignments.

The assignments initially made in a single-pulse solid-state NMR spectrum can be confirmed by homonuclear or heteronuclear correlation experiments. The homonuclear correlation for nuclei with a spin of 1/2, other than protons, is actually effective when the NMR data are collected either by dipolar-based transfers or by J-based techniques.[27] Again, as in the case of solutions, 2D INADEQUATE MAS NMR experiments (e.g., on ^{29}Si nuclei) can show the interconnections between silicon atoms, thus providing unambiguous assignments in ^{29}Si MAS NMR spectra.

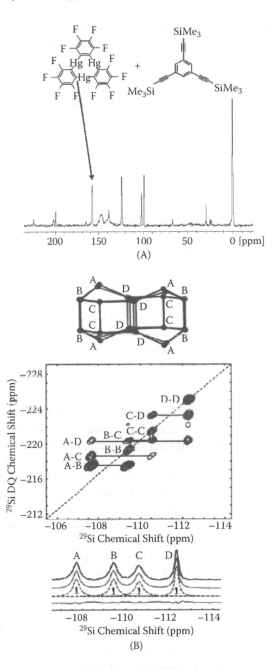

FIGURE 8.7   (A) Solid-state $^{13}C\{^1H\}$ CP MAS NMR spectrum recorded at a spinning rate of 10 kHz for a molecular system including a mercury complex. (B) The ^{29}Si double quantum MAS NMR spectra obtained in a sample of purely siliceous zeolite with assigned silicon atoms: 1D ^{29}Si DQ correlation spectrum (bottom) and 2D ^{29}Si DQ correlation spectrum (top).[27] (Adapted from Bakhmutov, V.I., *Solid-State NMR in Materials Science: Principles and Applications*, CRC Press, Boca Raton, FL, 2011.)

Figure 8.7B demonstrates how the signal assignments can be carried out in ^{29}Si MAS NMR spectra of zeolites investigated at the natural abundance of ^{29}Si nuclei. The four 1–1–1–1 resonances observed by 1D ^{29}Si NMR correspond to four structurally different silicon atoms, marked A, B, C, and D. However, it is then obvious that in the absence of Si–Si correlations their accurate assignment is impossible. The correlations can clearly be seen in the 2D ^{29}Si double quantum (DQ) NMR spectrum, which agrees well with the zeolite structure. It should be emphasized that only with reliable signal assignments can the zeolite structure be solved completely by, for example, use of the recoupling NMR technique to determine the corresponding silicon–silicon distances.

8.4.1 Proton–Proton Proximity and the Assignment of Signals in Solid-State ^{1}H NMR Spectra

As follows from Chapter 5, the proton–proton proximity, important for signal assignments in ^{1}H NMR spectra in solutions used for a structural analysis, can be found via nuclear Overhauser effect (NOE) measurements. In solids, this through-space information requires a more complex technique based on DQ experiments providing H–H proximities even in disordered and heterogeneous molecular systems. One such ^{1}H DQ MAS NMR experiment, performed on silica-based ZSM-12 material containing structural elements and modified by Zr species (Figure 8.8A), is shown in Figure 8.8B.[12] It is obvious that the single quantum ^{1}H MAS NMR spectrum should show all of the protons on the silica surface. Among them, the high-field resonances between 4.5 and 1.5 ppm can be easily attributed to the proton–silica units Si–H and Si–OH in accordance with their chemicals shifts. Also, the two low-field resonances observed at 10.1 and 12 ppm can be attributed to two zirconium hydride surface species; however, they cannot be assigned precisely in the absence of independent information on the proton–proton proximity (see the structure in Figure 8.8A) taken from a DQ experiment. In fact, the 2D DQ ^{1}H NMR spectrum (Figure 8.8B) shows clearly that only one ^{1}H resonance, appearing at 12 ppm, produces a diagonal peak on the DQ frequency axis. Thus, this resonance can be assigned to the ZrH_2 protons. The resonance at 10.1 ppm obviously belongs to the isolated ZrH proton showing a weak correlation at the DQ frequency of 14.5 ppm, and the resonance at 4.4 ppm belongs to the SiH fragment. Details regarding other approaches to proton–proton proximity, such as ^{1}H triple quantum (TQ) MAS NMR experiments to observe dipolar coupling for three protons, can be found in Reference 12.

8.4.2 Magnetic Shielding Tensors vs. Isotropic Chemical Shifts

There is no doubt that isotropic chemical shifts play a primary role in the identification of compounds and determination of their structure in solutions and the solid state. Nevertheless, there are situations when the $\Delta\delta_{iso}$ values of compared compounds are small and therefore are not conclusive in the structural context. In such cases, 3D chemical shifts can be measured as a source of additional data. As noted in Chapter 2, interactions giving rise to chemical shifts are anisotropic and can be characterized by chemical shift anisotropy $\Delta\sigma$ generally invisible in solutions

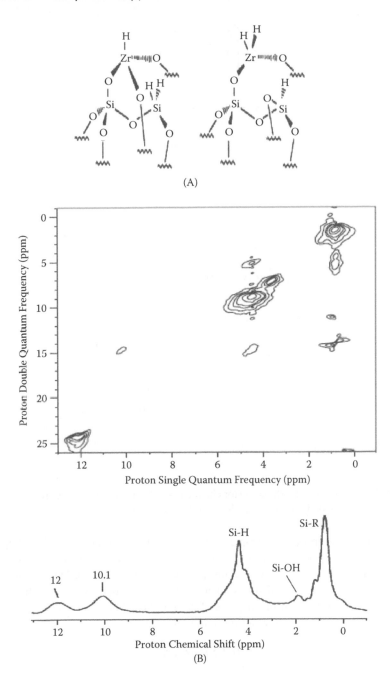

FIGURE 8.8 (A) Silica material containing Zr organic species on silica surface (Si-OH fragments on surface are not shown). (B) ¹H MAS and 2D double quantum MAS NMR spectra recorded in a silica sample at a frequency of 500 MHz and a spinning rate of 30 kHz; one rotor period of BABA recoupling has been applied for the excitation and reconversion of DQ coherence. (From Brown, S.P., *Prog. Nucl. Magn. Reson. Spectrosc.*, 50, 199, 2007. With permission.)

$$\ddot{\sigma} = \begin{bmatrix} \sigma_{xx} & \sigma_{xy} & \sigma_{xz} \\ \sigma_{yz} & \sigma_{yy} & \sigma_{yz} \\ \sigma_{zx} & \sigma_{zy} & \sigma_{zz} \end{bmatrix} \qquad 1$$

$$\delta_{ij} = \frac{\sigma_{iso,ref} - \sigma_{ij}}{1 - \sigma_{iso,ref}} \qquad 2$$

$$\ddot{\sigma}_{PAS} = \begin{bmatrix} \sigma_{11} & 0 & 0 \\ 0 & \sigma_{22} & 0 \\ 0 & 0 & \sigma_{33} \end{bmatrix} \qquad 3$$

(A)

$$\sigma_{11} \le \sigma_{22} \le \sigma_{33}$$

$$\sigma_{iso} = \frac{1}{3} Tr\{\ddot{\sigma}\} \equiv \frac{1}{3}\left(\sigma_{11} + \sigma_{22} + \sigma_{33}\right) \qquad 1$$

$$\Omega \equiv \sigma_{33} - \sigma_{11} \approx \delta_{11} - \delta_{33}$$

$$\kappa \equiv \frac{3\left(\sigma_{iso} - \sigma_{22}\right)}{\Omega} \approx \frac{3\left(\delta_{22} - \delta_{iso}\right)}{\Omega} \qquad 2$$

$$\left|\sigma_{zz} - \sigma_{iso}\right| \ge \left|\sigma_{xx} - \sigma_{iso}\right| \ge \left|\sigma_{yy} - \sigma_{iso}\right|$$

$$\left|\Delta\sigma\right| \equiv \sigma_{zz} - \frac{\sigma_{xx} + \sigma_{yy}}{2}; \quad \eta = \frac{\sigma_{yy} - \sigma_{xx}}{\sigma_{zz} - \sigma_{iso}} \qquad 3$$

(B)

FIGURE 8.9 (A) Magnetic shielding (or chemical-shift) tensor σ (see text). (B) The conventions commonly applied for reporting chemical shift tensors.

due to fast high-amplitude isotropic molecular motions. In contrast, line shapes in solid-state NMR spectra of static and spinning solids are strongly affected by the chemical shift anisotropy. This is a unique feature of solid-state NMR that allows measurement of 3D chemical shifts (or magnetic shielding tensors), which can then be used to characterize various aspects of compounds. The Hamiltonian describing anisotropic chemical shift interactions can be written as

$$\hat{H}(CS) = \gamma B_0 \sigma \hat{I} \tag{8.5}$$

where σ is the magnetic shielding tensor, symbol $\hat{I}$ represents the nuclear spin operator, and the magnetic shielding (or chemical-shift) tensor σ is expressed via a 3×3 matrix (1 in Figure 8.9A). In this matrix, each element corresponds to the i shielding component at field B_0, applied along the j-axis. In turn, this component is related to the chemical shift component (2 in Figure 8.9A) and measured experimentally. As is generally accepted, the principal components in the matrix labeled 3 in Figure 8.9A always correspond to the situation where $\sigma_{11} \leq \sigma_{22} \leq \sigma_{33}$ (Figure 8.9B). In other words, σ_{11} is always the smallest component. Following this terminology, the magnitude of the isotropic shielding σ_{iso} independent of the chosen coordination frame is defined by the equation labeled 1 in Figure 8.9B. Thus, this term is identical to the isotropic chemical shift δ_{iso} to account for $\delta_{iso} = -\sigma_{iso}$.

Generally, three conventional procedures are used to describe the chemical shift tensor values measured experimentally (Figure 8.9B). The first description uses the formalism that has been mentioned above. The second description adds two new terms: span Ω and skew κ. These terms describe the magnitude of the chemical shift anisotropy and a degree of axial symmetry of the chemical shift tensor. Thus, parameter Ω takes positive values measured in ppm. In turn, the κ parameter changes between +1 and −1. The shielding anisotropy ($\Delta\sigma$) and asymmetry parameter η are used in the third expression shown in Figure 8.9B. It should be noted that this terminology is applied more often in the NMR literature.

Figure 8.10A depicts the line shape typical of a non-axial symmetric shielding tensor observed in a static sample, where the values of the principle components are shown by two conventional symbols. This shape can be computed to obtain the chemical shift tensor parameters in the laboratory coordinate system.[29] However, one should remember that the principal (laboratory) coordinate system does not correspond to the molecular coordinate frame. In other words, the orientation of tensor components zz (33), yy (22), and xx (11) within a molecule remains unknown.

The measurement of chemical shift tensors is illustrated in Figure 8.10B, which shows the results of ^{199}Hg NMR experiments on $HgCl_2$ in spinning samples (15 kHz) and in static samples.[30] It is important to emphasize that the sideband pattern observed in the ^{199}Hg MAS NMR spectrum partially reproduces the static line shape. Both of the shapes, including the sideband pattern, can be computed by a standard software package for NMR spectrometers to obtain the following data: $\delta_{11} = -401$ ppm, $\delta_{22} = -442$ ppm, $\delta_{33} = -4034$ ppm, $\delta_{iso} = -1625$ ppm, and anisotropy $\Delta\delta = -2409$ ppm. The solid lines in the spectra correspond to the computer fittings. The principles of fitting procedures and the visualization of chemical shift tensors can be found in the recommended literature.

Knowledge of the shielding tensor parameters and particularly the chemical shift anisotropy ($\Delta\sigma$) obtained by solid-state NMR is also valuable for NMR in solutions. In fact, the chemical shift anisotropy ($\Delta\sigma$) allows evaluation of CSA contributions to total relaxation rates measured in solutions, particularly at very large $\Delta\sigma$ magnitudes. For example, ^{199}Hg T_1 relaxation of $HgCl_2$ in a DMSO solution is exponential, and the T_1 time is calculated as 34.3 ms in a magnetic field of 14.095 T (296 K). The T_1 value is dependent on the Larmor frequency and can be calculated on the basis of an anisotropy of 2409 ppm, thus illustrating the complete domination of the CSA contribution in the ^{199}Hg relaxation mechanism.

Tensor components can be accurately determined in the molecular coordinate frame only by solid-state NMR experiments on single crystals specially oriented in the magnetic field.[31] In contrast, for microcrystalline samples or amorphous and strongly disordered materials, the orientation of the chemical shift tensor components can only be proposed on the basis of the available elements of symmetry.

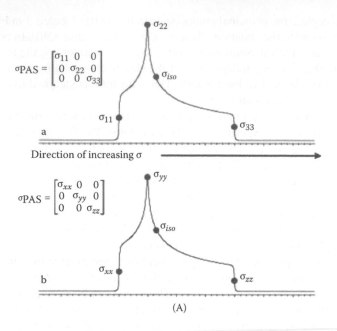

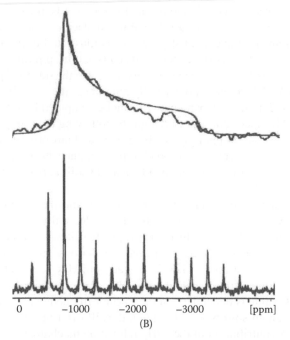

FIGURE 8.10 (A) Line shape typical of a non-axial symmetric tensor observed in a static sample and the terminology used for its description. (B) The ^{199}Hg solid-state NMR spectra recorded for spinning (bottom) and static (top) HgCl$_2$ samples.[30] (Adapted from Bakhmutov, V.I., *Solid-State NMR in Materials Science: Principles and Applications*, CRC Press, Boca Raton, FL, 2011.)

The tensors found in the molecular coordinate frame are directly associated with the electronic structures of the molecules under investigation. In fact, they can be interpreted in terms of electron distributions; however, it should be emphasized that the interpretation of shielding tensors in terms of the electron nature of molecules is not straightforward because of the complex nature of chemical shifts. For example, quantum chemical calculations performed for sets of similar molecules show that diamagnetic shielding components change insignificantly from one molecule to another. In other words, large changes in the total chemical shifts are actually connected with paramagnetic shielding components.

8.5 MULTINUCLEAR SOLID-STATE NMR APPLICATIONS

The great potential of the solid-state NMR technique can be explained by its ability to reveal resonances of any magnetically active nuclei in homogeneous and heterogeneous samples or even fast-relaxing nuclei, which are often invisible in solutions due to exclusively short relaxation times T_1 and T_2. In practice, however, the set of nuclei is still limited because of the following factors: (1) dependence of NMR sensitivity on the natural abundance of nuclei and their γ values, which dictate the duration of experiments; (2) availability of suitable NMR equipment; (3) complexity of experiments; and (4) the expectation of obtaining single-valued interpretations for the planned experiments. In this context, some nuclei are more attractive than others for researchers. They will be in focus here to demonstrate the most common principles of multinuclear NMR applied for solids and to show typical interpretations of these experiments. Because researchers working with NMR in solutions are generally familiar with ^{13}C, ^{19}F, ^{15}N, $^{11,10}B$, and ^{1}H NMR spectra, including their chemical shift ranges and the spectrum/structure relationships established for these nuclei, these topics are not considered in this section. Applications of these nuclei for studies utilizing solid-state NMR can be found in the recommended literature.

8.5.1 ^{31}P SOLID-STATE NMR

^{31}P nuclei are known to have spins of 1/2 and 100% natural abundance and to resonate at a relatively high frequency. These factors provide excellent NMR sensitivity, which is 390 times larger than that for ^{13}C nuclei. Therefore, ^{31}P nuclei are very attractive as targets in studies of individual phosphorus-containing compounds and complex materials or composites. ^{31}P chemical shifts in diamagnetic molecular systems cover a range between 500 and −200 ppm (Figure 8.11A), thus illustrating the large sensitivity of isotropic chemical shifts, $\delta(^{31}P)_{iso}$, to the nature of molecules and their structural features. To account for the large $\delta(^{31}P)_{iso}$ range, the simplest strategy in solid-state NMR studies consists of recording the highly resolved ^{31}P MAS NMR spectra to determine the number of different phosphorus sites. They are then characterized by isotropic chemical shifts to identify structural fragments or the full structure of a solid. The experiments can be performed with direct detection of ^{31}P nuclei or detection via cross-polarization, or both. The ^{1}H–^{31}P cross-polarization MAS NMR technique, particularly its kinetic version, is able to solve structural problems

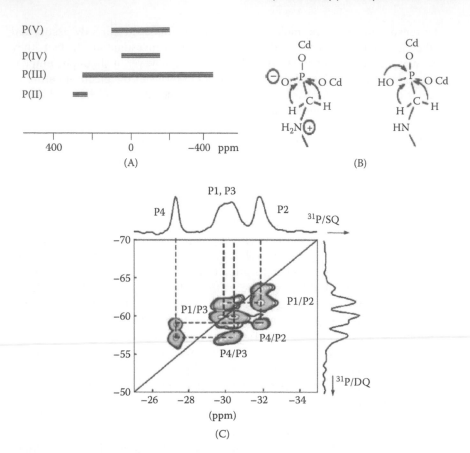

FIGURE 8.11 (A) ^{31}P chemical shifts in diamagnetic molecular systems compared to phosphoric acid. (B) Molecular (right) and zwitterionic (left) structure discriminated by proton localization via H–P cross-polarization. (C) The ^{31}P MAS-J-INADEQUATE NMR spectrum recorded in a sample of β-Ca(PO$_3$)$_2$. (From Hanna, J.V. and Smith, M.E., *Solid State NMR*, 38, 1, 2010. With permission.)

more precisely. For example, the molecular form and zwitterionic structure in Figure 8.11B can be discriminated by measuring the cross-polarization rates, which depend on the number of protons involved in the cross-polarization process.[32] In fact, three protons closest to ^{31}P nuclei (CH$_2$ and POH) are involved in magnetization transfer in the molecular form vs. two protons (CH$_2$) in the zwitterionic structure.

In addition to single-pulse ^{31}P MAS NMR experiments being used to measure isotropic chemical shifts, ^{31}P nuclei, due to their 100% natural abundance and high magnetogyric ratio, are very attractive for observing dipolar ^{31}P–^{31}P correlations in MAS NMR spectra. These correlations are very valuable in the assignment of signals in NMR spectra. For example, four resonances are clearly observed in the ^{31}P single quantum MAS NMR spectrum of the phosphate molecule β-Ca(PO$_3$)$_2$. The spectrum is represented as the corresponding projection in Figure 8.11C. However, accurate assignment, which is important in studies of bioglasses, for example,[33] is not possible, not even on the basis of chemical shifts obtained by first-principles calculations. In contrast, determination of the connectivity for PO_4 units and thus assignment of the resonances can be carried out by the MAS-*J*-INADEQUATE experiment shown in Figure 8.11C. The 2D ^{31}P NMR pattern exhibits cross peaks that directly show how the PO_4 units are connected in this solid.

8.5.1.1 [31]P Chemical Shift Tensors

The principle components of $\delta(^{31}P)$ tensors are associated with molecular structure and can be interpreted in terms of electron distributions. In addition to these rather theoretical aspects, in practice the chemical shift tensors are very useful for structural determinations. Some examples considered in this section illustrate the effectiveness of this approach. Layered metal phosphonates, attractive in chemistry as catalysts, represent a PO_3/metal inorganic core separated by the organic moiety bound to phosphorus. According to the generally accepted classification, PO_3 groups without protons on the oxygen atoms can have three different types of coordination or three types of connectivity. The (111) connectivity corresponds to the situation where each of the three oxygen atoms of the phosphonate unit is bonded to only one metal atom. For (112) connectivity, one of the oxygen atoms is bonded to two metal atoms and the other two atoms are bonded to only one metal atom. Finally, (122) connectivity corresponds to two oxygen atoms bonded to two metal atoms.[34] The ^{31}P isotropic chemical shifts referred to H_3PO_4 in a series of zinc phosphonates are 27.0 to 29.8, 32.4 to 35.9 ppm, and 35.9 to 37.7 ppm, respectively, for connectivity (111), (112), and (122). Because the chemical shift diapasons practically overlap, the [31]P δ_{iso} value is not a characteristic parameter. In contrast, the type of connectivity strongly affects the principal components of the ^{31}P chemical shift tensors, which dictate the shapes of the ^{31}P resonances in static samples of zinc phosphonates. As can be seen in Figure 8.12A, the signal shapes change dramatically as a function of connectivity.[34] The changes are very characteristic and can be used as a good structural criterion for phosphonates of Zr(IV), Al(III), Bi(III), and Cd(II).

Similarly, structural assignment can be made for the cadmium phosphonate hybrid system $Cd_2Cl_2(H_2O)_4(H_2L)$, synthesized from ethylenediamine-*N,N*-bis(methylenephosphonic) acid and named H_4L.[32] The ^{31}P{^{1}H} CP MAS NMR spectrum recorded in a sample of $Cd_2Cl_2(H_2O)_4(H_2L)$ shows a ^{31}P resonance with an isotropic chemical shift of 8.7 ppm. A very similar value of 10.5 ppm was measured in the ^{31}P{^{1}H} CP MAS NMR spectrum of initial phosphonic acid (H_4L). Generally speaking, the change in the $\delta(^{31}P)$ value is too insignificant to lead to a valuable structural conclusion. Even the formation of $Cd_2Cl_2(H_2O)_4(H_2L)$ itself cannot be confirmed. In contrast, the static ^{31}P NMR spectra shown in Figure 8.12B for both of the compounds differ strongly. In fact, the ^{31}P chemical shift tensor in the hybrid system $Cd_2Cl_2(H_2O)_4(H_2L)$ becomes axially symmetric with the components $\delta_{iso} = 8.7$ ppm, $\delta_{11} = -32$ ppm, $\delta_{22} = -32$ ppm, $\delta_{33} = 90$ ppm, and $\eta = 0.0$ vs. the asymmetric tensor in the initial compound: $\delta_{iso} = 10.5$ ppm, $\delta_{11} = -50$ ppm, $\delta_{22} = 3$ ppm, $\delta_{33} = 78$ ppm, and $\eta = 0.8$.

The principal axis values of δ_{11}, δ_{22}, and δ_{33}, obtained experimentally by ^{31}P solid-state NMR spectra, and the orientations of the chemical shift tensors, obtained by single-crystal experiments or computer calculations, can potentially provide more

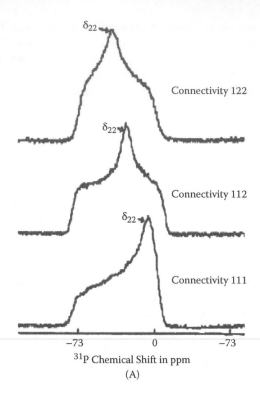

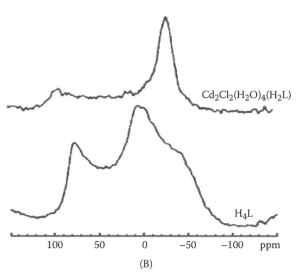

FIGURE 8.12 (A) ^{31}P NMR spectra in static samples typical of zinc phosphonates with different connectivity values.[34] (B) Static ^{31}P{^{1}H} CP NMR spectra of ethylenediamine-*N,N*-bis(methylenephosphonic) acid (H$_4$L) and the Cd$_2$Cl$_2$(H$_2$O)$_4$(H$_2$L) compound.[32] (Adapted from Bakhmutov, V.I., *Solid-State NMR in Materials Science: Principles and Applications*, CRC Press, Boca Raton, FL, 2011.)

structure details. In fact, these parameters are related to the local configuration, conformation, and coordination of the phosphorus-containing units. In many cases, even for unknown tensor orientations in the molecular coordinate frame, empirical correlations between structural parameters and $\delta_{ii}(^{31}P)$ values can be successfully applied for the determination of structure. It should be noted that such correlations are based on the assumption that the tensor principal axis system is related to the molecular coordinate system.

Figure 8.13A shows the CP MAS $^{31}P\{^{1}H\}$ NMR spectrum recorded in a sample of the compound $ZnAMP \cdot 3H_2O$ (where AMP = aminotris(methylene phosphonic acid)). The spectrum exhibits three well-resolved resonances corresponding to three phosphorus sites marked as P_A, P_B, and P_C in full agreement with x-ray diffraction (XRD) data.[35] The isotropic resonances are accompanied by the sidebands, and the pattern can be simulated to give the tensor parameters shown in the figure. It is of interest that these phosphorus sites are characterized by different principal tensor components. The ^{31}P δ_{11} values are practically the same, while the δ_{33} and δ_{22} components experience significant changes. Moreover, only one component, δ_{33}, obtained directly from MAS ^{31}P NMR spectra of $MAMP \cdot xH_2O$ compounds with different metals M, shows a correlation with the phosphorus sites (P_i) – terminal oxygen atoms (O_t) distances (Figure 8.13B), which were obtained independently by the x-ray diffraction method. The origin of the correlation is connected with the circular currents leading to the magnetic shielding contributions dependent on the interatomic distances. Thus, interpretation of chemical shift tensors requires data obtained by the complementary methods of NMR and XRD, and the correlations thus determined can be used in practice.

As in the case of other nuclei, *ab initio* methods (as complementary methods) can be applied for the calculation of magnetic shielding tensors for ^{31}P nuclei to compare theoretical values with experimental spectral frequencies. The calculations are generally carried out within the limits of the density functional theory (DFT) on the basis of the plane-wave pseudopotential method. For example, the combination of solid-state ^{31}P NMR data with *ab initio* modeling provides descriptions of the surface structure for phosphorus-containing materials.

8.5.2 ^{14}N SOLID-STATE NMR

^{15}N nuclei with spins of 1/2 are traditional targets in solid-state NMR studies of individual organic compounds, heterogeneous aggregates, and complex biomolecular systems. Depending on the nature of the objects, various ^{15}N NMR techniques and methodical approaches are available for their characterizations. For example, ^{15}N NMR describes the structures of membrane-associated peptides and proteins in the solid state.[36,37] Details regarding two-dimensional $^{1}H\{^{15}N\}$ heteronuclear correlation spectroscopy (HETCOR) NMR experiments on samples with natural ^{15}N abundance performed with indirect ^{15}N detection can be found in Reference 38. Also, a description and determination of ^{15}N chemical shift tensors by the $^{15}N–^{13}C$ REDOR technique and $^{15}N–^{1}H$ dipolar-shift CP MAS NMR experiments are provided in Reference 39. Structural problems solved for small organic molecules by solid-state ^{15}N NMR (e.g., application of ^{15}N chemical shifts to identify tautomeric forms of imidazole rings) are presented in Reference 40. In all of these cases, the

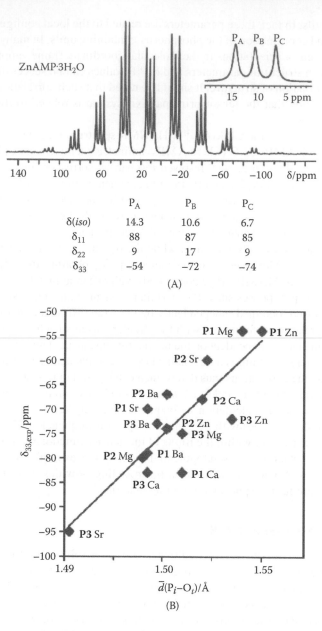

	P_A	P_B	P_C
$\delta(iso)$	14.3	10.6	6.7
δ_{11}	88	87	85
δ_{22}	9	17	9
δ_{33}	−54	−72	−74

(A)

(B)

FIGURE 8.13 (A) The $^{31}P\{^1H\}$ MAS NMR spectrum of compound ZnAMP in a sample spinning at a rate of 4 kHz and parameters of ^{31}P chemical shift tensors. (B) Correlation between bond length $d(P_i–O_t)$ and δ_{33} values obtained for MAMP·xH$_2$O compounds, where M is the metal. (From Weber, J. et al., *Inorg. Chem.*, 51, 11466, 2012. With permission.)

major inconvenience is the low sensitivity of ^{15}N NMR, which is particularly significant for solids. Therefore, such experiments often require ^{15}N-enriched samples. In contrast, the natural abundance of ^{14}N nuclei is high (99%) and the $^{14}N/^{15}N$ difference in resonance frequencies is not large. Therefore, at present, ^{14}N nuclei are attracting great interest among researchers investigating simple organic molecules, bioorganic systems, and materials to extract structural or dynamic information directly from the space where ^{14}N nuclei are located. The methodology of these studies is represented here by single-crystal experiments, stationary powder experiments, and MAS NMR experiments, where direct ^{14}N detection is in use.

8.5.2.1 Single-Crystal ^{14}N NMR Experiments

Single-crystal ^{14}N NMR experiments provide very precise measurements of the electric field gradients (EFGs) for ^{14}N nuclei, chemical shift (CS) tensor parameters, and dipolar couplings. In addition, they are unique because orientations of the $\delta(^{14}N)$ and EFG(^{14}N) tensors can be determined in the crystal frame. Although indirect detection in NMR experiments on ^{14}N nuclei can show third-order effects, only first- and second-order quadrupolar interactions are demonstrated with direct detection.[41] In the presence of magnetically non-equivalent ^{14}N nuclei in a single crystal, each set of the nuclei will be characterized by a common EFG and CS tensor to give two resonance lines, one of which corresponds to the fundamental spin transitions. Because the frequency distance between these resonances is determined by the first-order quadrupolar interactions, they are generally separated by several MHz (or less), while their mean frequency is altered by several kHz due to the second-order quadrupolar and CS interactions. Excitation bandwidths are limited by the power and duration of radiofrequency pulses; therefore, a number of such experiments should be performed to determine the signals. These signals will be broadened by only a few kHz due to heteronuclear dipolar couplings. Thus, the ^{14}N NMR spectra can be obtained at high signal-to-noise ratios with a spectral resolution satisfactory for analysis. Figure 8.14A shows the ^{14}N NMR spectra expected for a single-crystal compound containing a single nitrogen site, where the difference $\Delta\nu$ is controlled by first-order quadrupolar interactions. In the presence of protons, the lines will show dipolar nitrogen–proton couplings.

According to NMR theory, quadrupolar and chemical shift anisotropy interactions depend on the orientations of the EFG and CS tensors relative to the external magnetic field. Therefore, the resonance lines are shifted when the orientation of the crystal in the magnetic field changes. The behavior of the resonance frequencies with variations in the rotation angles around two orthogonal axes and treatments of these dependences to find EFG and CS give the principal components of the interaction tensors and their orientations in a molecular coordinate system. For example, in the early classical work of Griffin,[42] a complete ^{14}N single-crystal NMR study of N-acetyl-DL-valine allowed determination of the quadrupolar coupling constant for ^{14}N nuclei and asymmetry parameter as 3.21 MHz and 0.32, respectively. In addition to these data, the observed NH dipolar splitting corresponds to the orientation of the NH bond vector in this compound along the direction of the V_{yy} component of the quadrupolar tensor. One of the other components is perpendicular to the C–N–C plane, and the third component is orthogonal to these two directions. In addition to simple experiments, advanced NMR techniques can also be applied for single-crystal ^{14}N studies. Applications of PISEMA or 2D SLF experiments to discern the effects of quadrupolar and dipolar splitting into different dimensions can be found in Reference 41.

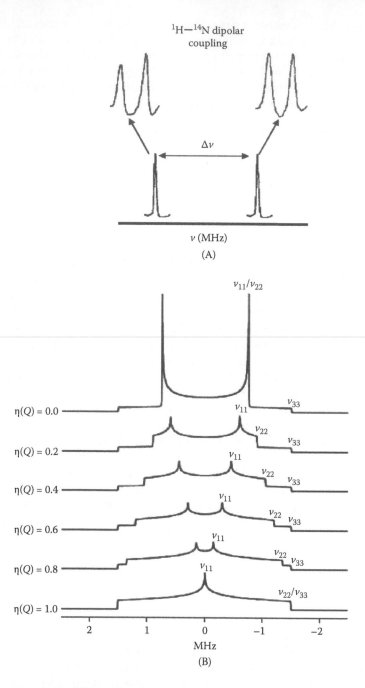

FIGURE 8.14   (A) The ^{14}N NMR spectra of a single-crystal compound with a single nitrogen site in the absence (bottom) and in the presence (top) of protons. (B) Evolution of the ^{14}N line shape in a powder sample with variations in asymmetry parameter $\eta(Q)$. (From O'Dell, L.A., *Prog. Nucl. Magn. Reson. Spectrosc.*, 59, 295, 2011. With permission.)

8.5.2.2 ^{14}N NMR in Static Powders

In contrast to a single crystal, even one nitrogen site in a powder sample of a rigid solid will show all of the possible EFG orientations, leading to the typical powder patterns shown in Figure 8.14B. Again, as in the case of other quadrupolar nuclei, the ^{14}N EFG tensor parameters can be easily obtained by NMR experiments on static powders, because the static line shape is very sensitive to quadrupolar coupling constants and asymmetry parameter η. For example, as seen in Figure 8.14B, the line shape changes greatly with the η value to show the features corresponding to the principal components of the EFG tensors. Because molecular motions reorient the EFG tensor, this shape is also sensitive to molecular dynamics on a very large time scale.

Technically, powder NMR spectra are easily recorded for nitrogen quarupolar coupling constants of ≤200 kHz, because standard RF pulses can uniformly excite the full frequency band in such ^{14}N powder patterns. This situation is typical of highly symmetrical nitrogen atoms, for example, in ammonium salts. Generally, the spectral patterns are obtained with regular quadrupolar echo pulse sequences and are treated by standard software to calculate quadrupolar coupling constant values and chemical shift parameters; however, this analysis is complicated by the presence of several nitrogen sites because the patterns represent overlapped lines, leading to a poor spectral resolution.

Experiments based on quadrupolar echo pulses and performed on solids with small ^{14}N quadrupole coupling constants can show the important features. Any motions on the T_2 relaxation time scale will result in dependence of the line shape on delay time between pulses. This effect is shown in Figure 8.15A by the ^{14}N NMR spectra of choline chloride. According to computer simulations, the line shape evolution corresponds to the fast 180° flipping choline cations.[41] Finally, the NMR patterns of ^{14}N sites with large quadrupole coupling constants (>200 kHz) can be recorded without distortions by a NMR technique based on the collection of several sub-spectra, the individual widths of which are determined by incrementing the RF frequency.

8.5.2.3 ^{14}N MAS NMR

When the first-order quadrupolar interactions at ^{14}N nuclei do not exceed the MHz magnitudes, ^{14}N MAS NMR experiments at spinning rates of ≤30 kHz can be successfully applied for the enhancement of spectral resolution. These experiments result in NMR spectra where the lines are broadened due to the presence of second-order quadrupolar interactions and are accompanied by wide patterns of spinning sidebands. Generally, the ^{14}N MAS NMR spectra show spectral resolution that is sufficient to resolve and characterize the EFG and chemical shift parameters corresponding to various nitrogen sites that would be invisible in static NMR spectra. However, it should be emphasized that the spinning sidebands and central resonance are very sensitive to the accuracy of the magic angle adjustment. Even a setting off by less than 0.01° can produce additional splitting and broadening.

As in the case of static NMR spectra, theoretical calculations of ^{14}N MAS NMR patterns can result in the EGF and chemical shift parameters. However, the computer fittings will be reliable only with high accuracy and stability of the spinning rates used in the MAS experiments. Modern MAS techniques can achieve high

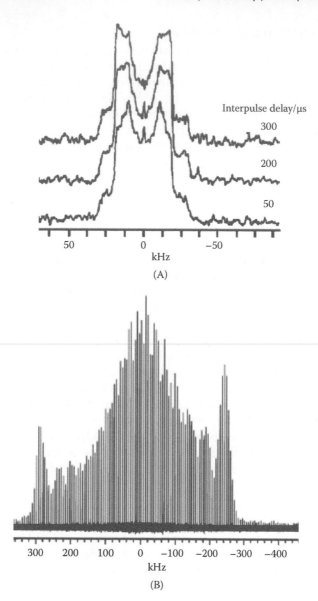

FIGURE 8.15   (A) ^{14}N NMR spectra in a static sample of choline chloride obtained at 48°C with the quadrupolar echo pulse sequence for variations in the delay time between pulses. The spectra can be simulated by a convenient program to correspond to two-site jumps. (B) The ^{14}N MAS NMR spectrum of potassium nitrate recorded at a spinning rate of 6 kHz. (From O'Dell, L.A., *Prog. Nucl. Magn. Reson. Spectrosc.*, 59, 295, 2011. With permission.)

spinning rates at a stability of around 0.1 Hz. For example, Figure 8.15B shows the ^{14}N MAS NMR spectrum obtained in a sample of solid KNO_3. The spectrum exhibits an exclusively wide spinning sideband pattern that covers a frequency range of ~600 kHz.[41] It is obvious that accurate fitting to such an experimental spectrum requires highly stable spinning rates to obtain the isotropic ^{14}N chemical shift δ_{iso} of 342.7 ppm, the quadrupolar coupling constant of 746 kHz, and the asymmetry parameter η of 0.02.

Another factor affecting the calculation of quadrupolar parameters from ^{14}N NMR spectra is the magnitude of quadrupolar coupling constants. It has been found that accuracy in the calculations decreases remarkably for quadrupolar coupling constants larger than 1 MHz. This problem can be partially avoided by application of so-called overtone spectroscopy, because the first-order quadrupolar broadenings are absent in these spectra. However, low inherent sensitivity and complex nutation behavior complicate these experiments.

8.5.3   ALKALI METAL IONS LI⁺, NA⁺, K⁺, RB⁺, AND CS⁺ FOR SOLID-STATE NMR

Alkali metal ions are good probes in NMR studies of chemical and biological systems and materials in spite of the pronounced quadrupolar character of the magnetically active isotopes of Li, Na, K, Rb, and Cs. Due to the ongoing development of solid-state NMR techniques, various methodologies are now available to obtain high-resolution NMR spectra for half-integer quadrupolar nuclei in solids.[43] Single-crystal NMR experiments provide complete information on the magnitudes of chemical shift/quadrupolar coupling tensor components and their orientations in the molecular coordinate system. When large single crystals are not available, then wide-line NMR can be successfully applied. However, because space orientations of the chemical shift and quadrupolar coupling tensors are generally different, a line-shape analysis to determine their parameters will be reliable if the NMR spectra are recorded at several magnetic fields.

In general, the quadrupolar coupling tensors of ^{23}Na, ^{87}Rb, and ^{39}K nuclei are significantly larger than their chemical shift tensors. Therefore, solid-state NMR of these nuclei is dominated by quadrupolar interactions even for the highest magnetic fields. In contrast, the quadrupolar couplings of ^{7}Li and ^{133}Cs nuclei are small; therefore, chemical shift tensor components of these nuclei can be easily determined directly from wide-line NMR spectra recorded during excitation of the central transitions. For comparison, the quadrupolar coupling constants in the solids $Li(CH_3COO) \cdot 2H_2O$, $NaNH_2$, and $Rb(CH_3COO) \cdot H_2O$ have been determined to be 0.15, 2.48, and 6.91 MHz, respectively. In addition to one-pulse experiments, high-quality powder NMR spectra can be achieved with spin echo pulse sequences that minimize the NMR probe ringing.

In order to resolve multiple sites of alkali metal ions in powder samples, MAS NMR at the highest fields should be applied. MAS NMR greatly increases the resolution, particularly for excitation of the central transitions for nuclei that have small quadrupolar constants. Nevertheless, as earlier, this MAS resolution will depend on the second-order quadrupolar interactions at these nuclei. Because they cannot be completely eliminated by simple spinning, the resonances of the half-integer quadrupolar nuclei ^{23}Na, ^{87}Rb, and ^{39}K remain broadened even with high spinning rates. In many cases, this broadening is still larger than the chemical shift dispersions. Advanced NMR techniques, such as dynamic-angle spinning (DAS), double rotation (DOR), multiple quantum magic-angle spinning (MQMAS), satellite-transition magic-angle

spinning (STMAS), and satellite transitions acquired in real time by magic-angle spinning (STARTMAS), completely remove these undesirable effects.[43] Solid-state NMR
of alkali metal ions can play a role complementary to x-ray crystallography to determine the structures of objects under investigation. Moreover, in contrast to x-ray crystallography, solid-state NMR spectroscopy can be applied to analyze powder samples
to extract information about the local electronic structure at ion binding sites in different systems, from inorganic to biological macromolecules. For example, K^+ and Na^+
ions are well observed in membrane-bound proteins and enzymes by solid-state NMR.
Here, NMR experiments are used to identify alkali metal ion binding sites and to probe
the properties of important binding sites and weakly bound alkali metal ions.

Solid-state NMR studies of alkali metal ions experiencing cation–π interactions
demonstrate the strategy for interpreting their chemical shifts. In fact, this strategy can
be applied for other studies. The cation–π interactions are non-covalent interactions
that play an important role in biomolecular structures and molecular recognition. It is
remarkable that they are also typical of many simple molecular systems, such as the
salts of tetraphenylborate, $M[BPh_4]$, where $M = Na^+$, K^+, Rb^+, or Cs^+ (Figure 8.16A).

According to solid-state NMR data,[43] the resonances of Na^+ and K^+ ions in $M[BPh_4]$
salts are strongly high-field shifted relative to those of free ions. The large shifts can be
attributed to high shielding environments around the metal centers due to the cation–π
interactions. The negative chemical shifts observed experimentally could be accepted
as being a signature of these interactions. However, this simple conclusion is in conflict
with the δ values of ^{23}Na and ^{39}K nuclei measured in crown ether complexes, where
the cation–π interactions are present *a priori*. In fact, the ^{23}Na and ^{39}K shifts have been
found within the normal chemical shift ranges. At the same time, the signals of nuclei
^{87}Rb and ^{133}Cs in $M[BPh_4]$ salts are also strongly high-field shifted, in agreement with
the ^{23}Na and ^{39}K shifts. Rationalization of this contradictory spectral behavior obviously requires knowledge of the paramagnetic and diamagnetic contributions to the
total chemical shifts that can be obtained by theoretical calculations performed for
model compounds with different Na–ligand interactions.

These calculations show that the paramagnetic contributions to the total ^{23}Na
shielding in the presence of specific ligands capable of cation–π interactions are
remarkably smaller than those found for other ligands. Thus, there is good agreement with the above statement that the negative chemical shift is proof of cation–π
interactions. However, this result does not confirm that the same tendency should
be expected for systems containing mixed ligands (not only π–ligands). In fact, the
crown ether complexes do not follow this tendency.

The diagrams in Figure 8.16B demonstrate how the exclusively large negative
chemical shifts inherent to molecules with cation–π interactions are related to the
benchmark chemical shifts of $M^+(g)$ free cations and $M^0(g)$ atoms in the gas phase.
In the case of $M^+(g)$ ions, only the diamagnetic contribution is present, and the $M^+(g)$
shielding values can represent the largest magnitudes that can be observed for M^+
ions in the condensed phase. In other words, the chemical shifts for $M^+(g)$ ions would
be the most negative shifts relative to all of the M^+ salts. As seen in Figure 8.16B, the
chemical shift tensor component δ_{33} (dashed lines) for M^+ ions in $M[BPh_4]$ salts is
very close to the δ values in $M^+(g)$ ions. Thus, the paramagnetic shielding contribution is close to zero when the magnetic field is applied along the δ_{33} direction. This

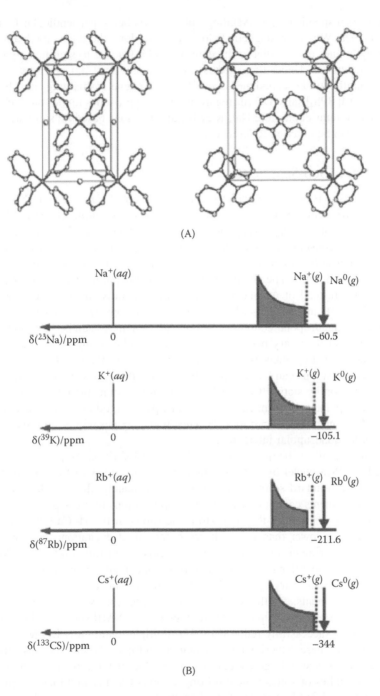

FIGURE 8.16 (A) The unit cell of M[BPh$_4$] along crystallographic axes: a-axis (left) and c-axis (right). (B) Relationship between the δ tensor components for M$^+$ ions in M[BPh$_4$] and chemical shifts for M$^+(aq)$, M$^+(g)$, and M$^\circ(g)$ (see text). (From Wu, G. and Zhu, J., *Prog. Nucl. Magn. Reson. Spectrosc.*, 61, 1, 2012. With permission.)

direction corresponds to the crystallographic c-axis being perpendicular to all four phenyl planes. When the magnetic field is applied along the a-axis (or b-axis), then the phenyl planes are parallel to the external field, leading to significant paramagnetic contributions. Under these terms, the chemical shift anisotropy determined experimentally will be a good measure of the paramagnetic shielding contribution in each of the $M[BPh_4]$ salts. Details regarding other approaches to the study of various molecular systems containing Na^+, K^+, Rb^+, and Cs^+ ions based on solid-state NMR techniques can be found in Reference 43.

8.5.4 ^{43}CA SOLID-STATE NMR

Ca^{2+} ions are invisible in ultraviolet spectra; therefore, ^{43}Ca solid-state NMR spectroscopy plays a major role in the study of solids containing calcium. The ^{43}Ca isotope has a spin of 7/2, a very low natural abundance (0.135%), and a negative gyromagnetic constant. In general, among seven possible spin transitions, NMR shows an intense line of central transitions, while other transitions are observed as low-intensity satellites. The chemical shift anisotropy parameters Ω and κ (used in the literature instead of $\Delta\sigma$ and η) found experimentally in many compounds by static ^{43}Ca NMR are not large. For example, anisotropy Ω reaches only 70 ppm. However, the quadrupolar coupling constants of ^{43}Ca nuclei can be significant. It has been found that they vary between 0 and 6 MHz. The strong quadrupole interactions broaden ^{43}Ca resonances. Due to the broadening effects, ^{43}Ca resonances of the central transitions can be of a low intensity, resulting in technical difficulties in their registration by static ^{43}Ca NMR experiments. As mentioned earlier, the first-order quadrupolar and chemical shift anisotropy interactions can be averaged by MAS NMR. Under these conditions, ^{43}Ca resonance shapes will be dominated by the second-order quadrupolar interactions.

In principle, all of the quadrupolar and chemical shift tensor parameters can be obtained for ^{43}Ca nuclei by MAS or static NMR spectra with further calculations performed by standard software packages. Unfortunately, the very low sensitivity in these experiments cannot be enhanced by applications of cross-polarization techniques. In fact, sometimes the signal-to-noise ratios in the H–Ca CP MAS NMR spectra are even lower than those observed for direct excitation; therefore, direct detection in experiments on ^{43}Ca-enriched systems is generally preferable. However, one should remember that the T_2 relaxation times of calcium are generally short while the T_1 times are long, varying from 3 to 30 s. The first factor complicates CPMG or spin echo experiments, and the second factor increases the experimental time.

The second-order quadrupolar broadenings in ^{43}Ca NMR spectra of labeled compounds can be successfully reduced by 2D multiple quantum (MQ) MAS experiments. In fact, these experiments can detect multiple Ca sites because the ^{43}Ca chemical shift range is large enough to do so. One of the most common tasks in ^{43}Ca NMR studies of solids is determining various Ca sites and their assignments, an aspect exemplified by studies of apatites, proton-containing molecular systems.[44] In ^{43}Ca-labeled samples, REDOR NMR or even CP MAS NMR can be applied to find the calcium units with different proton environments. The NMR spectra data obtained by various techniques in a hydroxyapatite sample are shown in Figure 8.17.

In this compound, one of the two crystallographic Ca sites, noted as Ca(2), is situated closer to group OH. The peak with the highest intensity in the 1D ^{1}H–^{43}Ca CP MAS NMR spectrum belongs to this site. The assignment is well supported by the

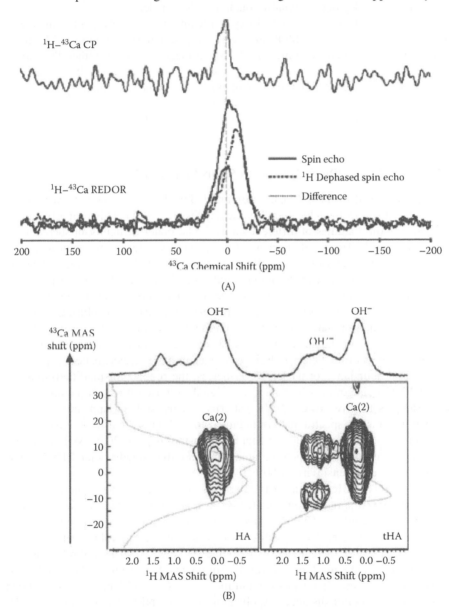

FIGURE 8.17 (A) The ^{1}H–^{43}Ca CP (top) and ^{43}Ca{^{1}H} REDOR MAS NMR (bottom) experiments performed on a sample of 60% ^{43}Ca-enriched Ca$_{10}$(PO$_4$)$_6$(OH)$_2$. (B) The 2D ^{1}H{^{43}Ca} HMQC NMR spectrum of 60% ^{43}Ca-enriched Ca$_{10}$(PO$_4$)$_6$(OH)$_2$ (left) and oxy-hydroxyapatite (right) phases. (From Laurencin, D. and Smith, M.E., *Prog. Nucl. Magn. Reson. Spectrosc.*, 68, 1, 2013. With permission.)

^{43}Ca{^{1}H} REDOR NMR experiment where the spin echo and ^{1}H dephased spin echo pulse sequences correspond to different signal intensities. In addition, the ^{1}H{^{43}Ca} transfer of population double resonance (TRAPDOR) spectrum shows independently that the OH group is actually coupled to ^{43}Ca nuclei.

The ^{43}Ca-enriched samples of hydroxyapatite and oxy-hydroxyapatite can be well characterized by ^{1}H{^{43}Ca} HMQC experiments. The pattern in Figure 8.17B shows that even weak dipolar couplings ^{1}H–^{43}Ca can clearly be seen. Other ^{43}Ca solid-state NMR techniques and methodical approaches are also available.[44] For example, manipulations of the satellite transition intensity to enhance the intensity of the central transition can be applied to probe various solids (e.g., natural and synthesized materials, minerals, glasses), the details of which can be found in the recommended literature.

8.5.5 ^{35,37}Cl, ^{78,81}Br, ^{127}I, and ^{17}O Solid-State NMR

Isotropic chemical shifts of ^{35,37}Cl, ^{79,81}Br, and ^{127}I nuclei are known to be very sensitive to variations in their local environments (Figure 8.18A). This property can be used in structural studies of different solids.[45] The major difficulty in such studies is detection of NMR signals because these nuclei experience strong quadrupolar interactions. The quadrupolar coupling constants of ^{35,37}Cl, ^{79,81}Br, and ^{127}I nuclei are extremely large and can reach magnitudes of 22, 49, or 30 MHz, respectively. Under these conditions, interactions between the quadrupole moments and the electric field gradients for these nuclei strongly perturb the Zeeman splitting, resulting in very large quadrupolar broadenings. Even the central transitions, generally better observed for most of the quadrupolar nuclei, will be broadened up to several MHz. The broadenings dramatically reduce sensitivity in NMR experiments in spite of a high natural abundance of nuclei. Nevertheless, the regular (or modified) quadrupolar CPMG pulse train, which increases signal-to-noise ratios, can be applied for static samples to give powder NMR patterns a sharp spikelet appearance. Figure 8.18A illustrates the spectral pattern obtained for ^{127}I nuclei in a static KIO$_4$ sample that corresponds to a quadrupolar coupling constant (C_Q) of 20.66 MHz at a zero asymmetry parameter. It should be noted that modified quadrupolar CPMG pulse trains can increase sensitivity by 80%.

The MAS NMR technique, particularly when applied at the highest magnetic fields and highest spinning rate of 70 kHz (a rate routinely reached with 1.3-mm NMR rotors), records ^{35,37}Cl, ^{78,81}Br, and ^{127}I NMR spectra where the isotropic resonances are not affected by any CSA interactions so the δ_{iso} values can be easily determined. Because there are currently numerous spectral structural relationships, the δ_{iso} values can be used for structural assignments. For example, the correlations between the chlorine chemical shift and the metal–Cl bond length can be applied to estimate interatomic distances. Applications of other NMR techniques (2D NMR and MQ MAS NMR) can be found in the recommended literature.

Because the ^{17}O chemical shifts cover a very large region between 1600 and 0 ppm (compared to liquid water), solid-state ^{17}O NMR (static or MAS) has great analytical potential in studies of oxygen-containing solids. In addition to isotropic

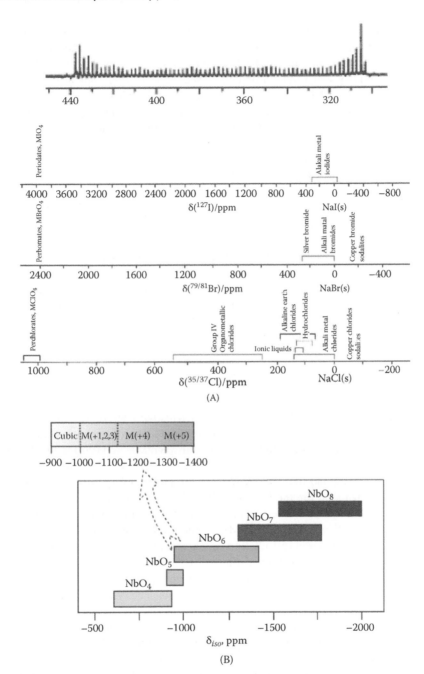

FIGURE 8.18   (A) From top to bottom: QCPMG ^{127}I NMR spectrum of static KIO_4 and isotropic chemical shift ranges for chlorine, bromine, and iodine nuclei. (B) The ^{93}Nb NMR isotropic chemical shift scale for NbO_x structural units. (Adapted from Bakhmutov, V.I., *Solid-State NMR in Materials Science: Principles and Applications*, CRC Press, Boca Raton, FL, 2011.)

chemical shifts, compounds under investigation can be characterized by the nuclear quadrupolar coupling constants and shielding tensors. The central transitions of ^{17}O spins do not show strong first-order broadenings; therefore, the detected lines are generally relatively sharp and dominated by second-order quadrupolar effects only.

The difficulties in detection of ^{17}O signals by solid-state NMR are connected with two factors: (1) very low natural ^{17}O abundance, and (2) relatively small amounts of oxygen atoms in samples. For this reason, NMR experiments performed on ^{17}O-enriched compounds are always preferable.

The problem of low sensitivity can sometimes be avoided by standard ^1H–^{17}O cross-polarization (see Figure 8.1C, for example). Other techniques applied for enhancement of ^{17}O sensitivity in the MAS spectra are based on fast amplitude-modulated pulses that lead to a rotor-assisted population transfer (RAPT). This approach can increase ^{17}O NMR sensitivity by a factor of 5.

Because spectral resolution in ^{17}O solid-state NMR spectra is directly connected with second-order quadrupolar interactions, it can be remarkably improved by ^{17}O MAS NMR experiments with strong magnetic fields or by complex mechanical motions of NMR rotors, such as by double rotation (DOR) or dynamic-angle spinning (DAS).[21] Alternatively, 2D multiple quantum ^{17}O NMR can be used to obtain well-resolved ^{17}O NMR spectra. This approach is often used to characterize various bridging oxygen sites in silicate glasses.

8.5.6 ^{51}V, ^{93}Nb, AND ^{181}Ta SOLID-STATE NMR

Solid-state NMR experiments on ^{51}V, ^{93}Nb, and ^{181}Ta nuclei are widely utilized in the inorganic chemistry and chemical industries to characterize molecular systems based on the vanadium, niobium, and tantalum oxides. In general, experiments based on direct detection are the most popular. $^{51}V(7/2)$, $^{93}Nb(9/2)$, and $^{181}Ta(7/2)$ nuclei in the external magnetic field produce energy-level diagrams typical of quadrupolar nuclei, and their NMR spectra are perturbed by first-order and second-order quadrupolar interactions. Quadrupolar coupling constants of ^{51}V nuclei are usually moderate and take values ranging from 2 to 6 MHz. In some compounds, however, they can reach 10 MHz. The ^{51}V δ_{iso} values in solid vanadium-based materials, compared to $VOCl_3$ as an external reference, vary within a range of 1200 ppm. However, the typical diapason for ^{51}V δ_{iso} is smaller (around 500 ppm). The ^{51}V chemical shift magnetic anisotropy is very large (≤ 1000 ppm) and strongly depends on the coordination environments. The typical values are within 100 and 500 ppm.[31]

Figure 8.18B shows isotropic ^{93}Nb chemical shifts, compared to a CH_3CN solution of $NbCl_3$, which change in a diapason of 4000 ppm for NbO_x units. The ^{93}Nb solid-state NMR spectra are also dominated by the quadrupolar interactions to manifest the characteristic powder patterns corresponding to central transitions from +1/2 to –1/2. It is known that the satellite transitions for nuclei with spins of 9/2 are located closer to the central transitions. Therefore, the solid-state ^{93}Nb NMR spectra can show not only central transitions but also several satellite transitions that are routinely observed.

In terms of signal detection, the NMR spectroscopic properties of [181]Ta nuclei are not ideal; therefore, they are not popular as probes for the solid state. In fact, these nuclei have a very large quadrupolar moment (317×10^{-30} m^2 vs. -5.2×10^{-30} m^2 for [51]V nuclei) and fast quadrupolar relaxation. Both of the effects result in the exclusively broad resonance lines. In addition, the gamma constant of [181]Ta nuclei is low, and the difference in energy-level populations is small, leading to poor NMR sensitivity. Finally, the [181]Ta isotropic chemical shift range covers a region of 3450 ppm referenced to solutions of $K[TaCl_6]$.

It should be emphasized that in order to perform NMR experiments on [51]V, [93]Nb, and [181]Ta nuclei in static and spinning samples, NMR spectrometers should meet the following minimal technical requirements: reasonably high magnetic fields of ≥ 9 T, spinning rates of ≥ 15 kHz, extremely short radiofrequency pulses (generally less than 0.5 µs), and very fast digitizing rates. Figure 8.19A illustrates the advantages of applying the highest magnetic fields to register [93]Nb nuclei in solid La_3NbO_7. The static powder NMR pattern of the compound, recorded at a magnetic field of 21.1 T, is much narrower than that recorded for a field of 9.4 T.

A large number of methodical approaches can be used to characterize solids by [51]V, [93]Nb, and [181]Ta NMR. Among them, the methods playing a major role are based on Hahn echo, solid echo, and CPMG (modified for quadrupolar nuclei as QCPMG) pulse sequences, particularly in NMR experiments on static samples. MAS NMR experiments can also be successful; however, one should remember that for [51]V nuclei, for example, which have a chemical shift anisotropy of ≤ 1000 ppm, a MAS NMR spectrum can be sideband free only at spinning rates greater than 40 kHz in combination with an applied magnetic field strength of 9.4 T. At lower spinning rates, the interpretation of very complex patterns will be difficult. A higher magnetic field (e.g., a field of 21 T) causes an increase in the chemical shift anisotropies such that a sideband-free MAS NMR spectrum could be obtained at a spinning rate of 90 kHz, which is technically impossible. Therefore, in reality, the MAS NMR spectra will always be complex and show wide sideband patterns. When magnetic shift anisotropies are relatively small and second-order quadrupolar effects are large, then the 2D multiple quantum MAS NMR method is very powerful for structural studies.[31] Figure 8.19B shows the [51]V 3Q MAS NMR spectrum obtained in a sample of a VO_x/ZrO_2 compound. As seen, the 1D projection of the spectrum (δ_{1d} axis in the figure) is well resolved; thus, eight to nine non-equivalent vanadium sites can be identified in this compound. Applications of heteronuclear correlation NMR spectroscopy, including [1]H–[51]V cross-polarization, double resonance techniques (e.g., SEDOR, REDOR, TRAPDOR, REAPDOR), and satellite transition spectroscopy, are also available and details can be found in the recommended literature and original articles.

REFERENCES AND RECOMMENDED LITERATURE

1. Kentgens, A.P.M., Bos, A., and Dirken, P.J. (1994). *Solid State NMR*, 3: 315.
2. Hartmann, S.R. and Hahn, E.L. (1962). *Phys. Rev.*, 128: 2042.
3. Kolodzieski, W. and Klinowski, J. (2002). *Chem. Rev.*, 102: 613.

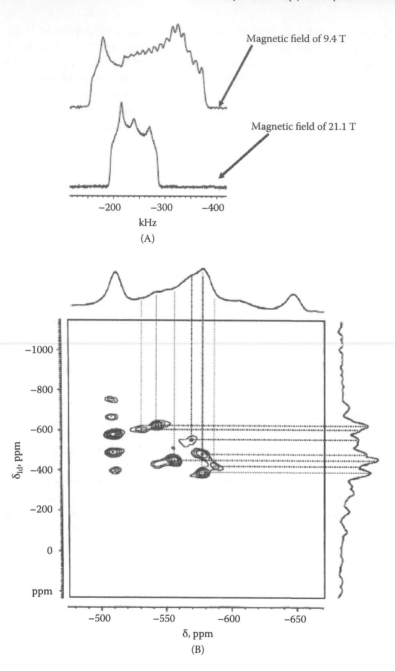

FIGURE 8.19 (A) The ^{93}Nb NMR spectra obtained in a static sample of the La$_3$NbO$_7$ compound at different magnetic field strengths. (B) The ^{51}V triple quantum MAS NMR spectrum obtained in a sample of VO$_x$/ZrO$_2$ at an external magnetic field of 9.4 T. (Adapted from Bakhmutov, V.I., *Solid-State NMR in Materials Science: Principles and Applications*, CRC Press, Boca Raton, FL, 2011.)

4. Bryce, D.L., Bernard, G.M., et al. (2001). *J. Anal. Sci. Spectr.*, 46: 46.
5. Bakhmutov, V.I. (2011). *Chem. Rev.*, 111: 530.
6. Smith, M.E. and van Eck, E.R.H. (1999). *Prog. Nucl. Magn. Reson. Spectrosc.*, 34: 159.
7. De Paul, S.M., Ernst, M., Shore, J.S., Stebbins, J.F., and Pines, A. (1997). *J. Phys. Chem. B*, 101: 3240.
8. Van Vleck, J.H. (1948). *Phys. Rev.*, 74: 1168.
9. Antzutkin, O.N. (1999). *Prog. Nucl. Magn. Reson. Spectrosc.*, 35: 203.
10. María, D.S., Claramunt, R.M., Vasilevsky, S.F., Klyatskaya, S.V., Alkorta, I., and Elguero, J. (2011). *ARKIVOC*, III: 114.
11. Fyfe, C.A., Lewis, A.R., Chezeau J.M., and Grondey, H. (1997). *J. Am. Chem. Soc.*, 119: 12210.
12. Brown, S.P. (2007). *Prog. Nucl. Magn. Reson. Spectrosc.*, 50: 199.
13. Gullion, T. and Vega, A.J. (2005). *Prog. Nucl. Magn. Reson. Spectrosc.*, 47: 123.
14. Ulrich, A.S. (2005). *Prog. Nucl. Magn. Reson. Spectrosc.*, 46: 1.
15. Lee, J.-A. and Khitrin, A.K. (2008). *Concepts Magn. Reson. Part A*, 32A: 56.
16. Ooms, K.J., Feindel, K.W., Willans, M.J., Wasylishen, R.E., Hanna, J.V., Pike, K.J., and Smith, M.E. (2005). *Solid State NMR*, 28: 125.
17. Amoureux, J.P., Fernandez, C., and Steuernagel, S. (1996). *J. Magn. Reson.*, A123: 116.
18. Jonas, J. (2007). High-pressure NMR. In: *Encyclopedia of Magnetic Resonance* (Harris, R.K. and Wasylishen, R., Eds.). Chichester: John Wiley & Sons.
19. Gerstein, B.C. (1981). *Philos. Trans. R. Soc. Lond. A*, 299: 521.
20. Lesage, A., Duma, L., Sakellariou, D., and Emsley, L. (2001). *J. Am. Chem. Soc.*, 123: 5747.
21. Ashbrook, S.E. (2009). *Phys. Chem. Chem. Phys.*, 11: 6892.
22. Madhu, P.K., Pike, K.J., Dupree R., Levitt, M.H., and Smith, M.E. (2003). *Chem. Phys. Lett.*, 367: 150.
23. Kinnibrugh, T.L., Ayi, A.A., Bakhmutov, V.I., Zo, J., and Clearfield, A. (2013). *Cryst. Growth Des.*, 13: 2973.
24. Maunu, S.L. (2002). *Prog. Nucl. Magn. Reson. Spectrosc.*, 40: 151.
25. Bohnsack, A.M., Ibarra, I.A., Bakhmutov, V.I., Lynch, V.M., and Humphrey, S.M. (2013). *J. Am. Chem. Soc.*, 135: 16038.
26. Palke, A.C., Stebbins, J.F., and Boatner, L.A. (2013). *Inorg. Chem.*, 52: 12605.
27. Lesage, A. (2009). *Phys. Chem. Chem. Phys.*, 11: 6876.
28. Cadars, A., Lesage, A., Trierweiler, M., Heux, L., and Emsley, L. (2007). *Phys. Chem. Chem. Phys.*, 9: 92.
29. Widdifield, C.M. and Schurko, R.W. (2009). *Concepts Magn. Reson. Part A*, 34A: 91.
30. Taylor, R.E., Carver, C.T., Larsen, R.E., Dmitrenko, O., Bai, S., and Dybowski, C. (2009). *J. Mol. Struct.*, 930: 99.
31. Lapina, O.B., Khabibulin, D.F., Shubin, A.A., and Terskikh, V.V. (2008). *Prog. Nucl. Magn. Reson. Spectrosc.*, 53: 128.
32. Bakhmutova-Albert, E.V., Bestaoui, N., Bakhmutov, V.I., Clearfield, A., Rodriguez, A.V., and Llavona, R. (2004). *Inorg. Chem.*, 43: 1264.
33. Hanna, J.V. and Smith, M.E. (2010). *Solid State NMR*, 38: 1.
34. Massiot, D., Drumel, S., Janvier, P., Bujoli-Doeuff, M., and Bujoli, B. (1997). *Chem. Mater.*, 9: 6.
35. Weber, J., Grossmann, G., Demadis, K.D., Daskalakis, N., Brendler, E., Mangstl, M., and Günne, J.S. (2012). *Inorg. Chem.*, 51: 11466.
36. Naito, A. (2009). *Solid State Nucl. Magn. Reson.*, 36: 67.
37. Davis, J.H. and Auger, M. (1999). *Prog. Nucl. Magn. Reson. Spectrosc.*, 35: 1.
38. Althaus, S.M., Mao, K., Stringer, J.A., Kobayashi, T., and Pruski, M. (2014). *Solid State Nucl. Magn. Reson.*, 57–58: 17.

39. Heise, B., Leppert, J., and Ramachandran, R. (2000). *Solid State Nucl. Magn. Reson.*, 16: 177.

40. Li, S., Zhou, L., Su, Y., Han, B., and Deng, F. (2013). *Solid State Nucl. Magn. Reson.*, 54: 13.

41. O'Dell, L.A. (2011). *Prog. Nucl. Magn. Reson. Spectr.*, 59: 295.

42. Stark, R.E., Haberkorn, R.A., and Griffin, R.G. (1978). *J. Chem. Phys.*, 68: 1996.

43. Wu, G. and Zhu, J. (2012). *Prog. Nucl. Magn. Reson. Spectrosc.*, 61: 1.

44. Laurencin, D. and Smith, M.E. (2013). *Prog. Nucl. Magn. Reson. Spectrosc.*, 68: 1.

45. Chapman, R.P., Widdifield, C.M., and Bryce, D.L. (2009). *Prog. Nucl. Magn. Reson. Spectrosc.*, 55: 215.

RECOMMENDED LITERATURE

Bakhmutov, V.I. (2011). *Solid-State NMR in Materials Science: Principles and Applications.* Boca Raton, FL: CRC Press.

Duer, M.J., Ed. (2002). *Solid-State NMR Spectroscopy: Principles and Applications.* Oxford: Blackwell Sciences.

Fitzgerald. J.J. (1999). *Solid-State NMR Spectroscopy of Inorganic Materials.* Oxford: Oxford University Press.

Günther, H. (2013). *NMR Spectroscopy: Basic Principles, Concepts and Applications in Chemistry.* Weinheim: Wiley–VCH.

Harris, R.K., Wasylishen, R.E., and Duer, M.J., Eds. (2009). *NMR Crystallography.* New York: Wiley.

MacKenzie, K.J.D. and Smith, M.E. (2002). *Multinuclear Solid-State Nuclear Magnetic Resonance of Inorganic Materials.* Oxford: Pergamon.

Schmidt-Rohr, K. and Spiess, H.W. (1994). *Multidimensional Solid-State NMR and Polymers.* London: Academic Press.

9 Molecular Dynamics and Nuclear Relaxation in Solids
Applications

Because some chemical and physical properties of solids, particularly polymeric systems or porous materials containing organic or inorganic fragments, depend on the mobility of molecules or molecular groups, the characterization of molecular dynamics is an important aspect of solid-state NMR studies. Generally, they are aimed at identifying motions to determine the type (motional model), frequencies, and activation energies of reorientations. It should be added that, in contrast to solutions, molecular motions in solids can be cooperative and give significant contributions to the mechanical properties of systems. Initially, studies of the dynamics in polymers, both crystalline and amorphous, considered rotational motions to be simple jumps about well-established axes, such as phenyl flips. However, later solid-state NMR and other methods have shown that the mechanism of the rotational motions is complex, such that large-angle fluctuations of polymers over rotational jumps play the important role. In other words, in reality, motions are cooperative, particularly in supramolecular systems. It is now generally accepted that conformational transitions in polymers involve rotations of individual groups through typically large bond angles (e.g., 120°) that occur via chain dynamics for angle fluctuations with small angular displacements, followed by large angle jumps due to conformational transitions of the chains.[1]

As mentioned in Chapter 4, fast motions in solids can be represented by high-frequency librations and low-energy internal rotations or even tunneling (e.g., for CH_3 groups), whereas reorientations of whole molecules or molecular groups, segmental motions in polymers, and translation diffusion are generally slow on the NMR time scale. In frequency terms, reorientations can be classified as very slow (frequencies from 1 Hz to <1 kHz), moderate (kHz frequencies), or very fast (frequencies of ~100 MHz). Motions on a millisecond or microsecond time scale can be observed by variable-temperature NMR spectra and analyzed by computer simulations of line shapes. These motions can be quantitatively characterized by T_2 and $T_{1\rho}$ time measurements. Slower motions from milliseconds to seconds can be monitored by magnetization transfer occurring due to chemical exchange (i.e., by application of exchange NMR spectroscopy). Finally, T_1 relaxation time measurements can characterize fast reorientations. This chapter illustrates the practice of dynamic solid-state NMR to demonstrate how to identify and describe molecular mobility in solids by applying various approaches and strategies in unusual and typical cases.

9.1 TEMPERATURE CONTROL AND CALIBRATION IN SOLID-STATE NMR

Because motions in solids can be characterized by activation energies, dynamic solid-state NMR experiments should be performed with variations in temperature. Such variable-temperature NMR experiments obviously require good accuracy in temperature control and calibration, particularly where a sample is located. Only under these conditions will quantitative interpretations of NMR data and determination of correlation times τ_C and E_a values be reliable and valuable. In fact, disparities between the temperature setting in variable-temperature units and the true temperature in NMR samples are well known, even for solutions.

In solutions, temperature calibration is not problematic and is well standardized. Generally, calibration is easily achieved by using the ^{1}H chemical shift difference, measured in neat methanol as parameter $\delta(CH_3)–\delta(OH)$ for a low-temperature diapason or in neat ethylene glycol as $\delta(CH_2)–\delta(OH)$ for a high-temperature diapason. The curves obtained are then used to determine the temperature required for the experiment. The situation is more complex in the case of solid-state NMR. Here, the design of the solid-state NMR probes (magic-angle spinning [MAS] or wide-line probes) and temperature control units dictates temperature stability and particularly the temperature gradient along a sample. It is generally accepted that temperature calibrations in solids should be carried out on the basis of the well-established temperature dependence of the ^{207}Pb chemical shift determined in lead nitrate. This shift is used as an internal NMR thermometer.

Measurements of the ^{207}Pb δ_{iso} values required for calibration can be carried out more precisely by ^{207}Pb MAS NMR. However, under MAS conditions with increasing spinning rates, the frictional heating effect significantly changes the local temperature in a spinning sample. This is valid for any MAS NMR experiment. Therefore, temperature calibration should also include the MAS rate. Clearly, this calibration is particularly important for variable-temperature NMR experiments on biological objects.

Currently, there is a protocol that is generally accepted for determination of the temperature in MAS NMR probes based on the empirical relationships shown in Equations 9.1 and 9.2.[2] It should be noted that these equations are applicable for BioMAS (temperature indicated by $T_S(b)$) and fast-MAS (temperature indicated by $T_S(f)$) NMR probes, respectively. In Equations 9.1 and 9.2, T_0 is the given set temperature and ω_r is the MAS rate, which typically changes between 8 and 35 kHz:

$$T_S(b) = 0.97T_0 + 1.34°C \exp(\omega_r/7.53 \text{ kHz}) - 0.77°C \tag{9.1}$$

$$T_S(f) = 0.98T_0 + 3.79°C \exp(\omega_r/(19.6) \text{ kHz}) - 3.49°C \tag{9.2}$$

For solid-state NMR, measurements of the frequency difference between the CH_3 and OH groups in the ^{1}H NMR spectrum of methanol can be also applied to determine the temperature in a sample. These measurements are generally carried out in a standard sample containing solid tetrakis(trimethylsilyl)silane, which is soaked in liquid methanol. Other calibration procedures are also available and can be found in the recommended literature.

9.2 UNUSUALLY FAST MOLECULAR DYNAMICS IN SOLIDS

The presence of extremely mobile molecular fragments that are very small in rigid solids, such as quantum tunneling methyl groups, is not surprising because their reorientations do not require a large space. It is not surprising also that these motions can contribute to NMR line shapes, second moments, and T_1, T_2 relaxation as a function of the motional frequency. However, some solids containing more massive molecular groups can also be very mobile.[3]

Figure 9.1A shows an organic molecule where the central unit can experience fast reorientations even in the solid state as is observed for CH_3 rotors. Variable-temperature spin–lattice relaxation experiments performed on a polycrystalline sample of this compound in a static wide-line NMR probe can explore the internal dynamics of this molecular rotor and quantitatively characterize these reorientations. In fact, magnetic fields oscillating due to this motion can cause effective dipolar T_1 relaxation. The ^1H T_1 times measured experimentally can be treated by the equation shown in Figure 9.1B to give the correlation times, τ_C. In turn, the variable-temperature measurements of τ_C times lead to determination of activation energy E_a.

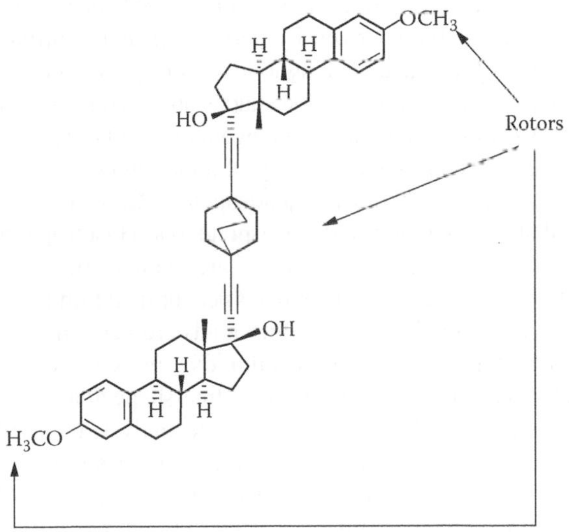

(A)

$$T_1^{-1} = C[\tau_C(1 + \omega_0^2 \tau_C^2)^{-1} + 4\tau_c(1 + 4\omega_0^2\tau_C^2)^{-1}]$$

$$C = (n/N)(9/40)[\mu_0/4\pi]^2\gamma^4\hbar^2/r^6$$

(B)

FIGURE 9.1 (A) Organic crystalline compound where three possible fast-rotating units (rotors) are present. (B) Equation used for treatments of the total relaxation rates T_1^{-1} for protons measured experimentally. (From Rodríguez-Molina, B. et al., *J. Am. Chem. Soc.*, 135, 10388, 2013. With permission.)

In spite of the simplicity of this approach, a problem complicates the interpretation of data obtained by static wide-line NMR experiments. In fact, there are three fast-rotating candidates—CH_3, OCH_3, and the central unit—in this molecule that can contribute to the total experimental relaxation rate, as shown in Equation 9.3:

$$1/T_1(total) = 1/T_1(CH_3) + 1/T_1(OCH_3) + 1/T_1(central\ unit) \qquad (9.3)$$

It should be noted that such a situation is widespread in the practice of solid-state NMR. This problem can be solved by accurate separation of the different relaxation contributions, using, for example, a model compound such as mestranol.[3] This model compound possesses only two rotors, OCH_3 and CH_3, one of which can be eliminated in the methoxy-deuterated analog. Due to the separation, it has been found that the central unit rotator has a very low energy barrier of 1.15 kcal/mol vs. 1.35 kcal/mol and 1.92 kcal/mol for CH_3 and OCH_3 rotors, respectively.[3]

9.3   MOLECULAR DYNAMICS IN SOLIDS DETERMINED BY FULL LINE-SHAPE ANALYSIS

As in the case of solutions, processes in solids occurring on the same time scale as the free induction decay (FID) affect line shapes in the NMR spectra. In fact, as follows from theory, any motions characterized by correlation times similar to T_2 times (or shorter than T_2 times) will average the magnetic anisotropic interactions and show, for the exchanging signals, an evolution of line shapes observed in static (or MAS) NMR spectra. Figure 9.2A illustrates a temperature evolution of line shapes that is typical of static ^2H NMR spectra. The gradual freezing of a motion upon cooling, shown from bottom to top, changes the line width and quadrupolar splitting of the static resonance. It is also obvious that a fast motion will correspond to a ^2H isotropic resonance in this spectra pattern. Because the resonance shapes are mathematically known, they can be simulated or iterated on the basis of semiclassical or quantum mechanical theory.

Deuterium resonances of ^2H-labeled compounds are generally accepted as being the most convenient for studies of molecular dynamics by variable-temperature static NMR spectra. The common algorithm of line-shape analysis applied for powder NMR spectra is presented in Figure 9.2A. The analysis is based on the Bloch–McConnell equation, which describes the vectors of the transverse magnetizations, $M(t)$, experiencing an exchange between different sites N.[4,5] The formal solution is represented by an exponential function, where Ω and K are the $[N \times N]$ diagonal matrices. The first matrix contains the anisotropic frequencies characterizing quadrupolar or chemical shift anisotropy (CSA) interactions, and the second matrix includes the exchange rates. Then, computer simulations based on the appropriate motional models and fittings to the experimental line shapes can recognize the types of motions and determine their correlation times.

This Bloch–McConnell approach can also be applied for variable-temperature ^2H MAS NMR spectra, where the Ω matrix is time dependent due to mechanical (macroscopic) sample rotation causing additional line broadenings. However, the MAS case requires more complex computer treatments, for which more information can be found in the literature.

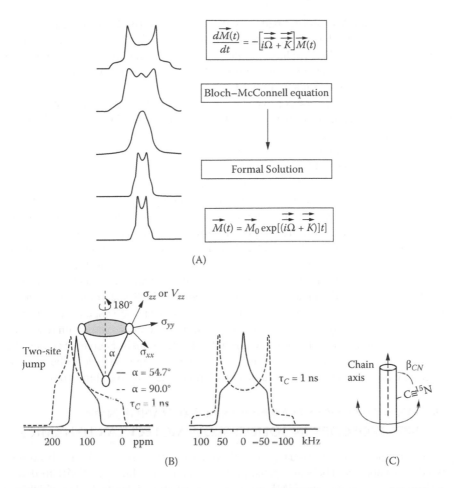

$$\frac{d\vec{M}(t)}{dt} = -\left[i\overleftrightarrow{\Omega} + \overleftrightarrow{K}\right]\vec{M}(t)$$

Bloch–McConnell equation

Formal Solution

$$\vec{M}(t) = \vec{M}_0 \exp[(i\overleftrightarrow{\Omega} + \overleftrightarrow{K})]t]$$

(A)

σ_{zz} or V_{zz}

180°

σ_{yy}

Two-site jump

σ_{xx}

— $\alpha = 54.7°$
-- $\alpha = 90.0°$
$\tau_C = 1$ ns

$\tau_C = 1$ ns

200   100   0 ppm

100 50 0 −50 −100 kHz

Chain axis β_{CN}

C≡^{15}N

(B)   (C)

FIGURE 9.2 (A) Temperature evolution of a static ^{2}H NMR resonance and the Bloch–McConnell formalism for its treatment.[5] (B) Powder line shapes for 1/2-spin nuclei (left) and ^{2}H nuclei (right) in a static sample in the presence of two-site jumps at a correlation time of 1 ns at various orientations of chemical shift anisotropy and EFG tensors, σ and V_{ZZ}, respectively.[4] (C) Rotational diffusion around the main chain in compound $[C_3H_2N]_n$–aPAN (see text). (Adapted from Bakhmutov, V.I., *Solid-State NMR in Materials Science: Principles and Applications*, CRC Press, Boca Raton, FL, 2011.)

In general, there are no limitations in the choice of target nuclei (with spins of 1/2 and quadrupolar) for studies of molecular dynamics by line-shape analysis as described above. However, in practice, ^{1}H, ^{13}C, ^{19}F, ^{15}N, and ^{2}H nuclei are the most popular. In this context, the ^{13}C- and ^{15}N-labeled organic and bioorganic compounds or materials will obviously be preferable.

In contrast to ^{2}H nuclei, line shapes of nuclei with spins of 1/2 are dominated by chemical shift anisotropy $\Delta\sigma$ and the asymmetry parameter η. When the correlation times (τ_C) of molecular reorientations (or motions of molecular groups) in the external magnetic field are larger than the inverse of the anisotropy factor (i.e., τ_C

$\gg 2\pi/\Delta\sigma$), then a NMR powder pattern will be typical of rigid solids. Fast isotropic reorientations will lead to an isotropic sharp resonance similar to that in solutions. Finally, in the intermediate exchange region at $\tau_C \sim 2\pi/\Delta\sigma$, the line shape will depend on correlation time τ_C as well as on the type of the molecular motions and geometry of the reorientations (i.e., on a motional model).

Figure 9.2B illustrates powder line shapes and their sensitivity to dynamic effects for two types of nuclei: nuclei with spins of 1/2 and ^{2}H nuclei. The shapes calculated for a two-site jump model at the τ_C value of 1 ns change remarkably with the geometry of the motions. Similarly, two-site jumps, three-site jumps, and other motional models can be recognized by line-shape analysis.

It should be emphasized, however, that in practice the situation can be much more complicated because different motional models can equally describe the shapes of lines observed experimentally. In other words, the computer simulation itself cannot discriminate between the different types of motions. In such cases, independent data are needed, which can be obtained by dynamic NMR experiments on two nuclear labels. An example of this approach is variable-temperature investigations of ^{2}H- and ^{15}N-labeled atactic poly(acrylonitrile), $[C_3H_2N]_n$–aPAN, where the labels are located in different places of the molecule.[4] Static ^{2}H and ^{15}N NMR spectra recorded in the temperature diapason between 10° and 140°C have shown line-shape evolutions that can be rationalized only as rotational diffusion around the main chain with restricted distribution of amplitudes (Figure 9.2C). Finally, in the exclusively complex cases, the concept of correlation time distributions (see Chapter 4) is applied to describe motions that cannot be characterized by a single correlation time at a given temperature.

9.4   ONE- AND TWO-DIMENSIONAL EXCHANGE NMR SPECTROSCOPY IN SOLIDS: SLOW MOLECULAR DYNAMICS

The general principles of exchange NMR spectroscopy in solutions have been discussed in Chapter 4. The same principles are valid for solid-state NMR to detect relatively slow motions on a NMR frequency scale from 1 Hz to 1 kHz which are typical of polymeric molecules and other systems experiencing segmental reorientations, conformation changes, etc. Details regarding one-dimensional (1D) and two-dimensional (2D) versions of exchange solid-state NMR spectroscopy, the pulse sequences applied, and technical details are reported in References 4 and 5 and in the recommended literature. Figure 9.3A shows the stimulated echo pulse sequence that is often used for recording solid-state 1D exchange NMR spectra. Pulse manipulation and delay times were used to evaluate the two- and multiple-time correlation functions required to describe molecular (or molecular group) rotational processes. The experiment included three sections. A 90° radiofrequency pulse in the first section was applied for excitation of nuclei and was followed by the time τ necessary for magnetization evolution by chemical shift anisotropy interactions or quadrupolar coupling. Then, in the second section, a storage pulse was used where the single quantum coherence converted into a term that does not evolve during the long mixing time, t_m. The third pulse in Figure 9.3A is the registering pulse used to collect FID data. Consideration of this pulse sequence shows that a decrease in the echo intensity depends on mixing time t_m. This feature allows determining the two-time

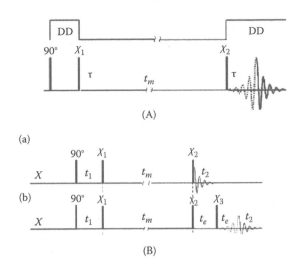

FIGURE 9.3 (A) Pulse sequence applied for stimulated-echo experiments giving 1D exchange NMR spectra.[5] (B) Two pulse sequences used for 2D exchange solid-state NMR experiments: (a) basic sequence with a single 90° excitation pulse; (b) modified pulse sequence with a single 90° excitation pulse and echo acquisition t_e. (Adapted from Bakhmutov, V.I., *Solid-State NMR in Materials Science: Principles and Applications*, CRC Press, Boca Raton, FL, 2011.)

correlation function for a two-site exchange process. In the case of multiple-center exchanges, the number of sites involved in the exchanges will also affect the echo intensity when the mixing time is long.

In addition to the stimulated echo shown in Figure 9.3A, many other modified 1D versions are currently used to identify the time scales and amplitudes of motions in organic polymers, inorganic materials, biopolymers, and proteins, including (1) 1D pure exchange (PUREX) NMR designed to suppress the resonances belonging to immobile molecules or groups; (2) one-dimensional exchange spectroscopy by side-band alteration (ODESSA) NMR aimed at the study of slow reorientations by analysis of spinning sidebands; and (3) center-band-only detection of exchange (CODEX) NMR, where the resonance intensities of the isotropic lines are also analyzed in high-resolution MAS NMR spectra. Further details can be found in the recommended literature.

A 2D exchange NMR experiment on static samples is represented by the pulse sequences shown in Figure 9.3B. The first 90° pulse excites the transverse magnetization, which develops during time t_1 due to the chemical shift anisotropy interactions. The second 90° pulse stores one component of the magnetization, situated along the Z-direction, which remains during the long mixing time, t_m. After time t_m, the registering pulse is applied to collect the NMR data during time t_2. Because a NMR signal now depends on times t_1 and t_2, double Fourier transformation results in a 2D NMR pattern that shows the intensity distribution as a function of the NMR frequencies and as a function of molecular motions occurring during the mixing time. Following the logic of this experiment, in the absence of motions the resulting 2D NMR pattern will be purely diagonal. However, in the presence of motions, the orientationally dependent frequencies will change and lead to the appearance of non-diagonal peaks. The second sequence in Figure 9.3B adds the echo acquisition time, t_e.[5] Again, as in the case of 1D NMR, various versions of 2D exchange NMR spectroscopy are available: (1) 2D PUREX NMR, (2) rotor-synchronized 2D-MAS NMR exchange, and (3) 2D CODEX NMR. They are also powerful tools for the characterization of correlation times τ_C and the geometry of motions in solids.

The practice of solid-state exchange NMR spectroscopy is represented by the probing dynamics in L-phenylalanine hydrochloride.[6] This compound is a good model of polymer systems and proteins where chains experience reorientations on a time scale from seconds to nanoseconds. It should be noted that the millisecond time scale is of greatest interest. Solid L-phenylalanine hydrochloride shows such motions, which represent the flip jumps of the aromatic ring. This motion can be studied by 2D exchange NMR, including quantitative analysis of cross-peak intensities.[6]

Figure 9.4A shows a 2D cross-polarization (CP) ^{13}C exchange NMR spectrum exhibiting cross peaks. The peaks corresponding to correlations between ^{13}C resonances δ_1 and δ_2 and ε_1 and ε_2 agree well with 180° jumps of the aromatic ring. Variations in the mixing time of the pulse sequence affect the cross-peak intensities, which can be calculated by convenient computer procedures to give a rate of jumps. Such 2D exchange NMR experiments are very useful for identifying the type of motions in solids.

As was already noted in the previous section, line-shape evolution in variable-temperature experiments, including experiments on L-phenylalanine hydrochloride, can be simulated to give the needed kinetic parameters of the phenyl jumps. One should remember that the line shape is very sensitive to the motional rate, which is comparable to the angular frequency characterizing the chemical shift difference for the two exchanging sites. In other words, interpretation of the data implies knowledge of the chemical shift difference between the two sites experiencing the exchange. In many cases, however, the isotropic chemical shift difference is not detectable experimentally. Under these conditions, R-CODEX (a modification of CODEX) NMR experiments can be applied. These experiments are based on dipolar couplings and do not require knowledge of the chemical shift differences, $\Delta\delta$. It should be emphasized that any motions that appreciably alter the direction of dipolar couplings can be investigated by this method.

The R-CODEX NMR data collected for solid L-phenylalanine hydrochloride are represented in Figure 9.4B. These data show the intensities of the ^{13}C signals δ_1, δ_2, ε_1, and ε_2 participating in the exchange, which decay with growing mixing time. The decays are exponential and can be treated by fittings to extract the necessary kinetic information.[6]

9.5 DYNAMICS IN SOLIDS BY CROSS-POLARIZATION NMR EXPERIMENTS

As noted earlier, cross-polarization $^1H–X$ is widely used in solid-state NMR to enhance the sensitivity required for observation of nuclei with low abundance (e.g., ^{13}C, ^{15}N, ^{29}Si). The CP pulse sequence can be easily incorporated into 2D exchange

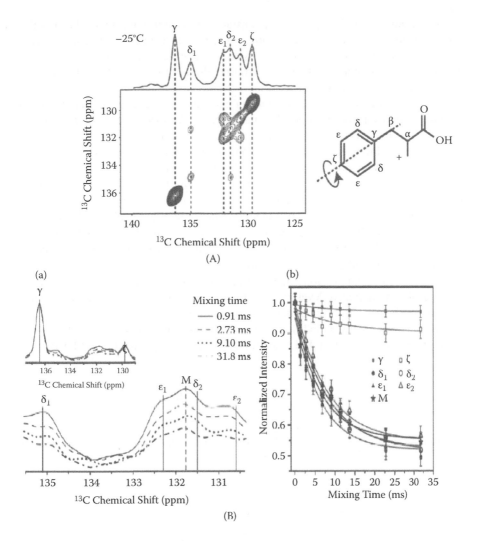

FIGURE 9.4 (A) The 2D exchange NMR spectrum recorded for ring carbons of crystalline L-phenylalanine hydrochloride undergoing flip motions. (B) (a) The ^{13}C R-CODEX NMR data obtained for the mobile ring sites of crystalline L-phenylalanine at different mixing times; (b) exponential decay fits for all peak intensities. (From Wenbo, L.I. and McDermott, A.E., *Concepts Magn. Reson. A*, 42A, 14, 2013. With permission.)

NMR for studies of molecular dynamics with reduced-duration experiments. At the same time, because the intensities of signals obtained by cross-polarization depend on molecular mobility (or motions molecular groups) via the T_{XH} constants and the ^{1}H $T_{1\rho}$ relaxation times in the rotating frame (see Chapter 8), the cross-polarization kinetics occurring during variations in the contact times allow the study of molecular dynamics in the solid state. The CP kinetics can be characterized by measurements of the T_{XH} and ^{1}H $T_{1\rho}$ parameters.[7]

For ^{13}C nuclei, the ^{1}H $T_{1\rho}$ and T_{CH} constants associated with molecular mobility are calculated from ^{13}C signal intensities measured as a function of proton–carbon contact times. These experiments are usually performed on fast-spinning samples because MAS conditions enhance spectral resolution, thus providing characterizations of different ^{13}C sites separately.

In general, the different ^{1}H $T_{1\rho}$ times in a sample can be averaged due to proton spin diffusion over distances of a few nanometers. For example, in rigid solids, the ^{1}H $T_{1\rho}$ times characterize large proton networks up to approximately several hundred nanometers, and they can give information on the homogeneity of a solid within the volume in the sub-nanometer range. One should remember that, if a sample is actually homogeneous, then all of the ^{1}H $T_{1\rho}$ times measured for different protons will be identical. In this case, a single proton spin reservoir can be considered for the whole sample. In contrast, in an inhomogeneous sample, there are different domains and the ^{1}H $T_{1\rho}$ values will be different for these domains because the sizes of the domains are larger than the spatial averaging by spin diffusion. All of these aspects play an important role in studies of composite materials or woods.

When solids contain regions characterized by different mobilities they can be distinguished at a qualitative level by wide-line separation (WISE) NMR experiments. Figure 9.5 shows the pulse sequence applied for WISE, which includes a regular cross-polarization section modified by incorporation of an extra time period, t_1.[4] After the regular step of ^{1}H–^{13}C polarization transfer, the ^{13}C magnetization will depend on

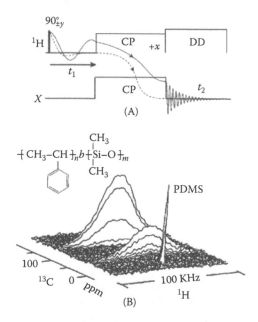

FIGURE 9.5 (A) The pulse sequence applied for 2D WISE experiments, including the standard cross-polarization section. (B) The 2D WISE NMR spectrum obtained for the rigid and mobile phases in the block copolymer PS-b-PDMS.[4] (Adapted from Bakhmutov, V.I., *Solid-State NMR in Materials Science: Principles and Applications*, CRC Press, Boca Raton, FL, 2011.)

the evolution of the ^1H magnetization during this time, t_1. If time t_1 is incrementing, then a 2D time matrix can be obtained to give a 2D NMR spectrum. Such 2D NMR spectra include high-resolution ^{13}C NMR MAS projections and wide-line spectra corresponding to ^1H nuclei. A WISE NMR experiments performed on a sample of the block copolymer polystyrene-b-poly(methylsiloxane) (PS-PDMS) is shown in Figure 9.5B. The experiment revealed the polystyrene fraction that is relatively rigid because it corresponds to very broad lines in the ^1H axis. In contrast, a relatively narrow ^1H resonance is clearly detected for the PDMS fraction that is obviously mobile.

9.6 MOLECULAR DYNAMICS IN SOLIDS BY NMR RELAXATION MEASUREMENTS: COMMON ASPECTS OF APPLICATIONS

Relaxation mechanisms were considered in Chapter 3, but this section provides a summary of the mechanisms important for solids. They can be represented by two categories. Nuclear relaxation in solids can be dominated by dipolar, quadrupolar, and chemical shift anisotropy interactions in the absence of paramagnetic centers. One should remember that the CSA contribution to total relaxation rates can be particularly significant in experiments at the highest magnetic fields. In the presence of paramagnetic centers, nuclear relaxation is governed by strong dipolar coupling by unpaired electrons and/or spin diffusion.

As in the case of solutions, dipolar or quadrupolar relaxation by isotropic molecular motions in solids results in temperature dependencies where the T_1 and $T_{1\rho}$ plots represented in semilogarithmic coordinates (see Figure 3.4 in Chapter 3) are V-shaped and show minima. In contrast, the T_2 time reduces consistently with decreasing temperature. The slopes of the T_1 and $T_{1\rho}$ curves correspond to the E_a values, which can be determined experimentally. In turn, the τ_C-independent T_1 minima are dictated by dipolar coupling (i.e., internuclear distances) or quadrupolar parameters directly calculated from the variable-temperature relaxation curves. It should be emphasized that the activation energies of motions in solids are much larger than in solutions (with the exception of fast CH_3 rotation or other small groups). Therefore, the T_1 minima in the relaxation curves move to a high-temperature zone and generally at room temperature $T_1 \gg T_2$. The T_1 and T_2 measurements allow the exclusion of correlation times τ_C from the well-known equations and determine the dipolar or quadrupolar coupling constants. In turn, these values can be applied to calculate correlation times τ_C and activation energies of motions from variable-temperature T_1 times. The same strategy can be used in the presence of relaxation by the chemical shift anisotropy mechanism.

Relatively slow molecular reorientations in solids correspond to the condition $T_{1\rho} \neq T_2 \neq T_1$. Thus, the $T_{1\rho}$ time is an additional important parameter characterizing molecular mobility in solids. For spin–lattice relaxation, governed by homonuclear coupling between spins I (for example, proton–proton) in the rotating coordinate frame, the relaxation rate can be expressed as follows:

$$1/T_{1\rho} = (1/5)\gamma^4\hbar^2 r^{-6} I(I+1)\left[3J^D(2\omega_1) + 5J^D(\omega_I) + 2J^D(2\omega_I)\right] \quad (9.4)$$

where the $T_{1\rho}(min)$ magnitude is shifted toward lower temperatures relative to T_{1min}.

All of the above considerations suggest that variable-temperature T_1, T_2, and $T_{1\rho}$ measurements can characterize isotropic molecular motions in solids via calculations of motional correlation times and activation energies only if the relaxation mechanism is well identified or the corresponding relaxation contribution is well evaluated. However, even under these conditions, the situation is complicated when molecular reorientations (or motions of molecular groups) cannot be described simply by a single correlation time at a given temperature. In these cases, the concept of correlation time distributions or other motional models can be applied.

9.6.1 NUCLEAR RELAXATION IN SPINNING SOLIDS

An important problem in the interpretation of nuclear relaxation measurements in solids is the fact that structurally different sites, particularly in amorphous systems, can show different motion correlation times. Then, instead of exhibiting exponential behavior, relaxation can become non-exponential corresponding to a multiple-exponential function or a stretched exponential, $\exp(-(\tau/T_1)^\beta)$. Moreover, the formal description of relaxation by a stretched exponential can lead to uncertainties, particularly for the β values changing in sets of similar compounds.[8] This effect is shown in Figure 9.6A. In such cases, the physical sense of the β parameter remains unclear. In the presence of paramagnetic centers as impurities, nuclear relaxation can again become exponential upon complete domination of the spin-diffusion mechanism. In fact, a single T_1 value will characterize all of the nuclei in a solid, even those that are not chemically equivalent, that are situated within the barrier radius effective for spin diffusion to paramagnetic centers. Because this fact is unknown *a priori*, additional experiments are necessary to clarify the relaxation behavior of nuclei used as probes.

It is generally accepted that MAS NMR is very effective for improving spectral resolution for the characterization of various sites in solids; however, the high spinning rates (currently up to 60 kHz) can affect relaxation behavior. For example, Figure 9.6B shows the ^1H T_1 time measured in spinning samples of talc. As can be seen, the dependence is significant for variations in the rates from 0 to 12 kHz. Moreover, it is not monotonic, leading to uncertainty regarding how to compare T_1 times available in the literature obtained at different spinning rates. The spinning effects obviously require explanation.[9]

In spite of the development of MAS techniques, rates applied for mechanical rotations of samples are still low relative to motional correlation times; therefore, generally, the spinning rates do not affect the T_1 times of relaxation caused by dipolar, quadrupolar, and chemical shift anisotropy interactions. However, theoretically, it has been shown that for high molecular mobility dipolar T_2 times (in contrast to dipolar T_1 times) can become dependent on spinning rates in accordance with Equation 9.5:

$$1/T_2^{DD} = (1/5)\gamma^4\hbar^2 r^{-6} I(I+1)\left[2\tau_C/\left(1+\omega_R^2\tau_C^2\right) + \tau_C/\left(1+4\omega_R^2\tau_C^2\right)\right] \quad (9.5)$$

where ω_R is the spinning rate. In fact, when $\omega_R^2\tau_C^2 > 1$, the dipolar T_2 time will be proportional to ω_R^2. In practice, this effect can be expected for a relatively long correlation time, τ_C. Simple calculations show that the correlation time is calculated as $>2 \times 10^{-5}$ s at a spinning rate of 10 kHz.

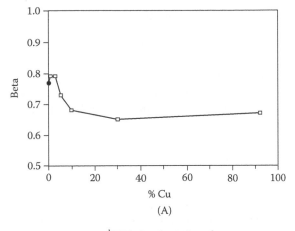

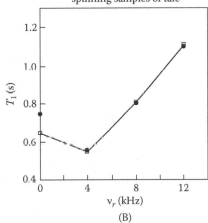

FIGURE 9.6 (A) Parameters obtained for treatments of the ^{29}Si T_1 relaxation curves for silica gels with the surface covered by copper ions at various concentrations.[8] (B) Dependence of the ^{1}H T_1 time in talc on spinning rates (room temperature, two experiments). (Adapted from Bakhmutov, V.I., *Solid-State NMR in Materials Science: Principles and Applications*, CRC Press, Boca Raton, FL, 2011.)

Nuclear relaxation occurring by the spin diffusion mechanism can also show a dependence on spinning rates. According to NMR theory and confirmed by a number of experiments, the spin diffusion coefficient, $D(\omega_R)$, reduces with spinning. Thus, the T_1^{SD} times depend on the MAS rates, ω_R, in Equation 9.6 via $D(\omega_R)$:

$$1/T_1^{SD} = (1/3)8\pi NpC^{1/4}D^{3/4}(\omega_R) \tag{9.6}$$

Thus, increasing the spinning rates elongates the T_1^{SD} times.[10] In turn, the T_2^{SD} times will also increase in spinning samples in accordance with Equations 9.7 and 9.8:

$$T_2^{SD} = (4/3)(\omega_R)^2 \tau^2 / \Delta v^2 \tau \qquad (9.7)$$

$$T_2^{SD} = (4/3)(\omega_R) \tau / \Delta v \qquad (9.8)$$

It should be emphasized that these expressions are valid when $v_R/\Delta v \geq 1$, where spinning rates v_R and line widths Δv are measured in Hz. Because, generally, in solids the dipolar spin–lattice relaxation times are independent of magic-angle spinning, the variable spinning rate experiments can potentially recognize the presence of the spin diffusion mechanism.

This test is very useful for the study of solids by relaxation methods. For example, spinning effects on ^{1}H T_1 values have been reported for solid adamantine and glycine. In high-density polyethylene, the ^{1}H spin diffusion reduces by 65% with an increase in spinning rates from 2 to 12 kHz. The ^{1}H T_1 time in a talc (Figure 9.6B) remarkably increases in spinning samples, clear proof of the spin diffusion mechanism being dominant in proton relaxation. In contrast, the ^{29}Si T_1 times do not change with spinning. This is reasonable because ^{29}Si–^{29}Si dipolar interactions are too weak to provide spin diffusion of ^{29}Si nuclei.

9.6.2 METHODOLOGY OF SOLID-STATE RELAXATION STUDIES

The methodology for relaxation studies of molecular dynamics in solids can be well represented by relaxation via dipole–dipole interactions in diamagnetic molecular systems. In such systems, the approaches used generally include the following:

- Variable-temperature or variable-resonance frequency measurements of T_1 times, where the number of minima observed in the temperature dependence or number of steps detected in the frequency dependencies will indicate the number of different molecular motions. In turn, the types of motions can be determined with the help of an appropriate mathematical formalism that gives correlation times τ_C and activation energies E_a.
- Measurements of the dipolar line widths and T_2 values, which can show magnitudes of the residual dipolar interactions averaged by molecular motions with correlation times much shorter than the T_2 times (generally on the order of 10^{-5} s).
- Measurements of temperature-dependent $T_{1\rho}$ times, because these times are sensitive to motions in the microsecond diapason.
- Measurements of times T_{1D} and $T_{1\rho\rho}$ (relaxation time in the doubly rotating frame), which are very sensitive to slow motions in the sub-millisecond to millisecond diapason.

It should be emphasized that when a motional model is unknown and nuclear relaxation is exponential, then the model-free approach, considered in Chapter 4, can be applied.

The first methodology is the one most commonly used in studies of molecular dynamics in the solid state. As has been shown theoretically in Chapter 4, for example, two motions can clearly show the presence of two minima in the

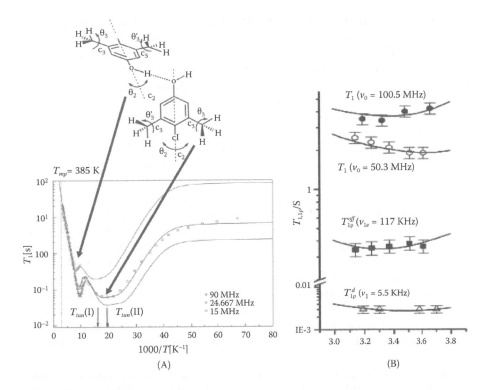

FIGURE 9.7 (A) Temperature dependence of ^1H T_1 time measured in a sample of compound PCMX at different frequencies (solid lines are theoretical curves obtained for the relaxation model with the shown motions. (From Latosiska, J.N. et al., *J. Phys. Chem. A*, 118, 2209, 2014. With permission.) (B) Temperature dependences of NMR relaxation times T_1, $T_{1\rho}$ (off-resonance), and $T_{1\rho}^d$ measured for carbons in a sample of solid polylysine. Solid lines show the results of fitting procedures.[12] (Adapted from Bakhmutov, V.I., *Solid-State NMR in Materials Science: Principles and Applications*, CRC Press, Boca Raton, FL, 2011.)

variable-temperature T_1 curves, if correlation times $\tau_C(1)$ and $\tau_C(2)$ differ by a factor of around 50 (see Figure 4.11 in Chapter 4). However, smaller differences in τ_C values mask the presence of two minima.

In practice, many systems comply with the condition formulated above; thus, T_1 times measured at different temperatures can provide detailed information on molecular dynamics. The compound PCMX with the structure shown in Figure 9.7A is such an example.[11] The ^1H T_1 times in a sample of PCMX were measured in a wide range of temperatures from 15 K to the melting point of the compound. The ^1H second moment, also describing molecular mobility, was obtained between 106 and 380 K. As follows from Figure 9.7A, the ^1H T_1 data collected at different frequencies clearly show two T_1 minima, corresponding to two types of motions. In addition, the curves demonstrate the different features near these minima. As can be seen, the high-temperature minimum is sharp and symmetric, while the low-temperature minimum is wide and asymmetric. The symmetric shape and position of the first minimum, depending on the spectrometer frequency, correspond to a classical motion with an activation energy of 10.5 kJ/mol found by the high-temperature slope. Density functional theory (DFT) simulations and an analysis of the second moment describe this motion as jumps by angle $\theta_2 = \pm15°$ (Figure 9.7A). The broad low-temperature T_1 minimum illustrates a very important property: Down to helium temperatures, the T_1 value is independent of the temperature. This effect is typical of tunneling. Thus, this T_1 minimum can be attributed to a hindered rotation of two methyl groups in combination with tunneling through the energy barrier estimated as <4 kJ/mol by the high-temperature wing of the minimum.

A second example, important from a methodology point of view, is ^{13}C relaxation in the solid polymer polylysine.[12] The dynamics of the polymer have been studied by determining the ^{13}C T_1, $T_{1\rho}$ (off-resonance), and ^1H-decoupled $T_{1\rho}^d$ times at different resonance frequencies and temperatures. The variable-temperature relaxation data are shown in Figure 9.7B. The character of these curves easily leads to accurate localization of the minimal relaxation times at different temperatures and different frequencies. Also, fitting to the experimental curves on the basis of various motional models determines the type of motions. In this case, the data have produced 15 to 25 relaxation parameters to describe the molecular dynamics as two motional processes with correlation times equal to 10^{-9} and 10^{-4} s. The main conclusion that can be reached on the basis of these two practical examples is that multinuclear relaxation experiments performed on nuclei located in different structural units of a compound can give more reliable and complete information on molecular dynamics, particularly if the target nuclei relax via different mechanisms.

Generally speaking, besides studies of molecular dynamics, relaxation times themselves are useful for planning advanced solid-state NMR experiments. Knowledge of relaxation times helps to optimize parameter settings and minimize the duration of an experiment to produce the best results. In addition, because different solids can show different relaxation behaviors, sometimes specific knowledge of a behavior is necessary for NMR studies of relative systems. For example, the design of new diamagnetic silica-based materials or zeolites is based on analysis of their solid-state NMR spectra, which are widely represented in the current literature. These compounds contain ^{27}Al and ^{29}Si nuclei in their matrices and have a relaxation behavior as summarized below:[13]

- ^{29}Si T_1 relaxation times typically take values of 5 to 30 s but can approach 100 s in some cases.
- ^{27}Al nuclei in the matrix of a material do not affect the ^{29}Si T_1 times, and *vice versa*.
- Spin diffusion is not effective in ^{29}Si T_1 and T_2 relaxation due to the large distances between neighboring silicon nuclei and their low natural abundance (both factors contribute to weak dipolar coupling).
- Molecular oxygen absorbed in the silica matrix strongly affects ^{29}Si T_1 times, whereas other paramagnetic impurities or water play a secondary role.
- ^{29}Si T_1 relaxation times of crystallographically different silicon sites are generally different.
- From crystalline to amorphous zeolites, the ^{29}Si T_1 times increase because many Si atoms in the amorphous samples are inaccessible to oxygen.

- The ^{27}Al T_1 relaxation times are relatively short (between 0.3 and 70 ms) because the relaxation is dominated by interactions with the crystal electric field gradients, which are modulated by translational motions of polar sorbate molecules.
- In zeolites containing organic components, the ^{29}Si T_1 times are generally longer.
- ^{27}Al T_1 times strongly increase upon dehydration of the samples.

All of the above points should be taken into account when planning NMR experiments and interpreting the data collected.

9.6.3 GENERAL COMMENTS ON NON-EXPONENTIAL RELAXATION IN SOLIDS

Exponential relaxation in solids, particularly in amorphous systems, is rather a surprising phenomenon, as non-exponential processes are detected more often. Nevertheless, in practice, researchers can be at a loss when they observe non-exponential relaxation. The relaxation behavior of nuclei in solids varies and depends on the nature of the nuclei, types of internuclear coupling, structural features of the molecular systems, and their nature. As a result, nuclear relaxation can be exponential, bi- or multi-exponential, non-exponential, or even more complex. However, in practice, determination of the type of relaxation is a result of computer fittings to magnetization decays obtained experimentally through the use of various model functions: exponential, double (or multiple) exponential, or stretched exponential. For all that, no physical sense is attributed to the function used; thus, the problem of non-exponential relaxation is not simple. In fact, even experimental conditions used for relaxation time measurements can affect the visible relaxation behavior. For example, ^{13}C spin–lattice NMR relaxation in simple organic solids follows a triple-exponential function when ^{1}H radiofrequency irradiation of a medium power is applied. In contrast, with strong ^{1}H field irradiation or in the absence of ^{1}H decoupling, the ^{13}C relaxation is exponential.

Therefore, the physical sense should play a major role when the experimental relaxation data are treated by different functions. For example, target nuclei having only two distinct environments in a solid or participating only in two different molecular motions are a good reason for a bi-exponential relaxation. This statement is quite valid for solid hexafluoroacetylacetone, where the ^{19}F T_1 relaxation is clearly bi-exponential due to the presence of two physically distinguishable CF_3 rotors characterizing different activation energies of the rotation.

In general, however, in the absence of an effective spin-diffusion mechanism, amorphous or other complex molecular systems show a non-exponential T_1, T_2 relaxation, often described by a stretched exponential function, $\exp(-(\tau/T_1)^\beta)$. Here, the β parameter can vary between 0 and 1, from a highly non-exponential relaxation to an exponential relaxation behavior. For example, non-exponential magnetic relaxation follows the stretched exponential in many disordered solids such as glasses, spin glasses, super-cooled liquids, liquid crystal polymers, dielectrics, magnetic systems, and amorphous semiconductors. If the relaxation function is well established, then the physical origin of the non-exponential decays can be discussed in terms of complex spin dynamics, adiabatic and non-adiabatic processes, anomalous diffusion, or models where the β parameter is taken as a measure of randomness. In addition, non-exponentiality is often connected with dynamic heterogeneities but not an intrinsic non-exponential behavior.

A special case of relaxation in solids is connected with the presence of paramagnetic centers. Here, interpretation of the stretched exponential function $\exp(-(\tau/T_1)^\beta)$ changes. As generally accepted, a stretched exponential with a β value of 0.5 corresponds to nuclear spin–lattice relaxation occurring via paramagnetic centers by dipolar interactions in the absence of spin diffusion. Intermediate values between 0.5 and 1.0 are discussed in terms of a spin-diffusion-limited mechanism for relaxation via random paramagnetic impurities. However, one should remember that in any case the β parameter in a stretched exponential relaxation should be determined reliably on the basis of a large number of delay times used in the relaxation experiments, which should be reproduced several times.

In addition, a non-exponential relaxation formally described by a stretched exponential function can also be approximated by a superposition of several exponential decays in the presence of smooth relaxation time distributions. For example, a β parameter of 0.67 corresponds to a Gaussian T_1 distribution. In other words, the situation is not single valued; therefore, independent information is needed to support either relaxation model.

Finally, in practice, it is not clear how to compare T_1 or T_2 times obtained for solids with different β values. One recommendation is to simplify the comparison of relaxation curves with different β values by using the parameter $T_1(50)$, which corresponds to the magnetization reaching 50% of the fully recovered value.

9.7 DYNAMICS IN HETEROGENEOUS AND DISORDERED SOLIDS: NMR SPECTRA AND RELAXATION DISPERSION

Disordered and heterogeneous solids are the focus of research into the mechanisms of chemical reactions occurring on the surface which can be valuable for heterogeneous catalysis. The solid matrix of such a system is usually filled with a liquid or a gas or small molecules as a function of pore size. Therefore, the dynamics of small molecules and the properties of the matrix, particularly their interconnection, are of great interest. Studies of self-diffusion in such systems using the pulsed field gradient (PFG) technique have been discussed in Chapter 4. Other aspects are considered in this section.

Liquids located in the matrix and on the matrix surface (i.e., at interfaces) are mobile, and their translational diffusion coefficients are similar to those of bulk liquids. Thus, the molecular mobility of liquids in heterogeneous solids is not solid like, and their local rotational/translational reorientations at interfaces are quite fast. Due to these motions, the various anisotropic magnetic interactions in the liquid spin system are significantly averaged and the dynamics of small molecules in heterogeneous solids can be probed by regular solid-state NMR techniques.

Figure 9.8A demonstrates the application of a simple qualitative approach based on line shapes in NMR spectra experiencing a temperature evolution. The room-temperature ^2H NMR spectrum recorded for a static sample of the silica-based material

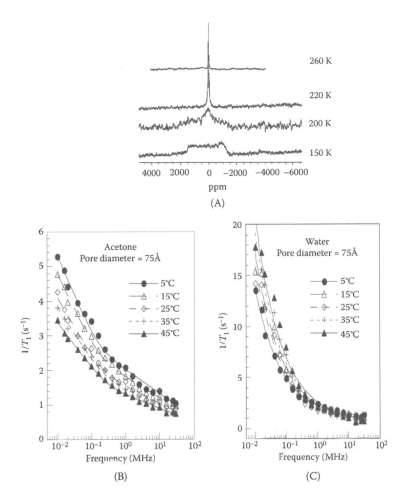

FIGURE 9.8 (A) Variable-temperature ^{2}H NMR spectra recorded in a static sample of silica-based material, SiO_2–Al_2O_3·^2H_2O (2 wt% Al). (B,C) The ^{1}H T_1 relaxation dispersions obtained for acetone and water in a sample of calibrated porous glass. (From Korb, J.P. and Bryant, R.G., *Adv. Inorg. Chem.*, 57, 293, 2005. With permission.)

SiO_2–Al_2O_3 with pores filled with ^2H_2O shows the sharp Lorenz-shaped resonance. This resonance, obviously corresponding to isotropic motions of the water in pores, remains sharp even at 220 K, insignificantly broadening at lower temperatures. Only at 200 K does the line transform toward a superposition of the broadened Lorenz-shaped signal and the typical rigid deuterium powder pattern, probably due to a slow exchange between moving and rigid water molecules.[14] Finally, at 150 K, this superposition is resolved to show a quadrupolar splitting of 148 kHz, which is typical of solid water. Similarly, the temperature evolution of solid-state ^{2}H NMR spectra characterizes the dynamics of other small molecules absorbed by porous solids, such as deuterated benzene and deuterated ethanol. Again, upon cooling, an isotropic resonance transforms to a powder pattern with quadrupolar splitting when motions are slow. Principally, applying line-shape analysis to these systems can characterize the dynamics of small molecules quantitatively by the correlation times of the motions.

Due to the high mobility of small molecules, the regular (solution-like) high-resolution ^{1}H and ^{1}H nuclear Overhauser effect spectroscopy (NOESY) MAS NMR approaches can be also used for characterization of their dynamics. For example, application of these techniques for studies of olefin molecules, such as 1-butene or 1-pentene, absorbed by zeolites has shown complex dynamics, including fast librations and overall reorientation–translation motions.[15] Details of these motions can be found in the recommended literature. In addition, due to the high spectral resolution of these techniques, they provide accurate measurements of the ^{1}H chemical shifts of the adsorbed molecules as a function of pore fillings. It is interesting that the chemical shifts of olefins at lower pore fillings are very close to those measured in the gas phase, while at higher pore fillings the chemical shift values approach those in bulk liquids.[15]

Even regular (solution-like) NMR relaxation experiments can completely describe the dynamics of small molecules in porous systems. For example, in the case of organic template molecules or liquids, ^{1}H T_1, ^{1}H $T_{1\rho}$, and ^{1}H–^{13}C cross-relaxation time measurements can be performed. The application of these solution-like approaches (see Chapter 4), especially ^{1}H T_1 measurements, plays a major role in the food and petroleum industries. The dynamics of water or oil at the surface can be characterized by diffusion coefficients, diffusion activation energies, the time of residence, and coefficients of surface affinity. Even the microporosity of ultra-high-performance concretes can be studied by these techniques.

If the molecular dynamics in heterogeneous solids containing pores filled with liquids are too complex to be described by regular relaxation time measurements, then the magnetic relaxation dispersion method can be used. As noted earlier, this method is based on the dependencies of the spin–lattice relaxation rates ($1/T_1$) on the strength of the magnetic field (i.e., resonance frequency). Technically these experiments can be performed on samples during their mechanical movement from one magnetic field to another one. On the other hand, rapidly changing the electric current in an available magnet can provide better results. Moreover, modern NMR spectrometers that use magnetic field switching are highly sensitive, and relaxation dispersion profiles can be obtained for different field strengths (even for the lowest magnetic fields) with practically the same signal-to-noise ratios in the NMR spectra of different nuclei. Nevertheless, in spite of such technical developments, variable-field experiments on ^{1}H nuclei are still the most popular. One should remember that the magnetic-field-switching method has an important limitation in that a rapidly decaying component in the spectra can be lost during the magnetic field switching time. In general, the modern technique is capable of measuring the $1/T_1$ magnitudes up to 2000 s^{-1}.

Relaxation dispersion experiments on protons can be divided into two categories: proton-poor objects (e.g., inorganic solids containing small ^{1}H fractions) and proton-rich samples such as organic solids. It should be emphasized that, in the case of inorganic objects, protons of liquid components and nuclei in the solid matrix are not coupled. In the absence of this coupling (or for weak coupling) an analysis of the relaxation dispersion profiles is remarkably simplified.

The main task of relaxation dispersion experiments performed on proton-poor porous systems and particularly on disordered glasses is determination of the liquid–surface interactions for various liquids. For example, the behavior of water and other solvents can differ when the molecules interact with the high surface area of chromatographic microporous silica glasses. Figures 9.8B and 9.8C illustrate ^{1}H NMR data collected with a field-cycling NMR spectrometer for acetone and water molecules in a calibrated porous glass sample.[16] In both cases, the ^{1}H T_1 profiles show frequency dependencies in spite of the temperature range, corresponding to fast molecular motions (see Chapter 3). However, these dependencies are different. As can be seen in Figure 9.8B, the field dependence observed for acetone consists of two logarithmic regions. In contrast, the water ^{1}H T_1 relaxation dispersion shows a power-law dependence on magnetic field strength. Moreover, the ^{1}H T_1 values of acetone and water are much different, as well as their temperature dependencies. The theoretical treatments are represented by solid lines, demonstrating a very good agreement with the experimental curves. The calculations give the correlation times and activation energies of motions in the surface.[16]

The situation is slightly complicated for proton-rich solids, in that the ^{1}H nuclei in liquid components situated on the surface and ^{1}H nuclei in solid components can be magnetically coupled. For a strong coupling, the ^{1}H T_1 relaxation process itself becomes complex and gives non-exponential relaxation curves. In spite of this feature, the mobility of liquids can still be characterized by the relaxation dispersion profiles, which are better interpreted compared to regular relaxation experiments.

Polymers and proteins, as proton-rich solids, are of greatest interest in the context of molecular dynamics; therefore, they are often the focus of ^{1}H spin–lattice relaxation dispersion experiments. Modern current-switched, field-cycling NMR instruments provide relaxation dispersion measurements for protons in proteins that are dry or especially treated and confined in gel. Even for such systems, dispersion profiles similar to those shown in Figures 9.8B and 9.8C can be obtained with great accuracy.

The relaxation dispersion profiles are different. They are dependent on the nature of the solids under investigation and can be approximated by various functions that correspond to the various relaxation models.[16,17] These models can be found in the recommended literature. It is remarkable that the ^{1}H relaxation dispersion profiles measured experimentally in proteins follow the power law shown in Equation 9.9:

$$1/T_1(\omega) = A\omega^{-b} + C \tag{9.9}$$

Here, constant b depends on the nature of the proteins; for example, for dry proteins, constant b takes a value of 0.78.

Researchers are showing a growing interest in relaxation dispersion studies. In fact, relaxation profiles describe molecular dynamics better than other methods, particularly for proteins and polypeptide molecules. For example, it has been found that variable-field proton relaxation rates do not depend on the dynamics of the side chains in a polypeptide molecule; however, they are controlled by the dynamics of the main chain itself. Physically, this means that the motions of the main chain modulate the time-dependent proton–proton dipolar couplings, which, in turn, cause proton relaxation.

9.8   DYNAMICS IN SOLIDS UNDER HIGH PRESSURE

High pressure affects the dynamic behavior of liquids, particularly hydrogen-bonded liquids, where the lifetime of hydrogen bond structures and rates of molecular motions change remarkably with pressure (see Chapter 7). Because heterogeneous disordered solids are relatively compressible, they are affected by the pressure applied during application of the high-pressure solid-state NMR technique (Chapter 8). At the present time, this technique routinely provides a pressure range between 1 bar and 5 kbar, which is sufficient for the characterization of molecular dynamics in layered compounds.[18] The pyridine–CdPS$_3$ system is one such compound, containing pyridine molecules intercalated into the gaps of the solid host. Figure 9.9A shows the structure of pyridine–CdPS$_3$, where the solid host layers are separated by two-dimensional van der Waals (VDW) gaps. In accordance with this structure, one can expect that the pyridine molecule will experience a reorientation motion and libration with a large and a low amplitude, respectively, as is shown in the figure. In fact, such motions are reasonable in structures where the pyridine molecules are oriented in the VDW gap so the molecular plane is perpendicular to the host layers and the C_2 symmetry axis is parallel to them.

This dynamics can be observed and characterized by static ^{2}H NMR spectra and a line-shape analysis applied for signals of ^{2}H-enriched pyridine. Generally speaking, an increase in the pressure can lead to a reduction in the van der Waals gap dimension, and the character of motions under high pressure can change. The ^{2}H NMR spectra in Figure 9.9B show the pressure effect observed at a temperature of 330 K, when the pressure changes from 10^{-3} to 4.5 kbar.

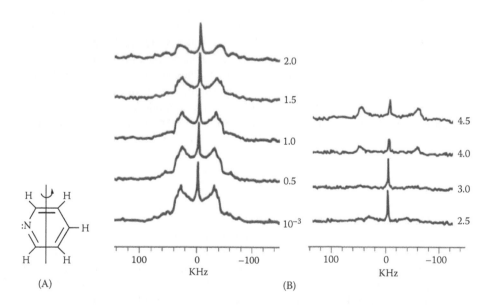

FIGURE 9.9   (A) Orientations of the pyridine molecule in the VDW gap of solid CdPS$_3$. (B) The ^{2}H NMR spectrum of ^{2}H-pyridine in solid CdPS$_3$ as a function of pressure. (From McDaniel, P.C. et al., *Physica A*, 156, 203, 1989. With permission.)

As can be seen, the ^{2}H NMR spectrum recorded at ambient pressure exhibits three resonance components showing quadrupolar splitting, in addition to the sharp resonance of pyridine molecules that experience fast isotropic motions (liquid-like). To put it simply, these solid-like components have been attributed to a threefold (or higher symmetry) rotational diffusion.[18] The data show that with a consistent increase in pressure the solid-like resonances broaden greatly and their intensity decreases. The latter can clearly be seen at a high pressure of 2.5 kbar. The effect of the intensity loss can be attributed to dephasing, which is typical of such systems. Then, at higher pressures of 3.0 and 4 kbar, the ^{2}H NMR spectrum shows only one rigid ^{2}H pattern, with quadrupolar splitting measured as 163 kHz. This value corresponds to the absence of any high-amplitude reorientations. Such a description of molecular dynamics, including a full line-shape analysis, can also be used to study biomolecular systems.[18]

In addition to the pressure effects observed as the evolution of line shapes in static NMR spectra, variable-pressure relaxation experiments can be performed to characterize molecular dynamics via motional correlation times. Spin–lattice relaxation measurements can be carried out for various nuclei, such as ^{65}Cu nuclei in a single crystal of $Sr_2Ca_{12}Cu_{34}O_{41}$[19] or ^{2}H nuclei in organic solids, particularly in polymers.[20] In fact, because the molecular dynamics contribute to the mechanical properties of polymers, knowledge of the high-pressure effects on nuclear relaxation in such systems is very valuable. The ^{2}H spin–lattice relaxation of methyl groups in a deuterated atactic polypropylene under pressures between 1 bar and 5 kbar is an example of such studies.[20]

The ^{2}H T_1 data obtained as the averaged relaxation times of methyl groups are shown in Figure 9.10. In general, the ^{2}H T_1 relaxation of CD_3 groups occurs due to their high rate of reorientations with the energy barrier, consisting of intra-chain

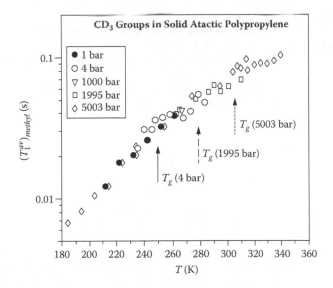

FIGURE 9.10 T_1 averaged times of methyl deuterons in atactic polypropylene as a function of temperature and pressure. (From Hollander, A.G.S. and Prins, K.O., *J. Non-Cryst. Solids*, 286, 12, 2001. With permission.

interactions. Reorientations have been found in the diapason between 10^9 and 10^{10} s^{-1}. As shown in Figure 9.10, the reorientation rate is affected only very little by the high pressure of 5 kbar. The low-field ^1H NMR relaxometry that occurs under high pressures is also applied for studies of nanoporous solids.[21] In this case, the pore spaces of porous materials have been determined by pressurized methane, so these studies are related primarily to liquids.

REFERENCES AND RECOMMENDED LITERATURE

1. Hansen, M.R., Graf, R., and Spiess, H.W. (2013). *Acc. Chem. Res.*, 46: 1996.
2. Guan, X. and Stark, R.E. (2010). *Solid State NMR*, 38: 74.
3. Rodríguez-Molina, B., Pérez-Estrada, S., and Garcia-Garibay, M.A. (2013). *J. Am. Chem. Soc.*, 135: 10388.
4. Azevedoa, E.R., Bonagambaa, T.J., and Reichertb, D. (2005). *Prog. Nucl. Magn. Reson. Spectrosc.*, 47: 137.
5. Krushelnitsky, A. and Reichert, D. (2005). *Prog. Nucl. Magn. Reson. Spectrosc.*, 47: 1.
6. Wenbo, L.I. and McDermott, A.E. (2013). *Concepts Magn. Reson. A*, 42A: 14.
7. Kolodzieski, W. and Klinowski, J. (2002). *Chem. Rev.*, 102: 613.
8. Alaimo, M.H. and Roberts, J.E. (1997). *Solid State NMR*, 8: 241.
9. Gil, A.M. and Alberti, E. (1998). *Solid State NMR*, 11: 203.
10. Kessemeier, H. and Norberg, R.E. (1967). *Phys. Rev.*, 155: 321.
11. Latosiska, J.N., Latosiska, M., and Tomczak, A.M. (2014). *J. Phys. Chem. A*, 118: 2209.
12. Krushelnitsy, A., Faizullin, D., and Reichert, D. (2004). *Biopolymers*, 73: 1.
13. Klinowski, J. (1991). *Chem. Rev.*, 91: 1459.
14. Bakhmutov, V.I., Shpeizer, B.G., Prosvirin, A.V., Dunbar, K.R., and Clearfield, A. (2009). *Micropor. Mesopor. Mater.*, 118: 78.
15. Bohlmann, W., Michel, D., and Roland, J. (1999). *Magn. Reson. Chem.*, 37: S126.
16. Korb, J.P. and Bryant, R.G. (2005). *Adv. Inorg. Chem.*, 57: 293.
17. Korb, J.P. (2001). *Magn. Reson. Imaging*, 19: 363.
18. Jonas, J. (2007). High-pressure NMR. In: *Encyclopedia of Magnetic Resonance* (Harris, R.K. and Wasylishen, R., Eds.). Chichester: John Wiley & Sons.
19. Fujiwara, N., Uwatoko, Y., Mori, N., Matsumoto, T., Motoyama, N., and Uchida, S. (2002). *J. Phys. Chem. Solids*, 63: 1103.
20. Hollander, A.G.S. and Prins, K.O. (2001). *J. Non-Cryst. Solids*, 286: 12.
21. Horch, C., Schlayer, S., and Stallmach, F. (2014). *J. Magn. Reson.*, 240: 24.

RECOMMENDED LITERATURE

Bakhmutov, V.I. (2011). *Solid-State NMR in Materials Science: Principles and Applications*. Boca Raton, FL: CRC Press.
Bohmer, R., Jeffrey, K.R., and Vogel, M. (2007). Solid-state Li NMR with applications to the translational dynamics in ion conductors. *Prog. Nucl. Magn. Reson. Spectrosc.*, 50: 87.
Duer, M.J., Ed. (2002). *Solid-State NMR Spectroscopy: Principles and Applications*. Oxford: Blackwell Sciences.
Huster, D. (2005). Investigations of the structure and dynamics of membrane-associated peptides by magic angle spinning NMR. *Prog. Nucl. Magn. Reson. Spectrosc.*, 46: 79.
McDaniel, P.L., Liu, C.G., and Jonas, J. (1989). High-pressure deuterium solid-state NMR study of the dynamics of pyridine intercalated $CdPS_3$. *Physica A*, 156: 203.
Ribeiro de Azevedo, E., Bonagamba, T.J., and Reichert, D. (2005). Molecular dynamics in solid polymers. *Prog. Nucl. Magn. Reson. Spectrosc.*, 47: 137.

Su, Y. (2012). *Structure and Dynamics of Membrane Peptides from Solid-State NMR*. Ann Arbor, MI: ProQuest.

Van Well, H.F.J.M. (1997). *Molecular Orientation and Dynamics in an Anisotropic Network: A Solid State NMR Study*. Maastricht, Netherlands: Shaker Publishing B.V.

10 Solid-State NMR
Special Issues

This book ends with a chapter addressing special issues in solid-state NMR. The principles, methodologies, and strategies discussed in the previous chapters can be applied for multinuclear solid-state NMR studies of various molecular systems— simple or complex, homogeneous or heterogeneous, amorphous or crystalline— where the experimental aspects, including registration of signals and methodology used, change only insignificantly with the changing nature of the samples. However, some solids and materials are exceptions to this versatility due to their complexity (proteins), porosity (diamagnetic porous catalysts with different pore sizes), or having a large amount of unpaired electrons (metallic systems and paramagnetic materials). Some of these materials are considered in this chapter to show how to modify NMR experiments, how to apply special methodological approaches, and what kind of information can be obtained.

10.1 SOLID-STATE NMR OF PROTEINS

When proteins cannot be crystallized for application of the x-ray method or have a low solubility in convenient solvents, solid-state NMR becomes the only method for accurate determinations of their structure. Just as for solutions (see Chapter 7), the structural analysis of solid proteins is based on primarily computerized treatment of numerous restraints, the parameters obtained by various solid-state NMR experiments. It is reasonable to expect that a greater number of experimental restraints will improve the precision in determinations of protein structures, a fact that has stimulated the development of a variety of solid-state NMR techniques.[1] For example, the first structure of the protein containing the SH3 domain was determined through solid-state NMR utilizing 286 carbon–carbon distances and 6 nitrogen– nitrogen inter-residue restraints. Due to further development of NMR techniques, the structures of other protein systems have been obtained by computer analyses of 643 restraints, providing the C–C and H–H distances that are necessary for formulation of the structure. Recently, the structure of GB1 protein was solved by computerized treatment of 880 interatomic distances and orientational restraints.[1]

As shown in Chapter 7, nuclear Overhauser effect (NOE) data (C–H or H–H) provide the most important distance restraints for structural studies of proteins in solutions. In solid proteins, the distance restraints can be obtained by modified NMR techniques such as proton-driven spin diffusion, dipolar-assisted rotational resonance, proton-assisted recoupling, and proton-assisted insensitive nuclei cross-polarization. Details regarding the pulse sequences applied for these experiments and their design can be found in Reference 1 and the recommended literature. One should remember that NOE experiments in solutions give short- and long-distance restraints, while the number of long-distance restraints in the solid state is obviously limited by the lower spectral resolution in the solid-state NMR spectra. As far as possible, this problem can be avoided by studies of protein molecules that are selectively labeled; however, in spite of the labeling, the protein size is still limited in the context of solid-state NMR measurements by small proteins containing fewer than 100 amino acids. For these small-size proteins, the precision of the structural determination can be enhanced by a factor of 2, when multi-dimensional (three- and four-dimension) solid-state NMR is applied to observe the dipolar line shapes. In fact, these shapes can potentially identify special orientations of individual chemical bond vectors.

The number of solid-state NMR restraints can be significantly increased by experiments performed on paramagnetic metalloproteins, which can be specially prepared, for example, by replacement of the diamagnetic metal ions in proteins with paramagnetic ions. In the NMR spectra of these proteins, the high magnetic anisotropies will cause pseudocontact (PC) chemical shifts directly connected with structural parameters. The PC shifts in combination with the NMR restraints obtained for the original diamagnetic metalloproteins provide enough data to determine the structure of proteins consisting of 159 amino acids. One of the structures obtained for the MMP-12 protein by solid-state NMR is shown in Figure 10.1A. It should be emphasized that treatments of combined types of restraints, including PC shifts, allow determination of the structure at a backbone root mean square deviation (RMSD) of 1.0 ± 0.2 Å. This is even better accuracy than that attained by single-crystal x-ray analysis (1.3 Å). Moreover, dipole–dipole interactions between unpaired electrons and nuclei in these proteins selectively enhance nuclear relaxation rates ($1/T_1$) and line broadenings ($1/T_2$). These additional paramagnetic effects, measured experimentally, give the metal–nucleus distance restraints that can also be used to determine the structure. Thus, in general, structural NMR studies of solid proteins are based on a combination of NMR spectroscopic data and measurements of relaxation times, the choice of which plays a very important role.

Another methodology useful for the structural characterization of complex molecular systems such as membrane proteins is represented by oriented solid-state (OSS) NMR spectroscopy, which measures dipolar couplings and chemical shift anisotropies in oriented samples. Generally, such samples are prepared in lipid bicelles.[2] Because the resonances observed in oriented samples are more dispersed due to anisotropic contributions to nuclear shielding, they show a larger dispersion of chemical shifts. This is an important advantage of OSS NMR, because the anisotropic interactions with the external magnetic field can be taken from the experiments and used to calculate the membrane protein topology.[2]

This advantage is amplified by the application of methods capable of enhancing spectral resolution and NMR sensitivity. Classical separated local field (SLF) experiments on membrane proteins are usually combined with sensitivity-enhanced (SE) experiments to give two-dimensional (2D) and three-dimensional (3D) NMR patterns. The pulse sequences applied in these experiments are complex, and the technical details are provided in Reference 2. The practical results obtained by ^{1}H–^{15}N SE

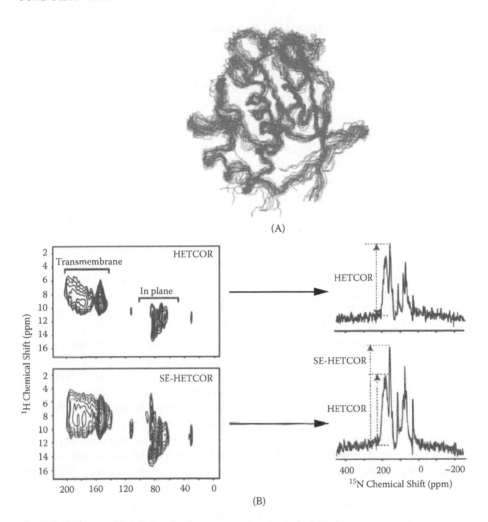

(A)

(B)

FIGURE 10.1 (A) Structure of the catalytic domain of MMP-12 protein obtained for a microcrystalline by treatment of the solid-state NMR data. (From Banci, L. et al., *Prog. Nucl. Magn. Reson. Spectrosc.*, 56, 247, 2010. With permission.) (B) ^1H–^{15}N HETCOR and ^1H–^{15}N SE HETCOR NMR spectra of the molecule U–^{15}N-SLN in aligned TBBPC bicelles and the corresponding 1D slices. (From Gopinath, T. et al., *Prog. Nucl. Magn. Reson. Spectrosc.*, 75, 50, 2013. With permission.)

heteronuclear correlation (HETCOR) experiments performed on the membrane protein [U–^{15}N]SLN in aligned bicelles are shown in Figure 10.1B. The 2D SE-HETCOR pattern indicates the excellent quality of the NMR spectra, where the cross peaks can be reliably assigned to solve the structure. The sensitivity of a SE-HETCOR experiment is remarkably improved compared to classical ^1H–^{15}N HETCOR experiments, as can be seen in the corresponding one-dimensional (1D) projections. In principle, these experiments can be performed on single crystals and powder samples.

10.1.1 CHEMICAL SHIFT TENSORS IN PROTEINS

In general, three-dimensional chemical shifts (chemical shift tensors) obtained by solid-state NMR are more informative than isotropic chemical shifts. In the context of solid peptides, chemical shift tensors can help with determination of local second-ary structures in the N–H···O=C hydrogen bond networks. In addition, knowledge of chemical shift tensors is very useful when the dynamics of individual residues are being studied. Usually, chemical shift tensors are determined by static solid-state NMR spectra, recorded in powder or polycrystalline samples of selectively labeled compounds, while their orientations in the molecular coordinate system can be deduced from the experiments on single crystals. Because the chemical shift tensors of ^{13}C, ^{15}N, and ^{17}O nuclei are particularly important for bioorganic molecules, they are represented here for peptides as a summary of some solid-state experiments.[3]

Magnitudes of ^{13}C chemical shift tensors measured for C=O groups in single-crystalline and polycrystalline peptides cover a wide range between 242 and 88 ppm. For example, the ^{13}C C=O tensor components in the twice-labeled peptide [1-^{13}C] Gly[^{15}N]Gly·HCl·H$_2$O have been determined as δ_{11} = 244.1 ppm, δ_{22} = 177.1 ppm, and δ_{33} = 87.9 ppm, where the δ_{22} direction deviates from the C=O axis by only 13°. The peptide Ac[1-^{13}C]Gly[^{15}N]GlyNH$_2$ shows δ_{11} = 243.0 ppm, δ_{22} = 184.9 ppm, and δ_{33} = 91.2 ppm, with the δ_{22} direction parallel to the C=O bond. These examples reflect a common tendency noted for C=O groups in peptides and proteins: $\Delta\delta_{22}$ > $\Delta\delta_{33}$ > $\Delta\delta_{11}$; that is, the δ_{22} values change remarkably from one compound to another, while the δ_{11} values remain very similar. The chemical shift anisotropy ($\Delta\delta$) does not change. It should be also added that the orientation of the principal component (δ_{22}) in the molecular coordinate system is not always collinear with the direction of the C=O bonds, although the δ_{33} direction is often perpendicular to the peptide plane.

An important element of ^{13}C chemical shift tensor studies in peptides and proteins is their sensitivity to conformational changes. It has been found that variations in the values of the principal tensor components as a function of conformational state are larger for δ_{22} and δ_{33} components, while δ_{11} components are practically constant. For example, the C=O group in a β-sheet conformation of peptide (Gly)$_n$-I shows the values δ_{11} = 243.0 ppm, δ_{22} = 174 ppm, and δ_{33} = 88 ppm, while a 3$_1$-helix conformation of (Gly)$_n$-II is characterized by δ_{11} = 243.0 ppm, δ_{22} = 179 ppm, and δ_{33} = 94 ppm.

The principal components of ^{15}N chemical shift tensors and their orientations in the molecular coordinate system have been found for many single crystalline and polycrystalline peptides and proteins. For example, the ^{15}N solid-state NMR spec-trum of the Gly-[15]Gly peptide shows chemical shift tensor values of δ_{11} = 204.7 ppm, δ_{22} = 56.5 ppm, and δ_{33} = 26.1 ppm, which change in peptide Val-[15]GlyGly: δ_{11} = 199.6 ppm, δ_{22} = 62.5 ppm, and δ_{33} = 19.9 ppm. ^{15}N solid-state NMR experiments on crystals have revealed tensor orientations in the molecular coordinate frame, where the δ_{11} component is generally oriented close to the N–H bond, deviating in some cases by 25°; the δ_{33} component is along the N–C(=O) bond; and the δ_{22} component is aligned perpendicular to the peptide plane. Similarly to ^{13}C tensors, the ^{15}N principal components are again sensitive to conformation states of peptides. For example, in Ala residues, the δ_{22} values are more downfield shifted (80 to 85 ppm) in β-sheets compared to those in α-helix conformations (74 to 80 ppm).

The low natural abundance of ^{17}O nuclei (0.037%) and large quadrupolar interactions hinder their detection in the solid state; thus, ^{17}O enrichment is required, and the application of high magnetic fields partially overcomes the problem of strong quadrupolar interactions. The ^{17}O chemical shift tensor components in the H-bonded peptides and polypeptides can be determined by solid-state NMR experiments performed for at least two different magnetic fields. Figure 10.2A shows the results of ^{17}O cross-polarization NMR experiments on a static sample of the labeled compound

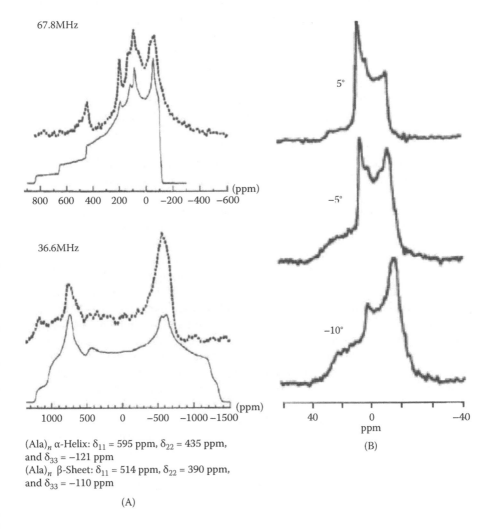

67.8MHz

(ppm)
800 600 400 200 0 −200 −400 −600

36.6MHz

(ppm)
1000 500 0 −500 −1000 −1500

5°

−5°

−10°

40   0   −40
ppm
(B)

(Ala)$_n$ α-Helix: δ_{11} = 595 ppm, δ_{22} = 435 ppm,
and δ_{33} = −121 ppm
(Ala)$_n$ β-Sheet: δ_{11} = 514 ppm, δ_{22} = 390 ppm,
and δ_{33} = −110 ppm

(A)

FIGURE 10.2 (A) ^{17}O cross-polarization NMR spectra recorded in a static sample of the helix form of a (Ala)$_n$ peptide molecule for different magnetic fields and ^{17}O principal tensor components as a function of conformational states obtained by computer simulations of line shapes (solid lines). (B) Variable-temperature ^{31}P NMR spectra recorded in a sample of egg phosphatidylethanolamine in the presence of decanol. (From Saita, H. et al., *Prog. Nucl. Magn. Reson. Spectrosc.*, 57, 181, 2010. With permission.)

[^{17}O](Ala)$_n$. The spectra vary widely due to the different contributions of quadrupolar and chemical shift anisotropy interactions to the line shapes. The line shapes can be simulated, and the fitting procedures give the corresponding ^{17}O chemical shift tensor components. As earlier, the tensor components can also be determined by ^{17}O magic-angle spinning (MAS) NMR experiments performed with high magnetic fields and fast spinning rates to enhance the spectral sensitivity.

Studies of peptides and proteins reveal common tendencies important for a structural analysis of bioorganic molecules. Again, as in the case of other nuclei, ^{17}O tensor components are sensitive to the conformation states of peptides and proteins. This effect can be seen in Figure 10.2A. With regard to the ^{17}O tensor orientations in the molecular coordinate system, the smaller quadrupolar coupling components and the most shielded chemical shift components are generally *normal* to the plane of peptides. The most de-shielded chemical shift components and largest quadrupolar coupling components are in the peptide plane.

In addition to providing statics information, chemical shift tensors are valuable for studies of molecular dynamics in solid peptides.[3] Backbone and side-chain dynamics in relatively small peptides can be characterized by variable-temperature ^{15}N solid-state NMR spectra recorded in static powder samples. The ^{15}N NMR line shapes of peptides consisting of Gly–Gly residues and the corresponding principal tensor components experience significant changes with temperature. For example, it has been shown for the Gly[^{15}N]Gly peptide that decreasing the temperature from 40° to –120°C leads to low and high field displacements of the ^{15}N δ_{11} and δ_{22} values respectively, while the ^{15}N δ_{33} component remains constant. An analysis of these data reveals the presence of librational motions around the δ_{33} axis which can be quantitatively characterized on the basis of a two-site jumping model.[3]

Similarly, the solid-like shapes of ^{31}P resonances and the ^{31}P chemical shift tensor components in phospholipids allow the study of their dynamics in biological cell membranes. They can also be used to characterize ligand–membrane interactions. The dynamics can be represented by polymorphic non-bilayer lipids, which are of great interest. In fact, lamellar-phase lipids can have different phase states. The hexagonal H_{11} and cubic phases depend on the type of lipids, their molecular shape, the presence or absence of other lipid molecules, the water content, and the temperature. When the lipid phase transforms from lamellar to hexagonal, the ^{31}P chemical shift anisotropy $\Delta\delta$ reduces remarkably and even changes its sign. The cylinders in this hexagonal phase are characterized by small radii, and the lateral diffusion around the cylinder axis averages the ^{31}P tensor components. Phase transitions under different conditions can be monitored by variable-temperature ^{31}P NMR spectra, where the line shapes of the ^{31}P resonances change strongly within the very limited temperature diapason (Figure 10.2B).

10.2   SOLID-STATE NMR IN METALS AND ALLOYS

As noted in Chapter 2, the large magnetic moments of unpaired electrons in metals, alloys, and metallic clusters give rise to strong local magnetic fields. These fields result in the exclusively strong NMR spectroscopic effects known as Knight shifts, the investigation and interpretation of which represent a special field of solid-state

physics. Solid-state NMR is traditionally applied for probing the electronic properties of metals and alloys, because the determination of resonance frequencies for
target nuclei and measurements of their relaxation rates allow characterization of
the density of states at the Fermi level and the character of the various s or d bands.
The theoretical expression of Knight shifts, K_{orb}, for electron components is given
by the equation provided in Figure 10.3.[4] At the present time, the electronic properties of small metallic particles and the metal cores in metal-cluster compounds are
attracting considerable interest because metallic systems containing small particles
of metals such as Pt, Pd, or Rh manifest important catalytic properties.

The electron effects in metals and alloys are so strong that they affect the choice
of methods for NMR signal registration. This statement is well illustrated by the
static [103]Rh NMR spectra shown in Figure 10.3 which were obtained for supported
metal catalysts. As can be seen, very pronounced electron–nucleus interactions
dramatically affect the [103]Rh resonance lines. The Knight shift is usually the main
parameter in such studies; however, as follows from Figure 10.3, its determination is
not trivial, as the [103]Rh resonances are too wide (particularly at small particle sizes)
and cannot be observed by standard NMR methods. In addition, the effect does not
depend on temperature. These extremely broad resonances can be recorded only
by point-by-point experiments, where the amplitude of a NMR signal is registered
by scanning the irradiation frequency (Figure 10.3). In spite of the very large line
widths of the observed resonances, the final [103]Rh NMR spectra in Figure 10.3 are

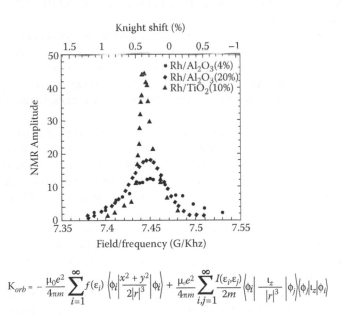

$$K_{orb} = -\frac{\mu_0 e^2}{4\pi m}\sum_{i=1}^{\infty} f(\varepsilon_i)\left\langle\phi_i\left|\frac{x^2+y^2}{2|r|^3}\right|\phi_i\right\rangle + \frac{\mu_0 e^2}{4\pi m}\sum_{i,j=1}^{\infty}\frac{I(\varepsilon_i,\varepsilon_j)}{2m}\left\langle\phi_i\left|\frac{l_z}{|r|^3}\right|\phi_j\right\rangle\langle\phi_j|l_z|\phi_i\rangle$$

FIGURE 10.3 Static [103]Rh NMR spectra of oxide-supported composites recorded by the
point-by-point method. The Rh–TiO$_2$ system containing 10% Rh has the largest particles,
with a diameter of 6 nm; therefore, the [103]Rh resonance line is strongly broadened. (Adapted
from Bakhmutov, V.I., *Solid-State NMR in Materials Science: Principles and Applications*,
CRC Press, Boca Raton, FL, 2011.)

very informative. In fact, they are highly sensitive to the nature of the components in the alloys. The shift of the ^{103}Rh resonance and its line-width change greatly moving from aluminum to titanium oxide.

It should be noted that the point-by-point method is universal and can be used to observe extremely wide lines directly in any solids. Technically, this method depends on varying the irradiation frequency for a single fixed magnetic field or varying the strength of the magnetic field while maintaining a constant NMR frequency. The latter approach has been used to record, for example, the exclusively broad ^{19}F NMR spectrum of crystalline CuF_2, where ^{19}F nuclei are direct neighbors of the paramagnetic copper ions.[5] The x-ray data obtained in crystalline CuF_2 show that the copper cation has two closest fluorine neighbors at different distances. Only the point-by-point method applied for variations in the NMR magnetic field corresponding to each point in the spectrum can allow observation of the two very broad ^{19}F resonances separated by 5000 ppm.

In rare cases, solid-state NMR spectra can be observed by regular detection of signals under MAS conditions due to the nature of alloys, such as the ^{125}Te solid-state NMR data obtained in doped PbTe, GeTe, SnTe, and $AgSbTe_2$ systems. In these systems, the ^{125}Te resonances can be affected by not only expected Knight shifts but also local bonding.[6] Understanding the effects of dopant or solute atoms on the ^{125}Te chemical shifts is obviously necessary in order to determine the composition of alloys, to assign the resonance lines, and to identify local chemical bonding. The strategy behind such studies can be demonstrated by ^{125}Te NMR of the three-component system $Pb_{1-x}Ge_xTe$.

The ^{125}Te MAS NMR spectra recorded by a long-term experiment performed on a sample of $Pb_{0.9}Ge_{0.1}Te$ at the lowest Ge concentration, spinning at a rate of 22 kHz, is shown in Figure 10.4A. It should be emphasized that these NMR experiments require minimization in the distortion of the peak intensities. It has been found that distortions are minimal when the carrier radiofrequency is close to the maximum of the spectrum.

The most intense resonance in the spectrum is centered at approximately −1860 ppm (referred to a $Te(OH)_6$ solution), which is close to the δ value observed in the two-component system Pb/Te. Thus, a Knight shift contribution to the resonance is negligible. This conclusion agrees with the long ^{125}Te T_1 relaxation time, measured as >5 s. Hence, as in the case of the well-established structure of PbTe, this signal belongs to the Te atoms that are bonded to six Pb atoms. The less intense resonances in the spectrum are remarkably shifted from the line observed in the two-component system PbTe. It has been found that their intensities increase with growing parameter x. The data provide reliable assignments of these signals to ^{125}Te atoms located in the GePb sublattice. In accordance with the assignments, the δ(^{125}Te) value changes with an increment of approximately +162 ppm per one Ge atom. Because the ^{125}Te T_1 relaxation times measured for all of the resonances are similar, even for growing parameter x, the Knight shifts are again negligible. In fact, in the presence of Knight shifts, the T_1 relaxation times of the shifted signals could be very short. In addition, the above assignments have been supported by density functional theory (DFT) simulations reproducing the line shapes shown in Figure 10.4A.

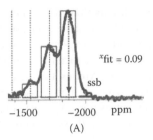

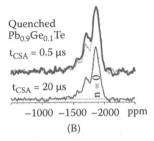

FIGURE 10.4 (A) ^{125}Te MAS NMR spectrum recorded at a spinning rate of 22 kHz in a sample of quenched $Pb_{1-x}Ge_xTe$ at $x = 0.1$ and computer fitting at $x - 0.09$ (including small spinning sidebands, SSBs) superimposed as a bar graph and dashed lines. (B) ^{125}Te MAS NMR data collected in a sample of $Pb_{0.9}Ge_{0.1}Te$ by recoupled CSA dephasing experiments. To make differential dephasing visible, the dephased spectrum is also shown scaled up and superimposed on the full spectrum (dashed line). (From Njegic, B. et al., *Solid State Nucl. Magn. Reson.*, 55–56, 79, 2013. With permission.)

The ^{125}Te chemical shift anisotropy has also been measured for each of the resolved NMR signals in the compound $Pb_{0.9}Ge_{0.1}Te$. The ^{125}Te NMR spectra obtained by CSA dephasing experiments at full (0.5 μs) and 20-μs CSA dephasing are represented in Figure 10.4B. The dephasing effects are in good agreement with the more symmetric environment of Te atoms by six Pb atoms for the line marked $n = 0$. The other signals (i.e., $n > 0$) give the ^{125}Te $\Delta\sigma(csa)$ values of ~200 ppm. Similar approaches can be applied for the characterization of other alloys and composites to clarify their structural details.

10.3 POROUS DIAMAGNETIC SOLIDS: POROSITY VIA NMR EXPERIMENTS

Porous organic or organic/inorganic molecular systems with a developed surface are widely used in different fields of chemistry, industrial chemistry, and materials science. In many cases, the design of such solids implies the creation of systems with specific pore sizes and channels, which can be well fitted to the field of practical applications. The latter is particularly important for catalysis or the storage of gases, where the selective uptake and retention of small gaseous molecules such as dihydrogen, small alkanes, or alkenes is necessary. The synthesis of new materials requires reliable control of the pore size and pore distributions throughout the volume of solids, which can be achieved by various physicochemical methods, such as gas adsorption, mercury porosimetry, neutron scattering, imbibation, and differential scanning calorimetry (DSC). It is obvious that complementary experiments will give more reliable and precise information on surface properties. Among the different methods available, NMR plays a very important role. Moreover, in many cases, NMR becomes the uniquely applicable method for measurement of porosity, particularly for porous systems containing partially filled pores or for dual-phase systems. Because NMR experiments collect time- and frequency-dependent data, there are two principally different NMR approaches to measurements of the porosity: NMR cryoporometry and NMR relaxometry. Both of these are focused on the NMR behavior of mobile small molecules, liquids or gases, situated within pores; thus, the techniques applied for measurement are related to solution NMR while the objects under investigation are solids.

10.3.1   NMR CRYOPOROMETRY

The NMR cryoporometric method is actually regular NMR spectroscopy applied for the simple registration of NMR signals to accurately determine their integral intensities. It is not a problem because the signals coming from small molecules, primarily liquids, situated in pore spaces are generally quite narrow. It should be noted that, because this approach is based on the melting-point suppression effect and ideologically close to DSC thermoporometry, both of them usually give similar pore size distributions.[7] In this context, two important advantages of NMR should be emphasized: (1) the signal intensity is directly proportional to the volume of liquids only in the filled pores, and (2) the temperature of a sample is not changed during the experiment but can be changed from one experiment to another.

Protons and deuterons are popular for the determination of pore spaces and pore size distributions in different porous systems. In such experiments, signal intensities coming from non-frozen liquid fractions are measured as a function of temperature. Then, the Gibbs–Thomson theory, the elements of which are represented in Figure 10.5A, is applied for quantitative interpretations of the variable-temperature NMR data. The melting-point suppression effect is expressed via equation (a) in Figure 10.5A, where the temperature constants T_0 and T are the melting points of the bulk and confined liquids, respectively. Signal intensity I is measured experimentally at different temperatures and is plotted as I vs. $X = 10^3/T$, shown as $1/T$. Then, the I parameter is expressed by equation (b), where N is the number of phases in a sample under investigation, and magnitudes I_{0i}, X_{ci}, and σ_i are the intensity, the inverse transition temperature, and the width of the temperature distribution curve, respectively, for the phase marked i. The term $erf(Z)$ in equation (c) is the function of errors. Finally, differentiation of equations (a) and (b) relative to the R and the X parameters gives equation (d), describing the pore size distribution as dI/dR vs. R.

Figure 10.5B shows the experimental data collected for perdeuterated benzene situated in pores of silica gels and a glass.[8] As follows from the data, the experimental curves I vs. $1/T$ are well fitted to equation (b), indicating a good agreement between the theory and experiments. The above experiments can be carried out with

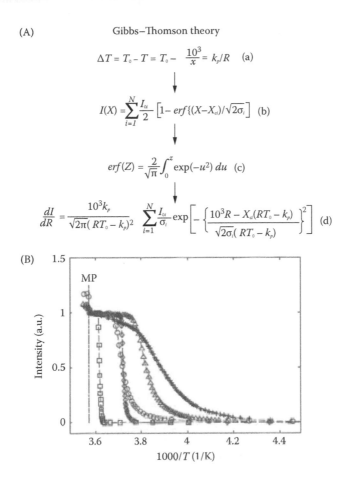

(A) Gibbs–Thomson theory

$$\Delta T = T_0 - T = T_0 - \frac{10^3}{x} = k_p/R \quad \text{(a)}$$

$$I(X) = \sum_{i=1}^{N} \frac{I_{0i}}{2} \left[1 - erf\{(X - X_{ai})/\sqrt{2\sigma_i}\right] \quad \text{(b)}$$

$$erf(Z) = \frac{2}{\sqrt{\pi}} \int_0^z \exp(-u^2)\, du \quad \text{(c)}$$

$$\frac{dI}{dR} = \frac{10^3 k_p}{\sqrt{2\pi}(RT_0 - k_p)^2} \sum_{i=1}^{N} \frac{I_{0i}}{\sigma_i} \exp\left[-\left\{\frac{10^3 R - X_{a}(RT_0 - k_p)}{\sqrt{2\sigma_i}(RT_0 - k_p)}\right\}^2\right] \quad \text{(d)}$$

FIGURE 10.5 (A) Elements of the Gibbs–Thomson theory. (B) Experimental curves for ^{2}H NMR signal intensity (arbitrary units) vs. temperature ($1/T$) collected for C_6D_6 molecules situated in pore spaces of 4-nm silica gel (noted as *), 6-nm silica gel (noted as $\triangle$), 10-nm silica gel (noted as $\bigcirc$), 7.9-nm glass (noted as $\diamondsuit$), and 23.9-nm glass (noted as $\square$).[8] (Adapted from Bakhmutov, V.I., *Solid-State NMR in Materials Science: Principles and Applications*, CRC Press, Boca Raton, FL, 2011.)

solution NMR. In contrast, the application of solid-state NMR techniques can lead to undesirable observations of the liquid and solid components simultaneously. The solution to this problem can be found in the recommended literature.

10.3.2 NMR Relaxometry

NMR relaxometry is based on relaxation or pulsed field gradient (PFG) experiments performed on samples of porous systems to characterize molecular motions or diffusion of pore-filling liquids. This application assumes limited liquid mobility in pores compared to bulk liquids. Elements of the theory describing the reducing motional

$$\frac{D(t)}{D_0} \approx 1 - A\sqrt{D_0 t}\,\frac{S}{V_p} \tag{a}$$

$$D(t) = \langle \Delta r(t)^2 \rangle / 6t \tag{b}$$

$$\frac{D(t \to \infty)}{D_0} = \frac{\phi}{T} \tag{c}$$

$$\langle \Delta r^2 \rangle \equiv \iint |r_1 - r_2|^2\, P(r_1, r_2, t) dr_1 dr_2$$
$$= \frac{1}{V_p} \iint |r_1 - r_2|^2 dr_1 dr_2 \tag{d}$$
$$= 2[\langle r^2 \rangle - \langle r^2 \rangle]$$

$$\langle r^2 \rangle = 3d^2 / 5 \tag{e}$$

(A)

$$T_{1,b} = \frac{1}{aT}\left(\frac{f_{g,b}}{b_g} + \frac{\omega^2}{\dfrac{f_{g,b}}{b_g} + c\omega\sqrt{T}} \right) \tag{a}$$

$$T_{1,s} = \frac{1}{aT}\left(\frac{f_{g,b}}{b_g} + \frac{f_{w,s}}{b_{w,s}} + \frac{\omega^2}{\dfrac{f_{g,b}}{b_g} + \dfrac{f_{w,s}}{b_{w,s}} + c\omega\sqrt{T}} \right) \tag{b}$$

$$f_{w,s} = \frac{V_{g,b}S}{4V} \tag{c}$$

(B)

FIGURE 10.6 (A) The theory of the dynamics valid for pore-filling liquids in porous systems. (B) The relaxation model used for the interpretation of T_1 times dominated by the spin–rotation mechanism.[12]

effect for diffusion are shown in Figure 10.6A.[9] According to this theory, the dynamics of the pore-filling fluid in porous media can be characterized by diffusion constant D. This constant is reduced relative to bulk liquids, because the molecules located close to the matrix surface will have a smaller free space. The effect is proportional to the surface-to-volume ratio in a porous system and is characterized by the measured diffusion coefficient in equation (a) in Figure 10.6A. Parameter D_0 in this equation corresponds to the bulk diffusion constant, S is the total surface area, V_p is the total pore volume, $D(t)$ is the diffusion constant at diffusion time t, and A is a constant that depends on dimensionality. At an isotropic diffusion, the $D(t)$

coefficient gives the vector displacement Δr in equation (b). When the diffusion distance is large relative to the pore size, then the second effect appears, expressed via the long time diffusion. Under these conditions, a decrease in the long time diffusion coefficient relative to constant D_0 will be connected with factor T in equation (c), which is a geometrical parameter determined by the pore structure and porosity Φ. If the system under investigation is closed, then the long time displacement (Δr^2) is limited by the size of the enclosing structure. At the limit of a long diffusion time (when the initial and final positions are not correlated), the mean square displacement is expressed by equation (d), where P is the diffusion propagator ($P = 1/V_P$ at long time and V_P is the pore volume). For spherical pores, characterized by diameter d, the Δr^2 factor is given in equation (e). Finally, because the diffusion displacement in NMR experiments is measured along one direction, this displacement is defined as $(\Delta r^2)/3$.

In practice, diffusion measurements in porous media can be performed with various NMR techniques, including 1D or 2D radiofrequency pulse sequences to determine spin–spin relaxation times T_2. Among the various pulse sequences that have been applied for such experiments, the 1D Carr–Purcell–Meiboom–Gill (CPMG) pulse sequence is generally the most popular pulse train. Here, the echo signal (M) is assumed to be the sum of multiple exponential decays with a distribution of times T_2 noted as function $F(T_2)$:

$$M(\tau_2) = \int e^{-\tau_2/T_2} F(T_2) dT_2 \qquad (10.1)$$

It has been emphasized that the $F(T_2)$ function can be described by signal amplitude $M(\tau_2)$ measured at different times (τ_2). Details regarding 2D T_2 NMR experiments based on the application of PFGs and their mathematical treatments can be found in Reference 9, and the recommended literature describes a large number of diffusion measurements performed in different porous media.

The restriction of molecular motions in pore spaces obviously affects the T_1 and T_2 relaxation behavior of target nuclei. In turn, the enhanced relaxation rates observed for molecules situated at a pore surface allow the calculation of pore diameters when the T_1 and T_2 times are accurately measured. Generally, fast exchange is assumed between molecules at the surface and in pores.[10] Then, the relaxation rates $1/T_1$ and $1/T_2$ will be proportional to the surface (S)-to-volume (V) ratio in the porous media:

$$1/T_1 = \rho_1 S/V \qquad (10.2)$$

$$1/T_2 = \rho_2 S/V \qquad (10.3)$$

where ρ is the surface relaxation strength. Because the relaxation times due to the porous nature of compounds are distributed, fitting a distribution of relaxation times to the experimental T_1/T_2 relaxation data will result in estimations of a final pore size distribution. This approach is simple, but it is very sensitive to signal-to-noise ratios in the collected NMR spectra. Therefore, these measurements generally lead to distributions that are more broadened than the actual ones.

10.3.2.1   NMR Relaxometry for Gases

Pore space can be filled with gases, the relaxation behavior of which can also be used for porosity measurements. Nuclear relaxation of gas molecules absorbed in pores differs from the relaxation of liquids; therefore, the collection of NMR data for gases and their interpretation are considered in a separate section. Generally speaking, interpretations of relaxation data strongly depend on relaxation mechanisms operating in gases for target nuclei. Because ^{83}Kr nuclei are quadrupolar, their quadrupole moments affect ^{83}Kr relaxation and line shapes, and because these effects are highly sensitive they can be applied for the detailed description of macroporous systems. For example, hyperpolarized ^{83}Kr nuclei located close to the surfaces of porous polymeric and glass systems are good probes. It should be added that the ^{83}Kr T_1 times measured between 2 and 7 s strongly depend on the chemical composition of the surfaces in the vicinity of the krypton gas. They also depend on chemical treatments of the glass surfaces. Similarly to krypton, $^{129,131}Xe$ NMR is often applied for porosity measurements in different porous systems, such as zeolites, mesoporous silicas, and silica glasses.[11] In fact, the isotropic chemical shift, chemical shift anisotropy, line width, and T_1 relaxation time measured for xenon are very sensitive to surface properties.

The application of relaxation techniques for probing gases in porous systems is discussed here by using the example of ^{19}F relaxation in porous silica-based systems treated with CF_4 gas.[12] In Chapter 3, it was shown that small molecules in non-viscous media undergo fast rotations, resulting in domination of the spin–rotation (SR) relaxation mechanism. In turn, the SR relaxation rate is dependent on the molecular inertia moment, spin–rotational constant, and molecular motion correlation time τ_C. The rate increases with increasing temperature (this effect is a good test for the presence of SR relaxation). The NMR ^{19}F T_1 times for CF_4 gas in pore media are completely governed by spin–rotation interactions that are mediated by the molecular collision frequency. Due to the presence of pores, additional collisions of CF_4 molecules with the pore walls remarkably affect the ^{19}F T_1 times. As a result, the relaxation process becomes more complex, and quantitative treatments of relaxation data require adequate mathematical models.

The elements of a basic mathematical model are shown in Figure 10.6B. This model shows the relationship between spin–rotation interaction and molecular collision frequency. This relationship allows calculation of the ^{19}F T_1 times for a bulk gas ($T_{1,b}$) at a given pressure and temperature by fitting to curves obtained for various temperatures and pressures. The collision frequency of gas molecules in bulk, noted as $f_{g,b}$, can be calculated using the Lennard–Jones collision theory, the Clausius state equation, and a specific term that shows how the molecular velocity changes with density. It should be emphasized that the CF_4 gas density in pores is greater than that of the bulk gas.

When the molecular collision frequency is known or found independently, then the bulk gas relaxation time, $T_{1,b}$, is expressed via equation (a) in Figure 10.6B, where parameter a is a coupling constant, ω is the Larmor frequency, T is absolute temperature, and c corresponds to thermally mediated intramolecular relaxation mechanisms in the absence of the collisions. The factor b_g represents the average number of wall collisions necessary for spin–rotation exchange.

An extension of expression (a), applied for collisions of gas molecules with pore walls, leads to time $T_{1,s}$ in equation (b) in Figure 10.6B,[12] where $f_{w,s}$ is the frequency of collisions with the wall and $b_{w,s}$ symbolizes the average number of wall collisions necessary for spin–rotation exchange. According to simple gas dynamics, the average wall collision frequency for molecules located in a pore space can be expressed via equation (c) in Figure 10.6B, where S is the surface area, $v_{g,b}$ is the average velocity, V is the pore volume, and $f_{w,s}$ is a factor proportional to the surface/volume ratio.

Thus, in practice, the ^{19}F T_1 times are accurately measured for CF_4 gas molecules in porous silica-based systems and the relaxation data are then fitted to the mathematical model. This approach represents a relatively rapid and precise way to measure surface/volume ratios from the increasing T_1 due to confinement in pores.

10.3.2.2 Low-Field NMR Relaxometry under High Pressure

Chapter 4 has shown that low-field NMR relaxometry is a powerful tool for characterization of molecular motions in solutions. In porous systems, due to a fast diffusion exchange between the fluid molecules at the surface layer and the molecules in the pores, the total relaxation rates are proportional to the pore surface-to-volume ratio (S/V) as was demonstrated in previous sections. Therefore, the T_1 relaxation times found by experiments with low magnetic fields can also be used as a measure for pore sizes.

In general, when porous solids investigated by NMR relaxometry are macroporous (i.e., their pore size is >50 nm), simple liquids (water, alcohols, etc.) remain liquid-like in the body of the macropores. In terms of nuclear relaxation, the T_2 times of liquids in such pores are only slightly smaller than the T_1 times. The situation changes in meso- and microporous systems. For example, micropores are so small that free liquid phases may be formed only in the center of the pore body. Due to van der Waals interactions with the pore walls, the molecules of liquids form an adsorbed phase at the matrix interface which dictates their complex relaxation behavior expressed as strong T_2 shortening. In such cases, a study of gas adsorption by low-field T_2 NMR relaxometry can be very effective. The interactions of methane—for example, with the inner surfaces of porous materials—are weak; therefore, methane can be considered a non-wetting pore fluid and is a convenient probe for ^1H relaxation.

The studies of gas adsorption for low magnetic fields under high pressures require special instrumentation for high-pressure adsorption. The center element of such a high-pressure adsorption setup is the NMR-compatible pressure cell.[13] T_2 relaxation measurements in these cells are performed with CPMG pulses, and the measured relaxation decays are transformed into relaxation time distributions by standard inverse Laplace transform software. One should remember that, for example, the ^1H T_1 time of methane in the gas phase increases with gas pressure. This effect must be taken into account when adjusting relaxation time delays in T_2 relaxation experiments at different pressures.

Figure 10.7 shows the results of measurements carried out for methane molecules located in the pores of a metal–organic material (ZIF-8) placed in a permanent magnet with $B_0 = 118$ mT.[13] At 5.6 and 10.8 bar (commonly below 40 bar), two T_2 relaxation time peaks characterize the ^1H T_2 relaxation time distributions. The right T_2 peak

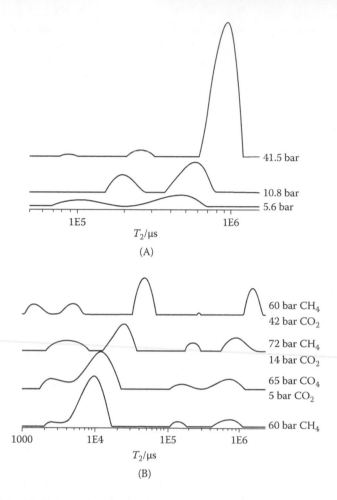

FIGURE 10.7 (A) T_2 relaxation time distributions obtained for methane in metal–organic framework (MOF) ZIF-8 material at different pressures. (B) Pressure effects on T_2 relaxation time distributions obtained for CH_4 molecules in MOF CuBTC material for two-component adsorption. (From Horch, C. et al., *J. Magn. Reson.*, 240, 24, 2014. With permission.)

experiences a remarkable shift from 350 ms to about 800 ms with increasing pressure. It should be noted that this value is much greater than those for free gas-phase methane relaxation times. In addition, the intensity of this peak strongly increases with pressure. Thus, the observed effect can be considered to be a fingerprint of the methane actually adsorbed in the micropores of the material. As noted earlier, an analysis of the signal intensities recorded by low-field high-pressure NMR experiments provides quantitative characterizations of microporous materials to obtain adsorption parameters. In addition, CH_4 T_2 relaxation time distributions can be probed in systems for pure methane and methane in the presence of CO_2. The data in Figure 10.7B show that the host–guest interactions lead to T_2 relaxation time distributions that can be used for determination of the CO_2/CH_4 adsorption separation factors important for porous systems.[13]

10.4 SOLID-STATE NMR AND PARAMAGNETIC MOLECULAR SYSTEMS

The theoretical aspects of NMR in the presence of paramagnetic centers were considered in Chapters 2 and 3, where the main attention was focused on interactions in electron–nucleus pairs. These interactions lead to strong spectral effects that are dependent on the nature of the nuclei, relaxation times of the electrons, strength and mechanism of electron–nucleus coupling, and distances between nuclei and electrons. The latter factor connects the paramagnetic effects with structural factors. Finally, the total spectroscopic effect depends on the number of paramagnetic centers.

In general, a synthesis of molecular systems and materials involving various organic and/or inorganic reagents can lead to products that contain paramagnetic impurities of a different nature. When concentrations of the paramagnetic impurities in solids are small and less than 0.5 wt%, the paramagnetic centers do not have any practical effect on NMR parameters measured experimentally, such as isotropic chemical shifts, magnetic shielding anisotropies, and line widths or shapes. Only relaxation times can be changed, insignificantly; therefore, in practice these systems can be accepted as diamagnetic objects. However, with increased concentrations of paramagnetic centers higher than 0.5 to 1 wt%, the solids can show specific NMR properties due to the presence of many paramagnetic centers located at various distances relative to the observed nuclei. Under these conditions, the effects of single electron–nucleus contacts should be summed to result in very large isotropic chemical shifts and magnetic shielding tensors with components coming from Fermi contact, pseudocontact, and dipolar electron–nucleus interactions. These pronounced effects are represented by ^{31}P NMR in solid paramagnetic phosphates with Fermi contact chemical shifts and ^{31}P line widths shown in Table 10.1 appearing at relatively short phosphorus–metal distances.[14] The data show that the ^{31}P isotropic chemical shifts reach extremely large magnitudes up to 17,000 ppm. Moreover, due to much shortening of the ^{31}P T_2 times in the presence of unpaired electrons, the resonances experience abnormal broadenings up to 1800 ppm, making their registration difficult. In addition, the bulk magnetic susceptibility (BMS) effect (B_{BMS}, discussed in Chapter 2), generally ineffective in diamagnetic solids,

TABLE 10.1
^{31}P Chemical Shifts and Line Widths (Δv), Determined as Full Width at Half Maximum and Reported for Solid Paramagnetic Phosphates at Room Temperature

Paramagnetic Compound	^{31}P $\delta(iso)$ (ppm)	Δv (ppm)
Heterosite ($FePO_4$)	5770	1217
Rhomb ($Li_3Fe_2(PO_4)_3$)	14,350	1839
Strengite	15,800	1078
Phosphosiderite	16,680	742
Lithium iron phosphate ($LiFeP_2O_7$)	14,398	2147
	8230	1460

becomes very pronounced in paramagnetic systems. According to theory,[15] this term causes line broadenings in static samples that are dependent on the shape of the NMR containers.

Thus, in solids containing high concentrations of paramagnetic centers, the orientation-dependent dipolar coupling, Fermi contact interactions, and BMS terms can produce strong broadenings of resonance lines in static samples which transform to intense sideband patterns in MAS NMR spectra at spinning rates comparable to or larger than line widths in static samples. Sometimes the effects can be so strong that the resonances can become spectrally invisible.

In practice, these paramagnetic effects lead to low NMR sensitivity and very low spectral resolution; therefore, study strategies or even the principles of signal registration should be altered. Strong dipolar electron–nucleus interactions can lead to very large chemical shift anisotropies that are even larger than the radiofrequency amplitudes. For example, if the anisotropy of a ^{1}H chemical shift reaches 500 to 1000 ppm, uniform excitation of a proton band covering 1000 ppm or 500 kHz in the frequency units at a magnetic field of 11.7 T can be realized only by a 90° radiofrequency pulse at a length of 625 ns! This is a significant technical problem that can be partly avoided by the application of short, high-power adiabatic RF pulses for population inversion.[16] The point-by-point method mentioned above can also be used to register NMR signals.

10.4.1 Spin Echo Mapping Technique for Detection of Invisible Nuclei

In general, conventional NMR does not allow direct observation of nuclei located closest to unpaired electrons in paramagnetic species due to the exclusively strong paramagnetic effects. For example, static samples of $CoAPO_4$-n aluminophoshate molecular sieves[17] show ^{31}P NMR spectra of low intensity with very broad lines, corresponding to the strong loss of intensity observed even at relatively low cobalt contents. In other words, a large portion of the nuclei neighboring cobalt ions cannot be observed. Under MAS conditions, the ^{31}P NMR spectra show wide sideband patterns which, however, are not sufficient to draw any structural conclusions. As follows from Figure 10.3, the point-by-point method of registration does not provide precise measurements of chemical shifts and accurate descriptions of line shapes. Therefore, another approach was needed to detect and describe the ^{31}P resonances in $CoAPO_4$-n molecular sieves.

Figure 10.8A shows a NMR experiment performed on static samples of the $CoAPO_4$-n molecular sieve and other paramagnetic solids that was based on a spin echo mapping technique involving Hahn echo ^{31}P static NMR spectra recorded at different carrier frequencies.[17–19] The final NMR spectrum, obtained by summing the sub-spectra, shows all of the lines within the broad spectral region corresponding to all of the ^{31}P nuclei in the sample. The first important element in these experiments is collection of the so-called whole echo. In fact, *a priori*, the resonance lines are expected to be very broad and echo delays (τ) must be short in order to avoid cutting off the free induction decay (FID). Short echo delays are better realized by NMR experiments on static samples. In contrast, MAS NMR experiments should be synchronized with spinning rates such that the echo delays will be automatically increased.

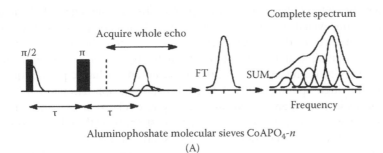

Aluminophoshate molecular sieves CoAPO$_4$-n

(A)

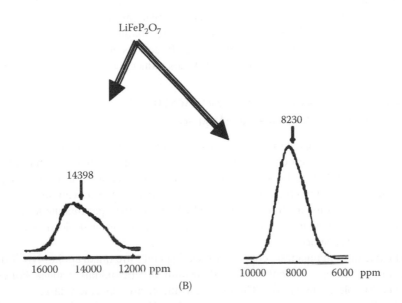

(B)

FIGURE 10.8 (A) Schematic representation of spin echo mapping experiments designed to record broad ^{31}P NMR spectra in static samples of paramagnetic aluminophoshate molecular sieves. Vertical dashed lines show dead time after the radiofrequency pulse. (B) Spin echo mapping ^{31}P NMR experiments performed on a static sample of paramagnetic material, LiFeP$_2$O$_7$, showing the presence of two phosphorus sites. (Adapted from Bakhmutov, V.I., *Solid-State NMR in Materials Science: Principles and Applications*, CRC Press, Boca Raton, FL, 2011.)

Another element is that spin echo mapping experiments require accurate tuning of the NMR probe for each carrier frequency to provide full detection of ^{31}P resonances for nuclei located closest to paramagnetic Co^{3+} ions even at high concentrations, where the phosphorus lines are particularly broad and their chemical shifts cover a range between –500 and 10,000 ppm.[20] Similarly, ^{31}P resonances in the paramagnetic iron phosphate FePO$_4$ and their lithium derivatives can be observed only by mapping experiments.[14] The solid-state ^{31}P NMR spectra recorded in static samples of FePO$_4$, Li$_3$Fe$_2$(PO$_4$)$_3$, and FePO$_4$·2H$_2$O exhibit extremely large ^{31}P Fermi contact shifts in a range of 5700 to 20,500 ppm, and their line widths reach magnitudes of

>4000 ppm. In spite of the strong paramagnetic effects, mapping ^{31}P NMR data allows identifying even different phosphorus sites within a compound. One of them, LiFeP$_2$O$_7$, is characterized in Figure 10.8B, where two phosphorus resonances are clearly observed at 8230 and 14,398 ppm. Alternatively, there is a technique for the detection of very broad resonance lines that is based on stochastic pulse sequences combined with high spinning rates.[21] Here, low-power, approximately 1° pulses operate stochastically and are 0° and 180° out of phase relative to one another.

Finally, because the paramagnetic effects depend strongly on the relaxation of electrons at favorable electron relaxation times (quite short, for example), even traditional MAS NMR experiments can be utilized for the study of paramagnetic solids, such as Cu(DL-Ala)$_2$ and Mn(acac)$_3$ molecules.[22] It has been demonstrated that they are well characterized even by cross-polarization and dipolar insensitive nuclei enhanced by polarization transfer (INEPT) MAS NMR spectra obtained for ^{13}C nuclei. Moreover [1]H–[13]C cross-polarization with insignificant modifications of pulse sequences increases NMR sensitivity, and relatively short [1]H T_1 times reduce the duration of experiments.

10.4.2 DETECTION OF NUCLEI AT PARAMAGNETIC CENTERS: PARAMAGNETIC METAL IONS

Metal ions are often used in the synthesis of molecular systems potentially active as catalysts; therefore, their chemical bonding and coordination in new compounds are of great interest. When the ions are paramagnetic (e.g., cations of manganese, copper, or nickel), studying them with solid-state NMR is neither trivial nor impossible. Generally, such ions are traditionally studied by electron paramagnetic resonance (EPR) to allow their direct observation and characterization if electron relaxation times are not too short.

The nuclei in manganese, copper, or nickel species are quadrupolar. NMR detection of these nuclei is difficult even for their diamagnetic oxidation states. For example, quadrupole moments of [63]Cu and [65]Cu nuclei are extremely large (–22.0 and –20.4 × 10^{-30} m^2/e, respectively). As a result, the [63]Cu(I) and [65]Cu(I) solid-state NMR spectra are dominated by strong quadrupolar interactions, and the breadth of their powder NMR patterns ranges from 760 kHz to 6.7 MHz. This was found in some organometallic complexes with different copper coordination environments.[23] The isotropic chemical shifts in these compounds range from 1050 to –510 ppm at magnetic shield anisotropies of 1000 to 1500 ppm. Quadrupolar coupling constants are between 22.0 and 71.0 MHz. Only special approaches are capable of reaching reasonable signal-to-noise ratios in such spectra. Figure 10.9A shows [65]Cu NMR spectra recorded in the diamagnetic Cu(I) compound (PPh$_3$)$_2$CuO$_2$CCH$_3$ by application of the wideband uniform rate smooth truncation/quadrupole Carr–Purcell–Meiboom–Gill (WURST/QCPMG) pulse sequence to increase excitation bandwidth. Incorporation of a high-power proton decoupling section into the above experiments helps with assignment of the signals. In fact, as seen in Figure 10.9A, the NMR pattern changes remarkably in the presence of decoupling. However, even this powerful technique will be ineffective with the appearance of an unpaired electron in these ions, where the use of EPR spectroscopy is preferable. At the same time, in rare cases, valuable solid-state NMR data can still be collected.

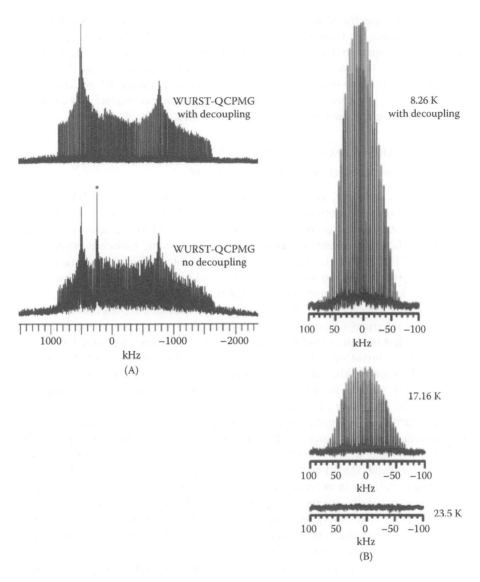

FIGURE 10.9 (A) ^{65}Cu NMR spectra recorded in a static sample of the diamagnetic complex $(PPh_3)_2CuO_2CCH_3$ (*, NMR signal from NMR proble). (B) Temperature-dependent ^{55}Mn NMR spectra obtained with CP/QCPMG pulse sequence in a sample of the complex $[Mn_2O_2(salpn)_2]$. (Adapted from Bakhmutov, V.I., *Solid-State NMR in Materials Science: Principles and Applications*, CRC Press, Boca Raton, FL, 2011.)

Figure 10.9B illustrates a case where the solid-state NMR data were obtained in a sample of the compound bis(μ-oxo)dimanganese(IV) $[Mn_2O_2(salpn)_2]$.[24] A signal in the ^{55}Mn solid-state NMR spectrum is observable at the magnetic field of 9.40 T only at the low temperature of 8.5 K. The resonance is complex (its interpretation can be found in the original article[3]) and centered between the signals observed in ^{55}Mn NMR spectra of species specially prepared and containing Mn(VII) and Mn(0). Increasing the temperature to 17.2 K leads to the ^{55}Mn NMR spectrum with a 70% loss in signal-to-noise ratio. The effect corresponds to electron–nucleus spin hyperfine interactions activated at a higher temperature. Then, at 23.5 K, the manganese signal completely disappears. This disappearance is explained by the fact that an only 2×10^{-5} percent chance of populating a paramagnetic state can cause too large broadening and the resonance becomes invisible.

10.4.3   NMR SPECTRA OF PARAMAGNETIC SOLIDS: GENERAL ASPECTS AND STUDY STRATEGIES

Fermi contact coupling, electron–nucleus dipolar interactions, and BMS effects cause large changes in isotropic chemical shifts and chemical shift anisotropies connected with structural parameters; thus, they can be used for the analysis of paramagnetic solids in a manner similar to that for diamagnetic molecules when target nuclei (e.g., ^{1}H, ^{31}P, ^{13}C, ^{2}H, ^{29}Si, ^{7}Li, ^{6}Li, ^{51}V) are remote from paramagnetic centers. Various approaches and strategies can be directed toward the determination of different sites and structural units in paramagnetic systems.

No matter the target nuclei, there are common tendencies in their observation: (1) higher concentrations of paramagnetic centers cause stronger paramagnetic effects, which manifest in static samples as shortened relaxation times and broadened resonances; (2) the broad static lines transform to wide, intense sideband envelopes in MAS NMR spectra when the spinning rates are comparable with the line widths in static NMR spectra; and (3) even at short distances between nuclei and unpaired electrons and too-strong paramagnetic effects, the spectral invisibility of the nuclei is still informative. The last point requires an explanation. In such cases, the NMR spectra will show only remote nuclei at reduced total integral intensities of the signals. For example, the intensity loss is very large in the ^{31}P MAS NMR spectra of aluminophosphates doped with ions of Co, Ni, Fe, or Mn. This loss can be characterized quantitatively in samples containing a convenient internal standard by using the protocol generally accepted for quantification of invisible nuclei.[25] Because the signal intensity is declining for nuclei situated in the first and second coordination spheres of paramagnetic centers, quantification of the effect will be very valuable. At the same time, the loss of intensity itself cannot be predicted *a priori*.

Generally speaking, because electron relaxation times vary strongly, solid-state NMR spectra of the same host compound can depend on the nature of the paramagnetic centers. This effect is demonstrated in Figure 10.10A by the ^{29}Si MAS NMR spectra of zeolites prepared in the presence of different lanthanide cations having different electron relaxation times.[26] The spectrum of the diamagnetic zeolite (Na-Y) shows four well-resolved ^{29}Si resonances, which correspond to four silicon sites in the sample. These resonances broaden remarkably in the Sm and Nd systems and disappear in the Gd and Dy zeolites, with the spectral information being completely lost. The same effects have been reported for ^{27}Al MAS NMR in these materials. The practical conclusion in the context of NMR applications is that ^{29}Si MAS NMR studies of zeolites or other compounds containing Ce^{3+}, Pr^{3+}, Nd^{3+}, Sm^{3+}, or Eu^{3+} ions will be more successful[27] than those containing Gd^{3+} or Dy^{3+} ions. As an example of

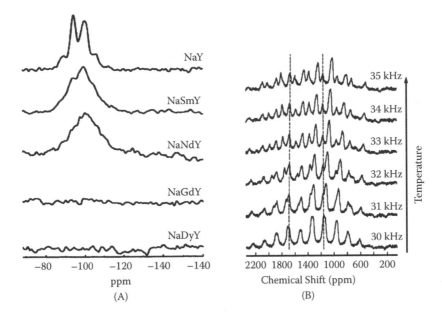

FIGURE 10.10 (A) ^{29}Si MAS NMR spectra of the zeolites (marked as Na-Y at a Si/Al of ~3 (diamagnetic)) NaSmY, NaNdY, NaGdY, and NaDyY obtained for similar concentrations of the corresponding lanthanide ions. (B) ^{31}P spin echo MAS NMR spectra in a sample of the paramagnetic complex αβ–[Co$_3$Na(H$_2$O)$_2$(P$_2$W$_{15}$O$_{56}$)$_2$]$^{17-}$ recorded at different spinning rates. The dotted lines show the positions of two different isotropic resonances. (Adapted from Bakhmutov, V.I., *Solid-State NMR in Materials Science: Principles and Applications*, CRC Press, Boca Raton, FL, 2011.)

the effect of the nature of an ion, isotropic high-field chemical shifts (up to –146 ppm) of a Fermi contact nature have been measured for a Nd-containing zeolite, whereas a very small and low-field shift of +6 ppm was observed in a Eu-containing zeolite.

An important step in the interpretation of chemical sifts measured in paramagnetic solids is determination of their origin, because Fermi contact, dipolar interaction, and pseudocontact shifts are differently connected with structure. Because the contact and pseudocontact isotropic chemical shifts are temperature dependent,[28] variable-temperature NMR experiments are helpful. At the same time, the character of the temperature dependence can identify contact and pseudocontact shifts because they change with temperature proportionally to factors T^{-1} and T^{-2}, respectively. To observe temperature effects, the NMR spectra obviously should be well resolved.

Figure 10.10B shows the ^{31}P spin echo MAS NMR spectra obtained in a sample of the paramagnetic complex αβ–[Co$_3$Na(H$_2$O)$_2$(P$_2$W$_{15}$O$_{56}$)$_2$]$^{17-}$ at spinning rates between 30 and 35 kHz. The spectra reveal the presence of two different phosphorus sites in the complex. It is also important that the isotropic chemical shifts of both sites (shown as the dotted lines) slightly change with the spinning rates. Because increasing the spinning rates from 30 to 35 kHz increases the sample temperature, the above change agrees with the Fermi contact nature of the ^{31}P chemical shifts. Similarly, the isotropic ^{13}C chemical shift measured in the MAS NMR spectra of the solid paramagnetic complex Cp_2Cr also changes, from -258 to -231 ppm, with increasing spinning rates from 3 to 15 kHz.[29] Significant temperature effects are also reported for ^{89}Y chemical shifts in MAS NMR spectra of mixed oxides.[30]

Two important statements can be made. First, temperature effects on isotropic chemical shifts δ_{iso}, often exceeding 20 ppm, are not negligible with regard to their interpretation. Second, the parameters obtained by NMR spectra in different paramagnetic compounds can be accurately compared when the data are collected at the same temperature.

Cross-polarization NMR experiments generally applied for enhancement of signal-to-noise ratios in NMR spectra of diamagnetic solids are very sensitive to the presence of paramagnetic centers. For example, ^1H–^{13}C cross-polarization experiments performed on the paramagnetic complexes Cp_2Cr, Cp_2V, Cp_2Co, and Cp_2Ni have demonstrated the ineffectiveness of cross-polarization at the regular contact times typical of these experiments.[31] Generally speaking, the ineffectiveness connected with short ^1H T_1 times can be used as a test for the presence of paramagnetic impurities. However, with reduced contact times, the cross-polarization still promotes the detection of ^{13}C signals necessary for signal assignments. Nevertheless, for paramagnetic solids, direct excitation of ^{13}C nuclei can even be preferable because the NMR experiments can be performed for short relaxation delays to decrease duration. The paramagnetic complexes Cp_2Cr, Cp_2V, Cp_2Co, and Cp_2Ni have been characterized by this method to show the ^{13}C isotropic chemical shifts and chemical shift anisotropies changing from -398 to $+1594$ ppm and from 828 to 2640 ppm, respectively, as a function of the metal center.[31]

In addition to direct excitation and cross-relaxation with short contact times, paramagnetic solids can be probed with ramped-amplitude cross-polarization (RACP).[32] For example, this type of cross-polarization from ^1H nuclei has been applied to create transverse ^{13}C magnetization at a contact time of 0.5 ms to characterize free organic radicals in the solid state by NMR.[33] Figure 10.11A shows the structure of the nitronylnitroxide radical and its ^{13}C RACP MAS NMR spectrum recorded at a spinning rate of 12 kHz. In spite of strong paramagnetic effects, the spectrum is still quite qualitative and shows differently broadened ^{13}C resonances at $\delta(^{13}C)$ values covering a very large range, from $+750$ to -750 ppm. Such resonances can be accurately assigned and interpreted in structural terms.[33]

10.4.3.1 NMR Spectra of Quadrupolar Nuclei in Paramagnetic Solids

Strong quadrupolar interactions in combination with paramagnetic properties can result in a situation where solid-state NMR data are difficult to collect and interpret. Because these effects depend on the nature of nuclei, their choice plays a very important role. Therefore, a common tendency in studies of paramagnetic solids is to choose nuclei with the smallest quadrupolar moments possible. For example, the quadrupolar moments of ^7Li and ^6Li nuclei are relatively small at -4×10^{-30} m^2/e and -0.08×10^{-30} m^2/e, respectively. They are even smaller than 14.66×10^{-30} m^2/e for ^{27}Al nuclei. In addition, the natural abundance of ^7Li and ^6Li nuclei is high (92.4% and 7.6%, respectively). Therefore, ^7Li and ^6Li nuclei are quite convenient for observing NMR spectra, even for paramagnetic solids. These nuclei are sensitive to their environments; for example, the ^7Li and ^6Li static and MAS NMR spectra recorded in

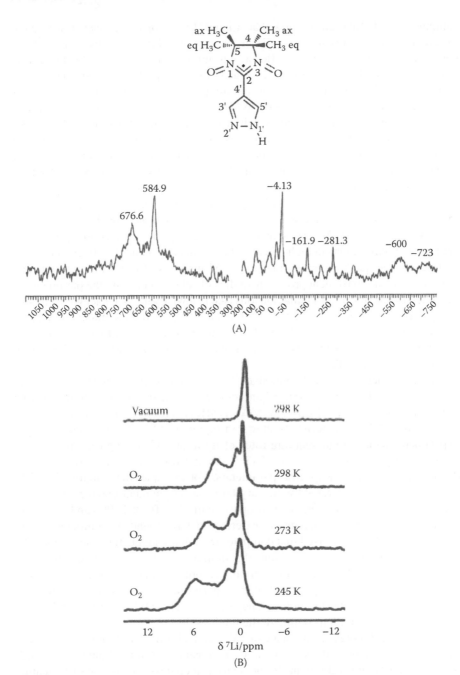

FIGURE 10.11 (A) The ^{13}C RACP MAS NMR spectrum of the nitronyl nitroxide radical recorded at a spinning rate of 12 kHz. (From María, D.S. et al., *ARKIVOC*, III, 114, 2011. With permission.) (B) The ^{7}Li MAS NMR spectra for the zeolite Al-ZSM-5 at various temperatures. (Adapted from Bakhmutov, V.I., *Solid-State NMR in Materials Science: Principles and Applications*, CRC Press, Boca Raton, FL, 2011.)

samples of the LiNiO family show that ^7Li and ^6Li isotropic chemical shifts change by greater than +700 ppm due to Fermi contact and dipolar interactions with the electron spins on the Ni centers.[34] Because lithium atoms are easily added to porous materials, ^7Li and ^6Li nuclei can serve as convenient probes to study them using solid-state NMR techniques. Even fine paramagnetic effects can be investigated. For example, the resonances in ^7Li MAS NMR spectra belonging to extra-framework lithium cations in porous solids are sensitive to the presence of paramagnetic gases.[35] The NMR signals of the nuclei located close to paramagnetic gaseous species show paramagnetic shifts and experience broadenings, and their relaxation times shorten.

The effect of oxygen is shown in Figure 10.11B,[36] where the ^7Li MAS NMR spectra were obtained in samples of the zeolite Al-ZSM-5. The ^7Li signal is sharp and practically symmetric in the ^7Li MAS NMR spectrum of the oxygen-free zeolite. However, in the presence of O_2, the line shape of the ^7Li signal changes strongly to show anisotropic paramagnetic shifts due to interactions with oxygen. Upon cooling, the accessibility of lithium cations in high-silica zeolites is improved and the effect increases.

Because the quadrupolar character of ^{27}Al nuclei is also not pronounced, they are often used as convenient probes in studies of solids containing the paramagnetic metal ions Mn^{2+} or Co^{2+} or lanthanide ions (see below). Even ^{51}V nuclei with a moderate quadrupole moment of -5.2×10^{-30} m^2/e can be applied for solid-state ^{51}V NMR studies of paramagnetic solids. It should be noted that vanadium atoms are important structural units in many molecular systems that are active in catalysis, so their investigations are of particular interest.

The principles of applying solid-state ^{51}V NMR techniques can be demonstrated by the study of the paramagnetic 3d and rare-earth vanadates.[37] Due to the quadrupolar and paramagnetic effects, ^{51}V MAS NMR spectra generally show very wide and intense sideband patterns, as shown in Figure 10.12A. It is obvious that these experiments require high spinning rates of up to 30 kHz or higher. These conditions simplify the analysis of ^{51}V MAS NMR spectra by computer simulations to determine the necessary spectral parameters. For example, the compounds $FeVO_4$ and $CeVO_4$ are characterized by the very large vanadium quadrupole coupling constants of 1.7 and 3.5 MHz and big isotropic chemical shifts of 17,000 and –559 ppm, respectively (usually in reference to external $VOCl_3$), found by computer simulations. It should be emphasized that computer treatments should take into account the second-order quadrupolar interactions that dominate in the ^{51}V NMR spectra.

The ^{51}V MAS NMR spectra are generally recorded with single or spin echo pulse sequences. Single pulse experiments can be performed with short relaxation delays (0.1 to 3 s) and very short radiofrequency pulses, such as 0.7 μs corresponding to a 16° pulse. These conditions optimize the final results.

Figure 10.12B represents the scale of ^{51}V chemical shifts valid for vanadium paramagnetic solids. This scale is much larger than that reported for diamagnetic solids, with a region of only 1200 ppm. In the analytical context, this is a positive feature. However, in practice, the initial search of ^{51}V NMR signals for unknown compounds is not simple. The signals can be localized by step-by-step changes in the values of the carrier frequency with accurate tunings of a NMR probe. Similar principles can be used for the study of other quadrupolar nuclei in paramagnetic solids, the details of which can be found in the recommended literature.

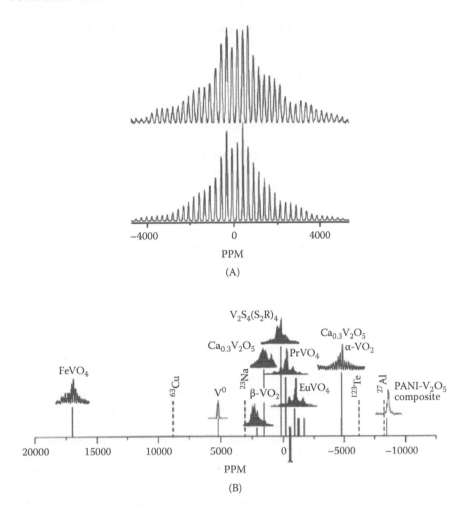

FIGURE 10.12 (A) Experimental (top) and calculated (bottom) ^{51}V MAS NMR spectra of the vanadate $YbVO_4$ recorded at a spinning rate of 30 kHz. (From Shubin, A.A. et al., *Catal. Today*, 142, 220, 2009. With permission.) (B) Chemical shift scale for ^{51}V nuclei in various paramagnetic solids. (Adapted from Bakhmutov, V.I., *Solid-State NMR in Materials Science: Principles and Applications*, CRC Press, Boca Raton, FL, 2011.)

10.4.3.2 Common Strategies in Structural Studies of Paramagnetic Solids

As noted earlier, the effects of paramagnetic centers in compounds under investigation are visually manifested in MAS NMR spectra as broadened resonances accompanied by intense sidebands to form very wide spectral patterns. This behavior is typical of zeolites, silica-based materials, aluminophosphates, and other systems containing paramagnetic Mn^{2+}, Co^{2+}, Ni^{2+}, and lanthanide ions and can be observed on ^1H, ^{27}Al, ^{29}Si, ^{31}P, and other nuclei in MAS NMR spectra. When the resonances corresponding to isotropic chemical shifts are localized by experiments at different spinning rates and thus the number of structurally different sites is established, then sideband patterns can be simulated to obtain chemical shift anisotropies, while the nature of the paramagnetic shifts can be evaluated by variable-temperature studies and quantum chemical calculations. These treatments can be quite accurate when one of the paramagnetic effects is dominant.

A common strategy in such investigations can be represented by ^{31}P MAS NMR spectra of the complex $K_{14}(H_3O)_3[Er(\alpha_1\text{-}P_2W_{17}O_{61})_2] \cdot 4KCl \cdot 64H_2O$ containing the paramagnetic europium ion (Figure 10.13A).[38] The ^{31}P MAS NMR spectra recorded at different spinning rates clearly identify two phosphorus sites in the complex. Theoretically, the NMR parameters of these sites can be calculated on the basis of the point–dipole approximation, where the common Hamiltonian describing ^{31}P nuclei in the presence of a paramagnetic center is represented by Equation 10.4:

$$H = H_{Zeeman} + H_{RF} + H_{DIP}^{Paramagn} + H_{CSA} + H_{FC} + H_{DIP} \tag{10.4}$$

The Hamiltonian includes the Zeeman interaction (H_{Zeeman}), the contribution from the radiofrequency field (H_{RF}), the electron–nucleus dipolar interaction ($H_{DIP}^{Paramagn}$), the chemical shift anisotropy (H_{CSA}), the Fermi contact interaction (H_{FC}), and the nuclear dipolar interaction (H_{DIP}). The tensor expressing the electron–nucleus dipolar coupling is shown in Figure 10.13A. Here, r is the electron–nucleus distance, $\delta_{\alpha\beta}$ is the Kronecker delta function, and e_α and e_β are the (x, y, z) components of the electron–nucleus dipolar vector. In turn, the Hamiltonian expressing the total electron–nucleus dipolar interactions can be obtained by summing up the dipolar matrixes that correspond to the individual electron–nucleus spin pairs.

Assuming complete domination of the term $H_{DIP}^{Paramagn}$, the theoretical chemical shift anisotropies for the ^{31}P NMR spectra can be calculated to account for the paramagnetic centers located within 100 Å relative to phosphorus atoms. At good agreement between theoretical and experimental values, the ^{31}P–paramagnetic ion distances can be determined. The results of the calculations carried out for the compounds in Figure 10.13B show that the solid-state NMR data agree well with the x-ray analysis.

In general, such domination of one paramagnetic contribution, which simplifies calculations, is rather a rare phenomenon requiring independent confirmation. In addition, evaluation of different contributions by analysis of sideband patterns, such as those caused by dipolar and BMS interactions, is practically impossible in the absence of accurately determined coordinates of paramagnetic centers. This circumstance complicates the quantitative analysis of solid-state NMR data.

As mentioned above, the presence of paramagnetic ions in proteins can improve accuracy in their structural studies by solid-state NMR in spite of paramagnetic effects. Here, the strategy of NMR studies is based on the appropriate choice of proteins. In general, proteins containing paramagnetic metal ions with large magnetic susceptibility anisotropy, such as low-spin Fe(III), high-spin Fe(II), cobalt(II), or lanthanide ions, provide measurements of the pseudocontact shifts for nuclei surrounding the paramagnetic centers. These shifts can be used as additional parameters in a structural analysis.

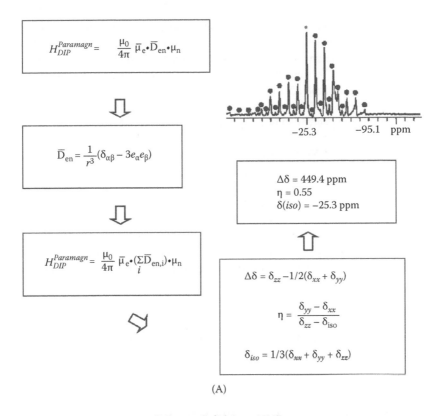

(A)

X-Ray vs. Solid-State NMR

Compound	X-ray distance (Å)	NMR distance (Å)
$K_{14}(H_3O)_3[Nd(\alpha_1\text{-}P_2W_{17}O_{61})_2] \cdot 4KCl \cdot 64H_2O$	P1: 4.67 P2: 6.23	P1: 4.96 P2: 6.30
$K_{14}(H_3O)_3[Er(\alpha_1\text{-}P_2W_{17}O_{61})_2] \cdot 4KCl \cdot 64H_2O$	P1: 4.56 P2: 6.03	P1 P2: 6.56
$K_{14}(H_3O)[Nd(H_2O)_3(\alpha_2\text{-}P_2W_{17}O_{61})_2] \cdot 4KCl \cdot nH_2O$	P1: 4.44 P2: 8.09	P1: 4.77 P2: 6.51

(B)

FIGURE 10.13 (A) The ^{31}P MAS NMR spectrum of the compound $K_{14}(H_3O)_3[Er(\alpha_1\text{-}P_2W_{17}O_{61})_2] \cdot 4KCl \cdot 64H_2O$ recorded at 15 kHz (the symbols show two sets of ^{31}P signals with different intensities) and elements of the theory used for calculating electron–nucleus dipolar interactions shown for one of the ^{31}P sites. (B) Distances between phosphorus atoms, marked as P1 and P2, and lanthanide ions determined by x-ray and NMR methods. (From Huang, W. et al., *J. Am. Chem. Soc.*, 130, 481, 2008. With permission.)

In such cases, ^{13}C nuclei, even those located close to paramagnetic centers, can be observed by direct detection or other standard techniques due to the relatively small effects of paramagnetic relaxation. For example, in solutions, Curie relaxation greatly broadens the ^{13}C resonances, but in solids this effect is remarkably suppressed. Thus, for example, 2D proton-driven spin diffusion (PDSD) NMR experiments generally lead to ^{13}C–^{13}C correlation spectra with good spectral resolution.[39] The signals in these spectra can be assigned on the basis of NMR spectra in solutions or by N–C(α)–C(β) (or other) correlation experiments in the solid state.

Two-dimensional PDSD NMR experiments demonstrating excellent spectral resolution are shown in Figure 10.14.[39] Here, the ^{13}C pseudocontact shifts are measured in the solid state as particularly valuable restraints for structural determinations. The spectra are obtained for the catalytic domain of the matrix metalloproteinase 12 (MMP12). MMP is a zinc protein for which the diamagnetic Zn(II) ions can be easily exchanged with paramagnetic ions Co(II). The remarkable pseudocontact shifts are clearly observable by comparison of the data collected in samples of the proteins Co-MMP and Zn-MMP. The technical details regarding these experiments can be found in Reference 39.

10.5   NUCLEAR RELAXATION IN PARAMAGNETIC SOLIDS: APPLICATIONS

Studies of paramagnetic solids by relaxation measurements are based on the following theoretical consequences. The decay of macroscopic transverse and longitudinal magnetizations measured experimentally follows the exponential law for complete domination of the spin diffusion mechanism. Then, a single T_1 (or T_2) time characterizes all of the nuclei in rigid solids, even those that are not chemically or structurally equivalent. More precisely, these nuclei are situated within the barrier radius effective for spin diffusion to paramagnetic centers. Under these conditions, the relaxation does not give valuable structural data. The spin diffusion mechanism is typical of systems with nuclei experiencing strong mutual dipolar interactions, such as protons or phosphorus or ^{19}F nuclei at relatively small concentrations of paramagnetic centers. For other nuclei with weak dipolar coupling and a high concentration of paramagnetic centers, relaxation is not exponential following a stretched exponential, $\exp(-(\tau/T_1)^\beta)$ or $\exp(-(\tau/T_2)^\beta)$. Recent approaches to theoretical interpretations of stretched exponential relaxation can be found in Reference 40. In general, a β value of 0.5 corresponds to relaxation via direct electron–nucleus dipolar interactions without spin diffusion, while intermediate values between 0.5 and 1.0 can be attributed to a diffusion-limited mechanism. Finally, the comparison of T_1 and T_2 values obtained in different samples will be reasonable at equal (or similar) β values found experimentally. As in the case of diamagnetic molecules, NMR relaxation studies of paramagnetic solids are numerous and various and concern different aspects of their properties. However, in spite of this, one can formulate a general methodology to demonstrate how to set up a relaxation experiment and how to interpret the data using, as an example, porous solids doped with paramagnetic metal ions.

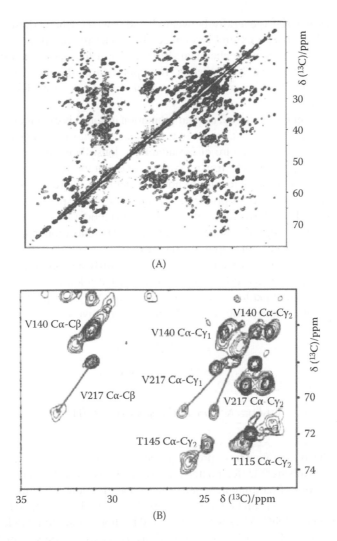

(A)

(B)

FIGURE 10.14 PDSD NMR experiments represented as a superimposition of the NMR spectra of the diamagnetic protein Zn-MMP (dark gray contours) and the paramagnetic protein Co-MMP (light gray contours), where (A) is the aliphatic region of the two spectra and (B) is the expansion with corresponding assignments. (From Bertini, I. et al., *Dalton Trans.*, 3782, 2008. With permission.)

10.5.1 STRATEGY FOR RELAXATION STUDIES OF PARAMAGNETIC SOLIDS

In general, preparation of catalytically active porous systems is based on the sol/gel synthesis of the matrix in the presence of metal ions. The final step of the synthesis is calcination at high temperatures to incorporate the metal ions into the matrix. An important element in the study of these porous solids is the distribution of paramagnetic ions through their volume. Porous systems prepared in this way are usually active in catalysis by small concentrations of metal ions incorporated into the matrix of the material (generally between 2 and 5 wt%). Under these conditions, their MAS NMR spectra exhibit resonances for nuclei remote from paramagnetic ions. Therefore, paramagnetic chemical shifts, as a test for incorporation, are practically invisible or absent. Nevertheless, incorporation of the metal ions can still be determined by relaxation measurements performed even on nuclei remote from paramagnetic ions.

As follows from NMR theory, nuclear relaxation in solids is extremely sensitive to the presence of paramagnetic centers. In turn, T_1 and T_2 relaxation times will depend on concentrations and the nature of paramagnetic ions, their distributions through the volume of samples, and electron relaxation times. At the same time, *a priori*, the significance of all of these factors is unknown; in addition, the relaxation mechanism can be different. These initial problems should be solved experimentally (or theoretically) to provide the basis for a suitable relaxation model for the interpretation of relaxation data. Five important statements based on ^{29}Si relaxation in the silica-based material $SiO_2Al_2O_3MeO$ can help in this context.

First, the rates of spin–lattice relaxation via spin diffusion and dipolar electron–nucleus interactions depend differently on the concentration of paramagnetic centers, as is shown in Equations 10.5 and 10.6, where they are proportional to the concentrations of paramagnetic centers N_p or N_p^2, respectively:

$$1/T_1^{SD} = (1/3)8\pi N_p C^{1/4} D^{3/4}$$
$$C = (2/5)\gamma_I^2 \gamma_S^2 \hbar^2 S(S+1) \tag{10.5}$$

$$1/T_1 = (16/9)\pi^3 N_p^2 (2/5)\gamma_I^2 \gamma_S^2 \hbar^2 S(S+1) r^{-6} \tau_e \big/ \left(1 + \omega_I^2 \tau_e^2\right) \tag{10.6}$$

Varying the N_p parameter is probably the simplest approach to evaluation of relaxation mechanisms.[41] Unfortunately, these equations are valid only for homogeneous distributions of paramagnetic ions throughout the sample volume. Therefore, in practice, the situation can be more complex. For example, the ^{29}Si $1/T_1$ rate in the silica-based material $SiO_2Al_2O_3NiO$, dominated by dipolar electron–nucleus interaction, increases linearly with Ni^{2+} concentrations for small amounts of nickel. Then, it reaches a maximum and decreases again at higher Ni^{2+} concentrations.[28] It is most likely that such systems violate the condition of homogeneity, or the nature of the nickel ions (oxidation states) changes with higher nickel loadings. The latter can be probed by independent physical measurements.

Second, the presence of the spin diffusion relaxation mechanism can be confirmed or refuted by ^{29}Si T_1 measurements in static and MAS NMR experiments at different spinning rates. For example, inversion–recovery and saturation–recovery experiments performed on a static sample of solid $SiO_2Al_2O_3MnO$ and on a sample spinning at rates between 2.5 and 14 kHz have produced ^{29}Si T_1 stretched exponential curves with β parameters between 0.6 and 0.7. In addition, they are independent of spinning rates.[28] Thus, spin diffusion is ineffective, and the ^{29}Si spin–lattice relaxation is completely governed by direct dipolar coupling between electrons and nuclei.

Third, when concentrations of paramagnetic metal ions are independently defined by convenient methods their locations relative to the matrix of a system can be found, but doing so requires another strategy in relaxation studies. Decreasing T_1 and T_2 relaxation times alone cannot be regarded as an indicator for the presence of paramagnetic ions in the matrix, because mechanically mixing the ions and compounds also reduces relaxation times. Semiquantitatively, this problem can be solved by relaxation experiments performed on two different groups of nuclei, one of which is knowingly located within pores. This is the case for ^1H and ^{29}Si nuclei in porous silica-based systems ($SiO_2Al_2O_3NiO$) prepared with different nickel loadings.[42] Here, T_1 MAS NMR relaxation measurements can be performed for ^1H and ^{29}Si resonances belonging to water molecules in pores and the silica matrix, respectively. Because both of the nuclei relax via direct dipolar interactions with nickel ions, the relaxation rate ratios, $R_1(^1H)/R_1(^{29}Si)$, found experimentally can be compared with the ratios calculated theoretically at equal distances $r(Si\cdots Ni)$ and $r(H\cdots Ni)$.[42] The experimental ratios obtained in these systems were significantly larger than the theoretical ratios; thus, the nickel centers accumulated within pores but not in the matrix.

Fourth, when nuclear relaxation can be well approximated by a bi-exponential function, a simple relaxation model can be applied for the localization of paramagnetic ions, as shown in Figure 10.15, where the ^{29}Si T_1 relaxation in the paramagnetic porous solid $SiO_2Al_2O_3MnO$ (3.8 wt% manganese) is clearly bi-exponential (Figure 10.15A). It should be noted that the partially relaxed ^{29}Si MAS NMR spectra (Figure 10.15B) recorded at different τ values and scaled to the sideband intensities show that all of the sidebands relaxed identically but remarkably faster than the central isotropic line. These facts support the relaxation model shown in Figure 10.15C. Within the framework of this model, the ^{29}Si MAS NMR spectrum can be represented by a superposition of resonances from the ^{29}Si nuclei located in the second and third coordination spheres of Mn^{2+}, while the ^{29}Si nuclei from the first coordination sphere, Si(1), in the SiOMn are spectrally invisible. The nuclei from the second sphere, Si(2), show a resonance accompanied by intense sidebands, while a resonance belonging to the more remote nuclei situated in the third coordination sphere, Si(3), exhibits less intense sidebands. Following this model, the main contribution to the total relaxation rates measured for the sidebands in the spectrum comes from the ^{29}Si nuclei located in the second coordination sphere. The central resonance, relaxing as a sum of ^{29}Si magnetizations Si(2) and Si(3), shows two components with short and long relaxation times measured as 0.76 s and 0.0097 s, respectively. These results correspond to the presence of paramagnetic ions that are chemically bonded to the matrix of the system.[43] Moreover, it is a relatively simple matter to conclude that the thickness of the silica walls is relatively small and created by three to five layers of condensed silica units and that ions of manganese are situated on the silica surface.

Fifth, due to the very strong paramagnetic effects, the $^{29}Si(1)$ nuclei in Figure 10.15C are obviously invisible, while the more remote nuclei $^{29}Si(2)$ and $^{29}Si(3)$ will show a very wide line in a static ^{29}Si NMR spectrum, as shown in Figure 10.16A. The angle-dependent term ($3\cos^2\theta-1$) and diverse $Si\cdots Mn$ distances controlling dipole–dipole electron–nucleus interactions produce such as line shape. With fast spinning, the wide line transforms to a broad sideband pattern where the central

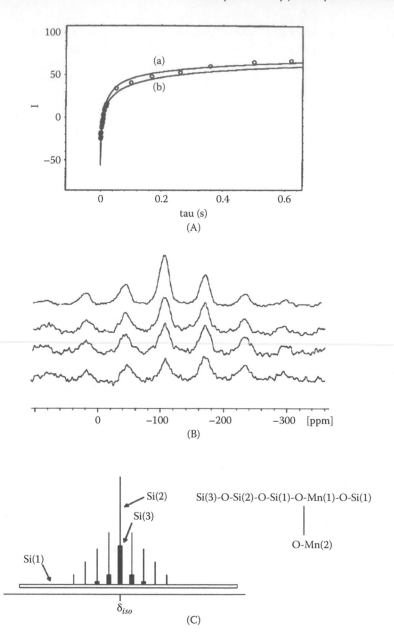

FIGURE 10.15 (A) The ^{29}Si inversion–recovery curve obtained for the central (isotropic) resonance in a sample of the compound $SiO_2Al_2O_3MnO$ containing 3.8 wt% manganese. The intensity (in arbitrary units) vs. delay time τ (in s) is fitted to a stretched exponential function (solid line b) and to a bi-exponential function (solid line a). (B) Completely and partially relaxed ^{29}Si MAS NMR spectra of $SiO_2Al_2O_3MnO$ obtained at τ values of 0.02, 0.015, and 0.012 s (from top to bottom). (C) Relaxation model used for interpretation of the relaxation data. (Adapted from Bakhmutov, V.I., *Solid-State NMR in Materials Science: Principles and Applications*, CRC Press, Boca Raton, FL, 2011.)

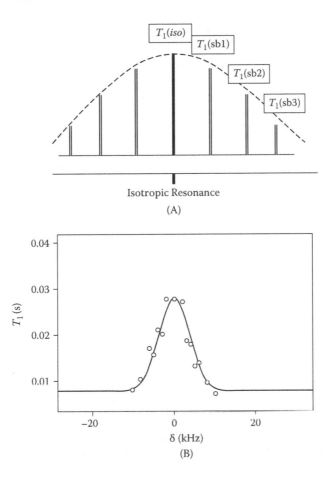

FIGURE 10.16 (A) Static NMR spectrum (dashed line) of a paramagnetic solid which transforms to the MAS NMR pattern with spinning, where T_1 times of the sidebands and the isotropic resonance are different. (B) ^{29}Si T_1 times experimentally measured for the sidebands and the isotropic (central) resonance (0 kHz). The experiments were performed on $SiO_2Al_2O_3MnO$ (2.5 wt% Mn and 2 wt% Al) spinning at different rates. Solid line corresponds to a Gaussian distribution.[28] (Adapted from Bakhmutov, V.I., *Solid-State NMR in Materials Science: Principles and Applications*, CRC Press, Boca Raton, FL, 2011.)

line is an isotropic resonance. It should be noted that, in many cases, the T_1 NMR data obtained in spinning paramagnetic solids indicate that the relaxation times of isotropic resonances are longer than those of sidebands. Formally, this situation corresponds to the presence of T_1 distributions observed for various nuclei. For the ^{29}Si nuclei in Figure 10.15C, this situation is quite reasonable, as the ^{29}Si nuclei are at different distances from the paramagnetic ions. The lines remote from the isotropic signal belong to nuclei that experience stronger dipolar electron–nucleus interactions and their relaxation occurs more rapidly. According to the Solomon theory of electron–nucleus dipolar interactions (Chapter 3), relaxation times can be quantitatively analyzed in terms of electron–nucleus distances r, when electron relaxation time τ_e is accurately determined independently. For example, relatively long τ_e times can be easy measured by advanced EPR techniques.[43] In the absence of EPR data, the τ_e times can be determined by accurate T_1 and T_2 NMR measurements, which can be used to exclude unknown distances r from the Solomon equations to calculate τ_e times. For example, applying this method to the solids $SiO_2Al_2O_3MnO$ and SiO_2MnO, prepared with different Mn^{2+} contents, has given relaxation times for electrons in Mn^{2+} ions of between 0.6×10^{-8} and 0.8×10^{-8} s.[28] Here, the ^{29}Si R_1 and R_2 NMR relaxation rates can be treated via Equation 10.7:

$$R_2^D / R_1^D = 2\tau_{1e}^2 \omega_I^2 + 1.5 \tag{10.7}$$

where ω_I is the Larmor frequency of the ^{29}Si nuclei. Because the τ_e time is now known, Equation 10.8 can be applied for calculation of the $Mn \cdots Si$ distances:

$$1/T_1 = 43.5/(0.1r(Si \cdots Mn))^6 \tag{10.8}$$

where T_1 and $r(Si \cdots Mn)$ are measured in seconds and Å, respectively. Equation 10.8 can be obtained from the Solomon equations by adding the well-known factor $(\mu_0/4\pi)^2$, a Mn^{2+} electron spin of 5/2, $\tau_{1e} = 1 \times 10^{-8}$ s, and the ^{29}Si resonance frequency of 79.46 MHz. Then, the $Mn \cdots Si$ distances obtained can be interpreted on the basis of the following criterion: Mn^{2+} ions are incorporated into the silica matrix when ^{29}Si T_1 times measured for isotropic resonances in ^{29}Si MAS NMR spectra correspond to the $Mn \cdots Si$ distances of 7 to 8 Å. These distances are typical of the second manganese coordination sphere. In general, the situation does not change greatly with the appearance of non-exponential relaxation approximated by a stretched-exponential function, where a T_1 time will be a representative value of whole spins. Even in this case, T_1 relaxation times measured in MAS NMR experiments for isotropic resonances and their sidebands can differ due to the presence of T_1 distributions.[44,45]

The pattern in Figure 10.16B shows the ^{29}Si T_1 time distribution in the solid $SiO_2Al_2O_3MnO$ observed by measurements at different spinning rates. The relaxation time consistently reduces from the isotropic resonance (0 kHz) to its sidebands, shifted to a low and high field. Moreover, this ^{29}Si T_1 pattern reasonably reproduces the shape of the static resonance, further supporting the above statements. Now, an electron relaxation time τ_e of 1×10^{-8} s opens the way to quantitative analysis of the shortest ^{29}Si T_1 components. The analysis can be based, for example, on Equation 10.6, where N_p (expressed as $(4/3)r_{AV}^3 \pi N_p = 1$) is the number of paramagnetic centers distributed homogeneously through the materials, and r_{AV} is the radius of the sphere around the paramagnetic center. This approach applied for $SiO_2Al_2O_3MnO$ systems with 2.5 and 4.8 wt% manganese gives r_{AV} distances of 8.1 and 7.6 Å, respectively. The value of 7.6 Å agrees well with Si atoms located in the second coordination sphere of Mn^{2+}. Chemically, this means that Mn^{2+} ions are incorporated into the matrix. Here, it is pertinent to emphasize again that the presence of T_1 distributions, in itself, is not a sufficient criterion for confirming the incorporation of paramagnetic ions in the matrix of materials, whereas quantitative analysis based on well-determined electron relaxation times can provide such confirmation.

The nature of the porous solids considered in this section is not special; therefore, the same ideas and similar methodologies and approaches to interpretation can be applied for relaxation studies of many molecular systems containing paramagnetic species. Finally, other solid-state NMR studies, such as zero-field NMR experiments that increase resolution in polycrystalline samples, solid-state NMR imaging applied for materials science, or NMR in suspensions, are also available and details can be found in the recommended literature.

REFERENCES AND RECOMMENDED LITERATURE

1. Banci, L., Bertini, I., Luchinat, C., and Mori, M. (2010). *Prog. Nucl. Magn. Reson. Spectrosc.*, 56: 247.
2. Gopinath, T., Kaustubh, R., Mote, K.R., and Veglia, G. (2013). *Prog. Nucl. Magn. Reson. Spectrosc.*, 75: 50.
3. Saita, H., Ando, I., and Ramamoorthy, A. (2010). *Prog. Nucl. Magn. Reson. Spectrosc.*, 57: 181.
4. Van der Klink, J.J. and Brom, H.B. (2000). *Prog. Nucl. Magn. Reson. Spectrosc.*, 36: 89.
5. Lee, C.E., Lee, C.H., Kim, J.H., Kim, K.S., and Kim, K. (1994). *J. Korean Chem. Soc.*, 38: 628.
6. Njegic, B., Levin, E.M., and Schmidt-Rohr, K. (2013). *Solid State Nucl. Magn. Reson.*, 55–56: 79.
7. Ishikiriyama, K. and Todoki, M. (1995). *J. Colloids Inter. Sci.*, 171: 103.
8. Aksnes, D.W. and Kimtys, L. (2004). *Solid State NMR*, 25: 146.
9. Song, Y.Q. (2009). *Prog. Nucl. Magn. Reson. Spectrosc.*, 55: 324.
10. Strange, J.H., Mitchell, J., and Webber, J.B.W. (2003). *Magn. Reson. Imaging*, 21: 221.
11. Bonhomme, C., Gervais, C., and Laurencin, D. (2014). *Prog. Nucl. Magn. Reson. Spectrosc.*, 77: 1.
12. Kuethe, D.O., Montano, R., and Pietra, T. (2007). *J. Magn. Reson.*, 186: 243.
13. Horch, C., Schlayer, S., and Stallmach, F. (2014). *J. Magn. Reson.*, 240: 24.
14. Kim, J., Middlemiss, D.S., Chernova, N.A., Zhu, B.Y.X., Masquelier, C., and Grey, C.P. (2010). *J. Am. Chem. Soc.*, 132: 16825.
15. Kubo, A., Spaniol, T.P., and Terao, T. (1998). *J. Magn, Reson.*, 133: 330.
16. Kervern, G., Pintacuda, G., and Emsley, L. (2007). *Chem. Phys. Lett.*, 435: 157.
17. Canesson, L. and Tuel, A. (1997). *Chem. Commun.*, 241.
18. Tuel, A., Canesson, L., and Volta, J.C. (1999). *Colloids Surf. Physicochem. Eng. Aspects*, 158: 97.
19. Mali, G., Risti, A., and Kaucic, V. (2005). *J. Phys. Chem. B*, 109: 10711.
20. Canesson, L., Boudeville, Y., and Tuel, A. (1997). *J. Am. Chem. Soc.*, 119: 10754.
21. Wilcke, S.L., Cairns, E.J., and Reimer, J.A. (2006). *Solid State NMR*, 29: 199.
22. Wickramasinghe, N.P., Kotecha, M., Samoson, A., Past, J., and Ishii, Y. (2007). *J. Magn. Reson.*, 184: 350.
23. Tang, J.A., Ellis, B.D., Warren, T.H., Hanna, J.V., Macdonald, C.L.B., and Schurko, R.W. (2007). *J. Am. Chem. Soc.*, 129: 13049.
24. Ellis, P., Sears, J.A., Yang, P. et al. (2010). *J. Am. Chem. Soc.*, 132: 16727.
25. Ziarelli, F., Viel, S., Sanchez, S., Cross, D., and Caldarelli, S. (2007). *J. Magn. Reson.*, 188: 260.
26. Nery, J.G., Giotto, M.V., Mascarenhas, Y.P., Cardoso, D., Zotin, F.M.Z., and Sousa-Aguiar, E.F. (2000). *Micropor. Mesopor. Mater.*, 41: 281.
27. Plevert, J., Okubo, T., Wada, Y., O'Keefe, M., and Tatsumi, T. (2001). *Chem. Commun.*, 2112.

28. Bakhmutov, V.I. (2011). *Chem. Rev.*, 111: 530.
29. Blumel, J., Herker, M., Hiller, W., and Koehler, F.H. (1996). *Organometallics*, 15: 3474.
30. Grey, C.P., Smith, M.E., Cheethem, A.K., Dobson, C., and Duptreet, R. (1990). *J. Am. Chem. Soc.*, 112: 4670.
31. Heise, H., Kohler, F.H., and Xie, X. (2001). *J. Magn. Reson.*, 150: 198.
32. Metz, G., Wu, X.L., and Smith, S.O. (1994). *J. Magn. Reson. Ser. A*, 110: 219.
33. María, D.S., Claramunt, R.M., Vasilevsky, S.F., Klyatskaya, S.V., Alkorta, I., and Elguero, J. (2011). *ARKIVOC*, III: 114.
34. Chazel, C., Menetrier, M., Croguenne, L., and Delmas, C. (2005). *Magn. Reson. Chem.*, 43: 849.
35. Accardi, R.J. and Lobo, R.F. (2000). *Micropor. Mesopor. Mater.*, 40: 25.
36. Terskikh, V.V., Ratciffe, C.I., Ripmeester, J.A., Reinhold, C.J., Anderson, P.A., and Edwards, P.P. (2004). *J. Am. Chem. Soc.*, 126: 11350.
37. Shubin, A.A., Khabibulin, D.F., and Lapina, O.B. (2009). *Catal. Today*, 142: 220.
38. Huang, W., Schopfer, M., Zhang, C., Howell, R.C., Todaro, L., Gee, B.A., Francesconi, L.C., and Polenova, T. (2008). *J. Am. Chem. Soc.*, 130: 481.
39. Bertini, I., Luchinat, C., Parigi, G., and Pierattelli, R. (2008). *Dalton Trans.*, 3782.
40. Almeida, R.M.C., Lemke, N., and Campbell, I.A. (2000). *Braz. J. Phys.*, 30: 25.
41. Scholz, K. and Thomas B. (1995). *Solid State NMR*, 4: 309.
42. Bakhmutov, V.I., Shpeizer, B.G., Prosvirin, A.V., Dunbar, K.R., and Clearfield, A. (2009). *Micropor. Mesopor. Mater.*, 118: 78.
43. Atsarkin, V.A., Demidov, V.V., Gaal, F.G., Moritomo, Y., Conder, K., Janossy, A., and Forro, L. (2003). *J. Magn. Magn. Mater.*, 258–259: 256.
44. Hayashi, S. (1994). *Solid State NMR*, 3: 323.
45. Bakhmutov, V.I., Shpeizer, B.G., and Clearfield, A. (2006). *Magn. Reson. Chem.*, 44: 861.

RECOMMENDED LITERATURE

Andrew, E.R. and Szczesniak, E. (1995). An historical account of NMR in the solid state. *Prog. Nucl. Magn. Reson. Spectrosc.*, 28: 11–36.
Bakhmutov, V.I. (2011). *Solid-State NMR in Materials Science: Principles and Applications*. Boca Raton, FL: CRC Press.
Blumel, J. (2008). Linkers and catalysts immobilized on oxide supports: new insights by solid-state NMR spectroscopy. *Coord. Chem. Rev.*, 252: 2410.
Demco, D.E. and Blumichm B. (2000a). Solid-state NMR imaging methods. Part I. Strong field gradients. *Concepts Magn. Reson. A*, 12: 188.
Demco, D.E. and Blumichm, B. (2000b). Solid-state NMR imaging methods. Part II. Line narrowing. *Concepts Magn. Reson. A*, 12: 269
Koehler, J. and Jens Meiler, J. (2011). Expanding the utility of NMR restraints with paramagnetic compounds: background and practical aspects. *Prog. Nucl. Magn. Reson. Spectrosc.*, 59: 360.
Koptyug, I.V. (2012). MRI of mass transport in porous media: drying and sorption processes. *Prog. Nucl. Magn. Reson. Spectrosc.*, 65: 1.
López-Cebral, R., Martín-Pastor, M., Seijo, B., and Sanchez, A. (2014). Progress in the characterization of bio-functionalized nanoparticles using NMR methods and their applications as MRI contrast agents. *Prog. Nucl. Magn. Reson. Spectrosc.*, 79: 1.

Concluding Remarks

In spite of its relatively small volume, this book clearly shows that NMR spectroscopy is the physical method most frequently applied in the various fields of modern fundamental and industrial sciences. The interdisciplinary nature of NMR is easily demonstrated by the yearly publishing of a vast number of scientific articles in numerous journals on physics, chemistry, geochemistry, biochemistry, chemical and biomedical engineering, biology, geology, and archeology, where key data describing an object under investigation have been obtained by NMR in solutions and in the solid state. In addition, articles regarding special NMR issues are often published in such journals as *Accounts of Chemical Research*, *Chemical Reviews*, and *Physical Chemistry Chemical Physics*. In this context, writing a complete NMR handbook does not seem to be possible or necessary because the NMR techniques, objects, and conditions of studies constantly change, affecting each other. Even the well-known ten-volume *Encyclopedia of NMR*, edited by R.K. Harris and R.E. Wasylishen, is constantly being updated by the addition of new NMR topics.

This text obviously does not claim to be a reference book, but it does demonstrate how to move from a solid or liquid sample via NMR experiments to NMR signals, how to understand the origin of their appearance and shape, and how to interpret such signals in terms of a formulated problem. Thus, this text underscores the spectroscopic logic that the assignments of signals in NMR spectra play a major role as the first and most important starting point in the analysis of data. It should be emphasized again that nuclear spins know nothing about the existence of molecules, and the connection between them is the product of interpretation. If a reader refers to this text to carry out NMR experiments, thus avoiding typical errors, and if it stimulates further reading, then the aim of the book has been achieved.

Index